AF594951

Multi-Agent Systems 2019

Multi-Agent Systems 2019

Editors

Andrea Omicini
Stefano Mariani

MDPI • Basel • Beijing • Wuhan • Barcelona • Belgrade • Manchester • Tokyo • Cluj • Tianjin

Editors
Andrea Omicini
Università di Bologna
Italy

Stefano Mariani
Università degli Studi di Modena e Reggio Emilia
Italy

Editorial Office
MDPI
St. Alban-Anlage 66
4052 Basel, Switzerland

This is a reprint of articles from the Special Issue published online in the open access journal *Applied Sciences* (ISSN 2076-3417) (available at: https://www.mdpi.com/journal/applsci/special_issues/Multi-Agent_Systems_2019).

For citation purposes, cite each article independently as indicated on the article page online and as indicated below:

LastName, A.A.; LastName, B.B.; LastName, C.C. Article Title. *Journal Name* **Year**, *Article Number*, Page Range.

ISBN 978-3-03943-046-8 (Hbk)
ISBN 978-3-03943-047-5 (PDF)

Contents

About the Editors

Andrea Omicini is Full Professor at DISI, the Department of Computer Science and Engineering of the Alma Mater Studiorum–Università di Bologna, Italy. He holds a PhD in computer and electronic engineering, and his main research interests include coordination models, multi-agent systems, intelligent systems, programming languages, autonomous systems, middleware, simulation, software engineering, pervasive systems, and self-organization. He has published over 300 articles on these subjects, in addition to having edited numerous international books, serving as Guest Editor of several Special Issues of international journals, and has held many invited talks and lectures at international conferences and schools.

Stefano Mariani received his PhD degree in Computer Science from the University of Bologna, Bologna, Italy, in 2016. He is currently Assistant Professor of Computer Science with the University of Modena and Reggio Emilia, Reggio Emilia, Italy. He has been involved in the EU FP7 Project SAPERE, and in EU H2020 Project CONNECARE. His current research interests include coordination models and languages, agent-oriented technologies, pervasive computing, self-organization mechanisms, and socio-technical systems.

Editorial

Special Issue "Multi-Agent Systems": Editorial

Stefano Mariani [1,*,†] and Andrea Omicini [2,†]

1 Department of Sciences and Methods of Engineering, Università degli Studi di Modena e Reggio Emilia, 42122 Reggio Emilia, Italy
2 Department of Computer Science and Engineering, Università di Bologna, 47521 Cesena, Italy; andrea.omicini@unibo.it
* Correspondence: stefano.mariani@unimore.it
† These authors contributed equally to this work.

Received: 16 July 2020; Accepted: 25 July 2020; Published: 1 August 2020

Abstract: Multi-agent systems (MAS) are built around the central notions of agents, interaction, and environment. Agents are autonomous computational entities able to pro-actively pursue goals, and re-actively adapt to environment change. In doing so, they leverage on their social and situated capabilities: interacting with peers, and perceiving/acting on the environment. The relevance of MAS is steadily growing as they are extensively and increasingly used to model, simulate, and build heterogeneous systems across many different application scenarios and business domains, ranging from logistics to social sciences, from robotics to supply chain, and more. The reason behind such a widespread and diverse adoption lies in MAS great expressive power in modeling and actually supporting operational execution of a variety of systems demanding decentralized computations, reasoning skills, and adaptiveness to change, which are a perfect fit for MAS central notions introduced above. This special issue gathers 11 contributions sampling the many diverse advancements that are currently ongoing in the MAS field.

Keywords: multi-agent systems; agent-based modeling; agent-based simulation; decision support

---

1. Introduction

As intelligent systems pervade more and more our everyday life, the need for a coherent set of abstractions and technical tools to support their design, development, and maintenance keeps growing steadily. Multi-agent systems (MAS) nowadays represent the richest and most reliable source for such abstractions, given that they provide the components (the agents) to encapsulate essential features such as cognition and autonomy, as well as the notions required to put systems together (agent societies) and make them work in the real world (MAS environment) [1]. In addition, a few decades of intensive academic and industrial research on MAS, and their integration with the most recent advances in AI techniques and IoT technologies, have promoted the intense development and widespread diffusion of novel agent-oriented techniques, methods, and tools, and paved the way towards the acceptance of MAS as the forthcoming industrial mainstream for complex yet reliable intelligent systems.

Yet, the articulation of the MAS scenario is nowadays so overwhelming that the transition is going to make both researchers and practitioners busy for two more decades, at least—before all aspects and issues concerning MAS techniques and methods are fully understood and addressed within the many relevant application scenarios where MAS are required to operate. Providing a platform where MAS researchers can share their most novel and exciting findings and results is then crucial to support and promote the development and spreading of new MAS models and technologies: this is in fact the main motivation behind the special issue.

2. Overview

Before delving into the individual contributions gathered, a few general statistics and observations are useful to have an overview of the content and outreach of this special issue:

- 53 papers have been submitted for peer review, out of which 11 were finally published, resulting in an acceptance rate of $\approx 21\%$;
- the average time to publish, intended as the time passed from submission to online availability, is 41 days, with a standard deviation of ≈ 16 days—dates are publicly available on each paper web page, accessible from the special issue home page (https://www.mdpi.com/journal/applsci/special_issues/Multi-Agent_Systems_2019, last accessed 21 March);
- papers already generated an average of 538 downloads ($\approx$281 standard deviation)—we deem citations not worth considering yet, after just (less then) 1 year since publication;
- published papers have been co-authored by authors coming from 10 different countries, covering Europe, Asia, North and South America. Amongst these, Italy and China are the most represented, having 2 papers with more than one local author.

These numbers are in line with previous edition of the special issue [2], except for a lower acceptance rate, which reflects the more selective review process meant to increase the quality of the special issue and its potential impact on research and practice.

Figure 1 shows the wordcloud generated from the full text of the published papers.

Figure 1. Wordcloud generated from the full text of each publication of the special issue.

The most mentioned words are "agent" and "model", closely followed by "simulation" and "system", and then by "task" and "data". The former four words are not surprising and confirm our editorial for previous edition: MAS are well-known and widely adopted also outside the strict boundaries of computer science and engineering exactly for the purpose of modeling and simulating complex systems, in fields as diverse as bioinformatics, social sciences, network science, supply chain, and logistics. The latter two words may instead appear as rather novel, and point to the increasingly widespread exploitation of MAS for novel purposes, such as collecting, managing, and analyzing data to turn it into actionable knowledge, and support execution of tasks requiring peculiar capabilities such as reasoning, reactiveness to environmental conditions, compliance to complex inter-dependencies.

Other highly relevant words working as clues for relevant application areas and kind of systems are the following ones:

- "game", "role", "social", and "interaction", which point to the social dimension of agenthood;
- "robot", "environment", "action", and "time", which emphasize the situated dimension of MAS.

In our previous editorial we analyzed a similar wordcloud from the perspective of the topics that were subject of the publications, which were: agent-based modeling and simulation, situated systems, socio-technical systems, and semantic technologies. Except for the latter, relevance of the other is confirmed by the current edition. That said, this year we would like to take a different point of view, by answering the following question: what are MAS used for? In the following sections we classify the papers included in this special issue according to the following four usage destinations, and summarize their main contributions:

Decision support—papers gathered in this category exploit MAS, in particular their ability to perform distributed reasoning, to deliver insights about a certain topic, with the goal of enhancing humans' decision making processes and lower their cognitive overhead.

Modeling framework and methodology—in this category, what matters the most is the expressive power of the agent abstraction as a conceptual tool supporting engineering of complex systems featuring autonomous components.

Programming abstraction and simulation framework—complementary to the previous category, in this one MAS are mostly used for their operational features, as a software tool enabling development and execution of the complex systems already mentioned, especially in simulated scenarios.

Execution infrastructure—here, MAS are used as the backbone infrastructure executing the computations demanded by the application at hand, leveraging MAS themselves as an efficient and effective distributed computing platform.

3. MAS for Decision Support

Interestingly enough, the category reflecting the new entry w.r.t. previous editorial is also the most represented: 4 papers out of the 11 published exploit MAS to deliver decision support.

In [3], the authors exploit agent-based modeling and simulation to define a photovoltaic adoption prediction model based on self-reported behavior, then refined by a genetic algorithm looking at observed data. The goal is to help energy-related decision making by policy makers, by modeling and predicting households pondering whether to adopt photovoltaic energy solutions. Here, the agent abstraction is useful to model individual behavior driven by rational utility functions (such as economic savings), and the social dimension stemming from neighborhoods influencing each others' decisions.

In [4], an MAS is used as the operational backbone of a game-theoretic approach to task allocation under strict spatio-temporal constraints, applicable to deliver decision support in many critical scenarios such as disaster relief. Here the main motivation behind usage of an MAS lies in the preference of quickness over optimality as regards convergence to useful allocations, as the targeted scenarios do not mind optimal solutions if they do not come within a reasonable time. As such, an MAS is built to run a scheduling algorithm rooted in game theory in a decentralized fashion, improving convergence time while giving away optimality.

In [5], an MAS is proposed as a platform for instrumenting a collective of neural network based classifiers by adopting a crowdsourcing metaphor: each classifier is an agent, each classification is an opinion, and the overall prediction delivered by the system is the aggregation of the crowd's opinions. The goal is to improve prediction accuracy and transparency, by letting agents interact socially to exchange knowledge (e.g., new features), gain reciprocal trust, and change opinion when given enough evidence. The agent abstraction is then used mostly for its autonomy, and the MAS as an enabler of the sociality needed to improve transparency and accuracy through the exchange of information.

In [6], the authors target the green coffee supply chain with an agent-based decision support system devoted to planning production scheduling in face of fluctuating and peak demand. The modeled supply chain is rather complex, with plenty of interdependencies amongst activities and variables influencing the decision process at each step. An MAS is thus used to tame this complexity, by modeling all the different tasks and processes as autonomous agents, each undergoing its own reasoning to take decisions, while interacting with others upon need.

In spite of the heterogeneity of the application domains and the techniques adopted, all the described approaches leverage on MAS central notions to improve delivering of decision support functionalities, either by simulation [3,4] or as an operational platform [5,6].

4. Agent-Based Modeling and Methodology

As witnessed by the following papers, modeling complex systems of any sort within heterogeneous scientific disciplines is a staple in MAS application, either for observing such systems to devise out properties, patterns, and laws, or for crafting them in compliance with agent-oriented methodologies so as to obtain MAS non-functional properties—decentralization, reactiveness to change, etc.

In [7], an MAS is used in the context of social sciences to model residents in a smart city so as to study their social engagement during time (e.g., daylife vs. nightlife) and across space (city center vs. business district). The idea behind such a modeling is that activity of the residents are influenced by what others are doing and by environmental conditions, such as the presence of shops, events, etc., hence the social and situated nature of agents in an MAS is a perfect fit. Based on this modeling, the authors study various aspects of social and institutional engagement, such as mutual trust and trust in institutions.

The application context of [8] is instead totally different, as it deals with observation of emergent properties, in particular scale-free features, for robotic systems implementing swarming behaviors, such as collective foraging. The authors aim at testing whether scale-free attributes may also arise in artificial collective systems inspired to biological ones, such as ant colonies, and then whether such attributes have positive influence on the overall system performance. In such a context, the agent abstraction is particularly useful while modeling individual behavior of robots, which depends on environmental conditions (situatedness) and peers actions (sociality).

In [9], we are introduced to yet another research field exploiting the expressive power of the agent abstraction for modeling, while also considering a methodological perspective: psychology, in particular, educational games design. The authors describe a design process for educational games which heavily relies on the agent abstraction for modeling both human behavior and the software system engaging players, for instance, the admissible actions at each stage of the game, their effect on the system or the player(s), and the modalities of interaction between players and with the software control system. To further consolidate the agent metaphor, the authors also consider virtual avatars representing players and system characters, so as to leverage on more natural interactions.

5. Agent-Oriented Programming and Simulation

Complementing the modeling aspect discussed in previous section, the two following works exploit agent-oriented programming to deliver software tools enabling design and deployment of MAS and agent-based simulations.

In [10], a model-driven approach is proposed to reconcile all the different existing organizational models meant to let MAS designers operationally define the social dimension in an MAS. Organizational models respond to the need of guaranteeing correctness of the overall MAS behavior despite individual agents are autonomous entities, hence, as such, are able to choose their own course of actions in isolation—and while pursuing their own individual goals. Through these models and the corresponding software tools, MAS designers have ways of specifying co-operation protocols amongst agents, taming their individual behaviors and steering them towards a coherent system-level goal.

In [11], the focus of the contribution and the main novelty regard seamless deployment on simulated and production environments, with little modifications as possible to the agent logic. The authors propose a coherent and integrated Python development framework encompassing testing, simulation, validation, and deployment software production stages, as well as autonomy, reactiveness to environment events, and social ability facets of an MAS. The proposed framework, ARPS, revolves around a few crucial architectural components: the agent manager, agents themselves, a discovery service, and a discrete events simulator. Facilities for dealing with sensing and actuating in either simulated or physical environments are made available, and agent behavior as well as social interactions can be defined through policies dictating which actions correspond to which event.

Both the aforementioned contributions aim at providing general-purpose agent-based solutions to let other developers build their own MAS.

6. MAS as Execution Infrastructure

The last usage destination—that is, exploiting an MAS as the execution infrastructure for a given system—is quite common in MAS literature, as the agent abstraction is a general-purpose programming concept with applications in many business domains and for heterogeneous systems.

In [12], an MAS is used in the context of multi-robot formation: first, a distributed consensus algorithm is simulated on a multi-agent based simulation software to assess desired properties despite uncertainty of data and delay in communications, then such algorithm is implemented as an MAS and deployed on a real robotic platform comprising four mobile robots, further assessing effectiveness. In this work, the value added of the MAS lies in its natural predisposition to distribution and tight coupling with environment sensing and actuation, which are necessary features of multi-robot systems.

In [13], instead, an MAS is used as the platform for training unmanned surface vehicles: agents in the MAS correspond to vehicles' controllers and implement a distributed learning algorithm meant to achieve optimal coordinated behavior. Here, the agent abstraction is chosen for its capability to express adaptive behavior by learning new behavioral rules (likewise plans in BDI architectures) while operating.

Both contributions showcase the ability of MAS architectures to provide a suitable infrastructure for effective and efficient execution of heterogeneous tasks (consensus in the former, learning in the latter).

7. Conclusions

The large number of submissions to this second installment of the MAS special issue has made it clear that there is still a huge space that initiatives of this sort can help covering. In addition, the quality of the papers collected and published here testifies the effort that the scientific community is devoting to the development of novel MAS models, techniques, and methods. The breadth of the MAS-related topics faced by submitted papers (which for obvious reasons cannot be fully analyzed here) also witness the increasingly expanding reach of agent-based techniques and solutions.

This is mostly why this special issue on the one hand provides readers with a very representative picture of the state-of-the-art of MAS research, on the other hand is far from being conclusive under any possible viewpoint. The articulation and expansion of the MAS field leave the space open for many other initiatives like this special issue—so we expect to see many more of them in the next few years.

In the meanwhile, we are quite confident that the readers of Applied Intelligence will be able to understand the extent of the application scenarios that MAS are going to cover in the next decades, as they become the conceptual and technical foundation for the next generation of complex intelligent systems.

Author Contributions: Conceptualization, S.M. and A.O.; methodology, S.M.; software, S.M.; validation, A.O.; writing–original draft preparation, S.M.; writing–review and editing, A.O.; visualization, S.M. All authors have read and agreed to the published version of the manuscript.

Funding: This research received no external funding.

Acknowledgments: The guest editors would like to thank the Applied Sciences Editorial Office, in particular the reference contact Daria Shi, for the extreme efficiency and attention devoted to the handling of papers, from submission to publication, through the peer review process. We would also like to thank the many reviewers participating in the selection process (3 to 4 on average) for their valuable constructive criticism, often appreciated by the authors themselves. Last but not least, our gratitude goes to the authors who submitted their papers, and to the many readers who already generated citations and downloads.

Conflicts of Interest: The authors declare no conflict of interest.

References

1. Omicini, A. SODA: Societies and Infrastructures in the Analysis and Design of Agent-based Systems. In *Agent-Oriented Software Engineering*; Ciancarini, P.; Wooldridge, M.J., Eds.; Springer: Berlin/Heidelberg, Germany, 2001; Volume 1957, pp. 185–193. [CrossRef]
2. Mariani, S.; Omicini, A. Special Issue "Multi-Agent Systems": Editorial. *Appl. Sci.* **2019**, *9*, 954. doi:10.3390/app9050954. [CrossRef]
3. Borghesi, A.; Milano, M. Merging Observed and Self-Reported Behaviour in Agent-Based Simulation: A Case Study on Photovoltaic Adoption. *Appl. Sci.* **2019**, *9*, 2098. doi:10.3390/app9102098. [CrossRef]
4. Lim, M.C.; Choi, H.L. Improving Computational Efficiency in Crowded Task Allocation Games with Coupled Constraints. *Appl. Sci.* **2019**, *9*, 2117. doi:10.3390/app9102117. [CrossRef]
5. Grimes, S.; Breen, D.E. Woc-Bots: An Agent-Based Approach to Decision-Making. *Appl. Sci.* **2019**, *9*, 4653. doi:10.3390/app9214653. [CrossRef]
6. Pérez-Salazar, M.; Aguilar-Lasserre, A.; Cedillo-Campos, M.; Posada-Gómez, R.; del Moral-Argumedo, M.; Hernández-González, J. An Agent-Based Model Driven Decision Support System for Reactive Aggregate Production Scheduling in the Green Coffee Supply Chain. *Appl. Sci.* **2019**, *9*, 4903. doi:10.3390/app9224903. [CrossRef]
7. Olszewski, R.; Pałka, P.; Turek, A.; Kietlińska, B.; Płatkowski, T.; Borkowski, M. Spatiotemporal Modeling of the Smart City Residents' Activity with Multi-Agent Systems. *Appl. Sci.* **2019**, *9*, 2059. doi:10.3390/app9102059. [CrossRef]
8. Rausch, I.; Khaluf, Y.; Simoens, P. Scale-Free Features in Collective Robot Foraging. *Appl. Sci.* **2019**, *9*, 2667. doi:10.3390/app9132667. [CrossRef]
9. Ponticorvo, M.; Dell'Aquila, E.; Marocco, D.; Miglino, O. Situated Psychological Agents: A Methodology for Educational Games. *Appl. Sci.* **2019**, *9*, 4887. doi:10.3390/app9224887. [CrossRef]
10. Coutinho, L.R.; Brandão, A.A.F.; Boissier, O.; Sichman, J.S. Towards Agent Organizations Interoperability: A Model Driven Engineering Approach. *Appl. Sci.* **2019**, *9*, 2420. doi:10.3390/app9122420. [CrossRef]
11. Prado, C.; Bauer. ARPS: A Framework for Development, Simulation, Evaluation, and Deployment of Multi-Agent Systems. *Appl. Sci.* **2019**, *9*, 4483. doi:10.3390/app9214483. [CrossRef]
12. Wei, H.; Lv, Q.; Duo, N.; Wang, G.; Liang, B. Consensus Algorithms Based Multi-Robot Formation Control under Noise and Time Delay Conditions. *Appl. Sci.* **2019**, *9*, 1004. doi:10.3390/app9051004. [CrossRef]
13. Han, W.; Zhang, B.; Wang, Q.; Luo, J.; Ran, W.; Xu, Y. A Multi-Agent Based Intelligent Training System for Unmanned Surface Vehicles. *Appl. Sci.* **2019**, *9*, 1089. doi:10.3390/app9061089. [CrossRef]

Article

Merging Observed and Self-Reported Behaviour in Agent-Based Simulation: A Case Study on Photovoltaic Adoption

Andrea Borghesi * and Michela Milano

DISI, University of Bologna, 40136 Bologna, Italy; michela.milano@unibo.it

* Correspondence: andrea.borghesi3@unibo.it

Received: 2 April 2019; Accepted: 15 May 2019; Published: 22 May 2019

Abstract: Designing and evaluating energy policies is a difficult challenge because the energy sector is a complex system that cannot be adequately understood without using models merging economic, social and individual perspectives. Appropriate models allow policy makers to assess the impact of policy measures, satisfy strategic objectives and develop sustainable policies. Often the implementation of a policy cannot be directly enforced by governments, but falls back to many stakeholders, such as private citizens and enterprises. We propose to integrate two basic cornerstones to devise realistic models: the self-reported behaviour, derived from surveys, and the observed behaviour, from historical data. The self-reported behaviour enables the identification of drivers and barriers pushing or limiting people in their decision making process, while the observed behaviour is used to tune these drivers/barriers in a model. We test our methodology on a case-study: the adoption of photovoltaic panels among private citizens in the Emilia–Romagna region, Italy. We propose an agent-based model devised using self-reported data and then empirically tuned using historical data. The results reveal that our model can predict with great accuracy the photovoltaic (PV) adoption rate and thus support the energy policy-making process.

Keywords: simulation model; multi-agent systems; photovoltaic energy; parameter fine-tuning; self-reported behaviour; predictive model

---

1. Introduction

The European Union is deeply committed to curtailing its greenhouse gas emissions by at least 20% by 2020, w.r.t. 1990 levels, as stated in the sustainable growth strategy outlined in [1]. The path to achieve such a goal passes through an increase up to 20% of the share of renewable energy sources in final energy consumption and a 20% rise in energy efficiency. All EU members and regions should put an effort in this direction to contribute to these common objectives. For instance, Italy was supposed to reach a 17% share of final energy coming from renewable sources in 2014, a target that have been reached and slightly surpassed [2].

The complex task of enforcing these guidelines is shouldered by national and regional policy makers. Energy policies have a strong impact on sustainable development and they influence economy, society and environment. Policy makers have to devise plans targeting strategic objectives, e.g., cutting greenhouse emissions, with the goal of satisfying different constraints (i.e., limiting pollutant emissions, not exceeding a financial budget, etc.) and respect the EU guidelines. After having been devised, the plans need to be enforced with implementation instruments (from incentives to investment grants, passing through tax exemptions) [3–5]. One aspect that tends to be severely underestimated while planning energy policies is the strong influence of human behaviour together with social dynamics; it is often studied with the assumption that consumers are rational and guided only by financial and economic drivers [6–8] which severely affect the accuracy and realism of the study. In fact, the decision process of involved agents (i.e., private citizens) is deeply influenced by non-economical motivations, such as social influence, peer pressure, bandwagon effects, lack or wealth of knowledge, risk aversion, etc. [9–12]. Properly understanding the decision making process is critical to better influence the interested parties' behaviour and steering them toward good practices and policy objectives. In this context there is urgent need of appropriate and accurate models for enabling policy makers to design, evaluate and implement energy policies to satisfy strategic objectives and develop sustainable strategies that have a strong impact on economy, society and environment.

We propose to merge in the model definition two types of knowledge: (1) self-reported behaviour derived from large scale surveys and interviews, and (2) observed behaviour based on real data measuring the actual effect of the target energy policy. The models are used to bridge the gap between these two behaviours and enable a better understanding of private citizens decision-making processes. We claim that social and economic drivers and barriers can be extracted from quantitative analysis of survey data, whilst a deeper understanding of how these drivers operate and interact can be derived from interview findings. On the basis of these drivers and barriers, we build a parametric model, whose parameters can be empirically tuned so that the model reproduces the observed behaviour. We expect the parameter tuning to generate different outputs (i.e., parameter values for different drivers and barriers) for different private entities classes (private citizens, enterprises, etc.) and for different countries and geographical situations.

The final outcome of merging self-reported and observed behaviour is the creation of predictive models with the ability to forecast the stakeholder behaviour in the presence of specific energy policies, financial and economic situations. These predictive models can be inserted into simulations and used by policy makers in a what-if fashion, namely by proposing alternative scenarios and observing the emerging behaviour of consumers related to energy efficiency and overall cost.

In this work we focus on policies for promoting energy production from renewable energy sources and, in particular on photovoltaic (referred to as PV) power generation. We use, as a case study, self-reported and observed behaviour in the Italian region of Emilia–Romagna, where the majority of the total installed photovoltaic power is generated by small/medium panels installed by private citizens and enterprises. For this reason the regional policy makers cannot directly decide the total power installed, but they have to push the PV power generation through indirect means, usually in the form of incentives to the PV energy. The decision to install a PV panel is not driven exclusively by economical/technical considerations (although these aspects have clearly a significant impact), but it involves also different factors determined by the human behaviour and social interactions [13,14]. As observed behaviour, we employ the data regarding the historical yearly installation rate of new photovoltaic panels and the total amount of installed photovoltaic power reported by the national and regional governments. On these data we craft an agent-based model for simulating the adoption of photovoltaic panels. We consider individual households as the actors populating the simulation environment and deciding whether to install a PV panel or not. The behavioural rules of the agents are devised using self-reported data collected thanks to surveys and questionnaires conducted among private citizens. From these data we derive the drivers and the barriers that influence the adoption of a PV panel. The importance of each factor is decided during the following phase, when we use the observed data in order to fine-tune the parameters of the model. The model takes into account both geographical, economical and social aspects.

The validation and final evaluation of the proposed model has been performed over a period of 11 years by comparing the historical PV power installation trend in a certain period to the one generated by the agent-based simulators. The historical data collected over this time span is divided in two sub sets in order to achieve a two-fold purpose: (I) tune the agent-based model's parameters (combination of self-reported and observed behaviour) and (II) test the accuracy of the approach by assessing its predictive capacity. For this purpose, the first seven years were used for parameter tuning and for the remaining years we compare the historical data with the simulated behavior—a small discrepancy would mean a good accuracy of the model, otherwise, a large gap would indicate a model not really usable. The experimental results highlight that with our model it was possible to predict future trend of installed PV power; this information and predictive capability can greatly help policy makers in their task.

The structure of the paper is the following. In Section 2 related works are discussed. Section 3 provides a general overview of the proposed approach. Section 4 presents the surveys used to identify drivers and barriers governing people's decisions and the method to derive the model for the agents' behaviour. Then, Section 5 describes the proposed agent-based model. The method used for tuning the model's parameters is described in Section 6. Section 7 reports the evaluation of the proposed approach, validating the fine-tuned agent-based model and assessing its accuracy. Finally, Section 8 concludes the paper, summarizing the obtained results and suggesting future research directions.

2. Related Work

The adoption of renewable energy sources, such as photovoltaic panels, can be framed as an innovation diffusion problem, an issue that has been the subject of many research works. Several findings suggest that the diffusion of an innovation is a social process. A common methodology to deal with this problem is agent-based modeling and simulation, where the agents are connected to form a interconnected network; agent-based models are also referred in the literature (and in the rest of this paper) as ABMs. Agent-based modeling is a computational approach that provides a tool for researcher with the purpose of creating, analysing and experimenting with models composed of agents that interact within an environment. Agent-based models are a simplified representation of the reality that can be used to explore certain aspects that would be harder to study without the aid of computational experiments [15]. Agents are usually distinct parts of a program that are used to represent social actors/individuals, organizations such as firms and enterprises, or bodies such as nation-states. They are programmed to react to the computational environment where they reside; this "simulated" environment is a representation of the real environment where the social actors operate [16].

In particular, ABMs have been used to study how innovative technologies spread in the real world [17–20]. It has been noted that the adoption rate of innovation does not depend exclusively on economic factors (i.e., costs or available budget), but many other aspects can have a profound influence. For instance, Abrahamson et al. [21] describe a threshold ABM where the adoption rate of a new technology is influenced by the "bandwagon effect", with new adopters facilitating the spread of knowledge that in turn increases the adoption of the innovative technology by new agents. Similarly, Chatterjee et al. [22] consider that potential adopters can have precise information about the cost of a innovative technology but can only estimate its benefits and real value—hence the perceived worthiness is an important factor. The main idea is that the information about an innovative technology spreads among an increasing network of agent through communication with previous adopters—in this way the uncertainty about the innovation potential decreases. The PV technology diffusion can be cast as a problem of innovation adoption, hence these insights will be partially incorporated in the model proposed in this paper.

Extensive research has been devoted to investigating the PV technology via ABMs [23–26], with a special focus on rooftop PV panels—systems that are typically installed by private citizens and small enterprises. The rest of this section reviews some of the approaches proposed in the last few years.

Zhao et al. [27] describe an ABM for studying the diffusion of PV systems where agents are homeowners which decide whether to install a PV panel or not. They consider four main factors that affect the agents' decision: payback period, household income, neighbourhood and advertisement. The final decision of each agent is based on a linear combination of these four factors, called "desire level". If the desire level is above a certain threshold, the household will opt for installing a PV panel. Selecting the correct value for the threshold is not an easy task: it strongly impacts the decision algorithm of the agents but the authors do not offer a general method to compute it. Instead, the domain experts' knowledge is used to select a set of realistic values, which have to be tested and validated on test-case scenarios (without comparison with historical trends). With our approach we aim at finding the correct values for the ABM parameters through the fine-tuning process guided with observed data.

Extending the work of Zhao et al., Palmer et al. propose a different ABM [28]: again, the agent decision criterion is based on four different factors, but in this case these factors are weighted according to the agents' social class. Each agent (corresponding to a household) is associated to a specific social class; a small-world network connects all agents and households belonging to the same classes tend to be linked together. The model parameters are calibrated using the PV installation trend in Italy during the 2006–2011 period but all the data set is using for the training, therefore no validation is performed using new data. This poses a risk of overfitting the model parameters to the particular historical period taken into consideration. The risk is also increased due to the selected period: in 2006–2011 the PV installation rate was mainly governed by a set of incentives offered by the Italian government that changed considerably in 2012, as was described in detail by Borghesi et al. in [29].

The approaches listed so far discounted the geographical location of buildings. Robinson et al. [30] introduce this new element and integrate data coming from a geographic information system (GIS) with an ABM, with the goal of analysing the diffusion rate of PV panels. The addition of the actual topology of the target area permits to include the effects of solar exposure and population density on the diffusion of PV systems, thus improving the accuracy of the model. The parameters calibration is done using real data of the historical PV adoption in the city of Austin, Texas. While the results are interesting, no validation has been performed yet, i.e., all the data has been used to fine-tune the model, whose accuracy w.r.t. new observed behaviour has not been computed. Davidson et al. [31] take into account geo-spatial information as well as population demographics in order to forecast the photovoltaic adoption trends. Their goal consists in understanding the best predictors for the installation of PV panels. The analysis highlights that a relatively small subset of geo-spatial data can be used to obtain estimates (in terms of PV adoption trend) as accurate as those obtained with much larger and more comprehensive geo-spatial data sets. This work does not develop an agent-based model and it is mostly focused on understanding which are the geo-spatial factors with the major impact on the PV adoption and thus is not directly comparable with the one proposed in this paper.

Zhang et al. [32] outline an ABM to study the adoption (both at individual and community level) of rooftop PV panels, considering the San Diego county as a case study. The key point of the proposed approach is to learn a model for the behavior of individual agents using combined data of individual adoption characteristics and property assessment; the learned model is then integrated in the agents involved in the simulation. They also employ their system to evaluate different policy strategies targeted at fostering PV adoption. The proposed model is calibrated using observed PV adoption rates in the city San Diego, California. The authors also propose a preliminary validation of their model, comparing its prediction accuracy to the accuracy of baseline model (a model taking into account fewer factors than the presented one). The validation lacks a full comparison of the model predicted behavior and the observed one.

Macal et al. [33] describe an agent-based model (called BE-Solar), that incorporates a social and behavioral decision framework for understanding the technology adoption process. The main limitation of this approach is the lack of model calibration and validation. Rai et al. [34] present a empirically-driven agent-based model of technology adoption applied to residential solar photovoltaic. The variables describing the agents' behaviour are fine-tuned using historical data. In their work, they propose a theoretically-based framework and consider multiple validation criteria.

Agent-based models have also been employed to simulate policy scenarios and provide recommendations. For example, Lee et al. [35] proposed an ABM to model the decision-making process of homeowners while buying and installing energy efficient technologies in their homes. Homeowners' decisions are based on a simple additive weighting algorithm that estimates the utility values of different options, ultimately selecting the one with the maximum utility value. The utility values of different options are calculated based on a combination of empirical factors (derived from housing stock data), social factors, and policy regulations. Installations lead to altered energy demand and CO_2 emissions. The model was partially calibrated using observed data; due to the limited availability of historical data only a couple of technologies were subjected to calibration. Although this is a clear limitation, validation was not the main scope of the paper that instead was mostly focused on providing a tool for comparing different scenarios.

Johnson et al. [36] model households photovoltaic solar panels adoption following an approach where household agents initially make decisions based on their subjective beliefs, awareness, and attitude towards the technology. These factors determine the chance the homeowners meet with a photovoltaic installation company, at which point they become rational profit-maximising consumers, weighing up the costs and benefits (subsidies etc.) of installing solar panels. This model enabled the researchers to make recommendations to regional government on the potential impact of incentive policies, and how different policies compared in terms of costs, energy capacity installed, and participation rates. No validation technique for this model was proposed, as this work is more focused on studying the theoretical impact of different policies, rather than providing a predictive tool.

Adepetu et al. [37] employ an ABM to study the impact of realistic incentive mechanisms on the adoption and diffusion of PV-batteries. They observe that while many different types of incentives have been proposed in the last few years, those incentives did not have the same effect in different parts of the world, due to the underlying different conditions and contexts (referred to as jurisdictions). Then, in their work they propose jurisdiction-specific ABMs in order to find the best incentive policies. As a case study they consider two distinct jurisdictions, Ontario and Germany. The agents' decision process is partially based on questionnaires (self-reported behaviour) and the tuning of some of the model's parameters is performed using historical data. The proposed approach is interesting but it is explicitly focused on PV-batteries and do not consider PV panels on rooftops, hence it is not directly comparable to ours; moreover, the authors validate the models using historical data (looking for parameters values that lead to the best fit with observed data) but they do not evaluate the predictive capacity of the proposed ABMs—the main focus is the comparison of incentive policies among different jurisdictions. Furthermore, the proposed paper do not provide statistical, quantitative measures of the accuracy of the fitting method, hence the comparison is not possible.

Sinitskaya et al. [38] describe an agent-based model to explore the impact of installers of PV panels on the adoption rate of PV technology among residential households. PV installers and PV household buyers are modeled as agents and the explicit goal is to maximise the profits for panels installers—hence their work is not directly comparable to our methodology. However, a shared aspect is the use of interviews with the involved stakeholders (investors and homeowners) in order to devise the proper (realistic) decision algorithm of the model's agents.

Lee et al. [39] propose to combine ABMs with a logistic regression model to estimate the correct values for the model parameters. They apply this hybrid scheme to the case study of rooftop PV panels adoption in a neighborhood located in Seoul, South Korea. The agents in the ABM are buildings placed in a geographically accurate simulated world, thanks to a geographical information system (GIS). The house owners of the corresponding building decide whether to install a PV panel or not, depending on multiple factors (economical, social, geographic). The parameters of each agent are tuned using logistic regression and very fine-grained building-level data collected during multiple years. The validation is performed using the cumulative observed data (sum of all adopted PV systems in the neighbourhood). In this way, this work uses the observed behaviour to obtain realistic models (the logistic regression is guided by historical data), although the decision algorithm of the agents is not based on self-reported behavior. Moreover, the validation is performed on the same time period used for tuning the parameters (albeit at different granularity), and thus no indication is provided on the predictive capacity of the approach.

While there are recent works that strive to bridge the gap between self-reported and observed behavior (for instance [37,39]), they do not explicitly frame the problem in these terms, thus they only consider a partial aspect of the overall issue. For example, Lee et al. [39] do not consider self-reported behaviour. Furthermore, the majority of the approaches terminate their analysis at the first step of the validation phase: the ABM parameters are tuned using historical, observed data but no study on the prediction capability of the model is performed. In this way, these approaches are proven to be well suited for describing the observed data (a worthy task), but no guarantee is given about the usefulness for predicting future trends—which is a very important aspect for policy makers. On the contrary, our approach advances the state of the art in two ways: (I) it explicitly states the need of considering both self-reported and observed behaviour, as only via merging them it is possible to obtain accurate ABMs; (II) it is validated on a predictive task, that is the model parameters are tuned using a subset of the observed data while the test is performed using a separate subset (a different time period).

3. Methodology Overview

This section provides an overview of the proposed methodology. The rest of the paper is devoted to describe how the proposed approach has been applied to the photovoltaic adoption in the Emilia–Romagna region. Figure 1 depicts the scheme of the methodology. The main idea was to start from the self-reported behaviour (collected through questionnaires and interviews) and extract drivers and barriers influencing the stakeholders' decision process. These decision factors were encoded in an agent-based model via a set of parameters—the relative values of the parameters indicated the magnitude of the impact caused by the associated factor. This ABM was a template for the decision process inferred from the extracted drivers and barriers.

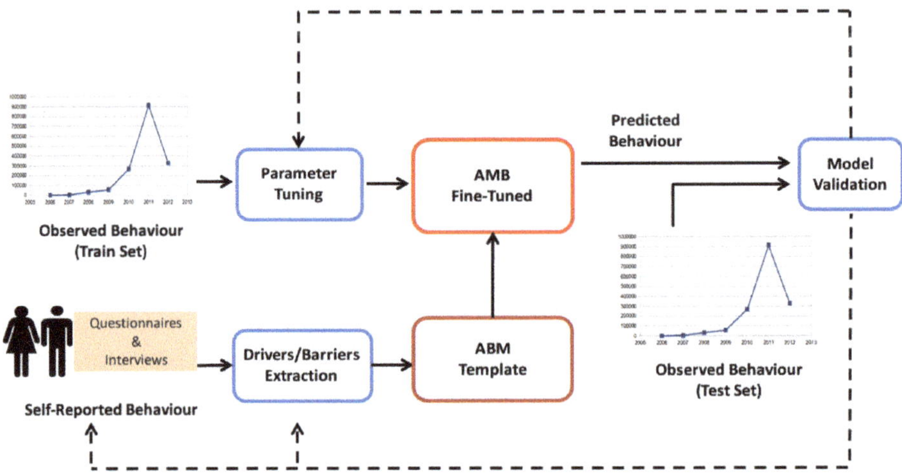

Figure 1. Methodology scheme.

A key part of an ABM was assigning a value to the model parameters. Since it was very hard to deduce appropriate parameter values only through self-reported behaviour, the model template underwent a fine-tuning phase where the parameters are empirically tuned using part of the observed behaviour. The observed behaviour was a time series describing the emergent overall behaviour that we want to simulate. Adopting standard machine learning terminology, we partitioned the observed behaviour into training and test sets. We trained the model (changing the parameters' weights) to make the output of the simulator as close as possible to the training observed behaviour and we validated it on the test set.

After the fine-tuning (also referred as calibration in the literature) the final outcome was a simulator that is able to predict with great accuracy the behaviour of interest. Now, the predicted behaviour and the test set (a subset of the observed behaviour) can be used to assess the quality of the approach; we defined this phase as model validation. The model validation stage can have different outcomes, depending on the accuracy of the predicted behaviour measured using the validation error (the difference between the predicted and the observed behaviour). If the accuracy was not high enough, it was possible to recalibrate the ABM parameters, for example using a different training set or improving the calibration method itself (i.e., letting the fine-tuning algorithms run for a longer time, thus exploring more parameters configurations). If the validation error is too big, a simple recalibration would probably not suffice; in this case the ABM template should be revisited in order to better encapsulate the drivers and barriers previously identified. If the validation results are very poor the best recourse would be a major overhaul of the ABM, possibly repeating the questionnaires and interviews phase with different questions, in order to get a deeper grasp of the agents decisional process. This second refinement should be performed by domain experts and it is out of the scope of this paper.

A key element of the proposed approach was the combination of agent-based models with automated fine-tuning techniques, to calibrate/validate the models. This way led to a method capable of filling the gap between observed and predicted data. We therefore merged methodologies belonging to both agent-based modelling and automated parameter tuning, moving towards a research area that has been very rarely explored so far. In our case, applying an automating tuning mechanism means exploring the configuration space of the ABM with the goal of finding the parameters with the best performance, that is those parameters that minimize the distance between the predicted behaviour and the observed one. As shown in Section 7, an ABM validated using historical data can be also used for predictive tasks with good accuracy, hence helping policy makers devising policies and strategies.

4. Driver and Barrier Extraction

The first challenge that needs to be addressed consists of the definition of the agents behaviour. The factors (drivers and barriers) influencing the decision-making process of the actors of the simulation model need to be identified. Drivers are those elements that lead to making a specific energy efficiency investment, while barriers are those elements against the investment. These factor were extracted from a large set of empirical data, gathered through questionnaires and interviews conducted in the Emilia–Romagna region as part of the ePolicy European research project [40]. The empirical data collection took place between March and August 2013. In this section we provide a partial overview of the methodology and the results of the surveys; for the complete and detailed discussion we refer to the work of Balke and Gilbert [41].

The tools used to collect the data are the following:

1. an online questionnaire concentrating on the general attitudes towards photovoltaic and its adoption (196 questionnaires were completed);
2. semi-structured interviews with both apartment block caretakers (building administrators) and employees of photovoltaic installation companies (11 interviews of average length between 60 and 90 min).

The results of both questionnaires and interviews indicated that a series of economic and social elements come into play when deciding to install a PV panel. Some of the factors involved in the decision process were purely economical (for example the profitability of the investment) or physical (such as the type of house and/or the availability of enough roof surface). Other elements reflected personal values and motivations (such as environmental awareness) and represented the social component of the decision process (for example the diffusion of PV-related information within peer networks). A very important aspect to be considered in the simulation is that many households pondering about PV installations are not able to make a clear costs-benefits calculation, but rather act on (other) perceived benefits of photovoltaic.

Some of the drivers identified are the following:

- high energy costs and potential saving after the PV panel installation,

- a general interest on renewable energy and sensitivity towards environmental concerns,
- the possibility to generate their own electricity,
- the knowledge regarding PV technologies (its costs, the incentives, etc.)—often acquired through social networks.

Conversely, these were some of the main barriers:

- misinformation about the cost of PV panels (often perceived as higher than the real one),
- uncertainty about the bureaucratic procedures,
- poor knowledge regarding the incentive mechanisms offered by government bodies,
- fiscal limitations (for example, the up front initial price of the panels),
- trust issues such as the stability of the price for solar-produced electricity in the long run,
- personal low risk propensity.

As illustrated in the following sections, the drivers and barriers identified via interviews and questionnaires form the basis of the decision model of each agent in our simulation. However, even when drivers and barriers have been identified, it is not clear how these parameters are weighted and which is their relative importance in understanding the decision making strategy of different house owners. For this purpose, the model parameters associated with the decision factors will undergo an empirical tuning phase exploiting the observed data, the historical installation rate of PV panels in the Emilia–Romagna region.

5. The Agent-Based Model

The process of extracting drivers and barriers from self-reported behaviours and embedding them in an agent-based model is not straightforward. Self-reported behaviour is not always easily quantifiable and, generally speaking, it cannot be used to directly infer a set of rules defining the decision algorithm of the simulated actors. In recent years, the authors of the current paper experimented with several ABMs with the purpose of better capturing the self-reported behaviour. In [42] a purely economic ABM was proposed, in order to understand the impact of national and regional incentives on the adoption of PV panels by residential homeowners. Each agent takes into considerations a series of economic factors that influence its decision to buy and install a PV panel. The model presented in [29] adds a preliminary social component: the behaviour of each agent is influenced by the decisions already taken by its neighbours and by the perception of PV technology possessed by each agent.

The knowledge gained with the previous works led to the methodology presented in this paper. For example, a very important lesson is that discounting non-economical factors generates ABMs fail to properly reflect the observed behaviour. The critical improvement of the currently proposed approach is the refinement of the agents decisional process and, most importantly, the fine-tuning strategy that lead to a model capable of predicting the future trends of PV panels installation rate. A preliminary version of the model discussed here (but without the optimal fine-tuning and predictive capability) was already presented in [43]. The model presented in this paper also adds an entirely new aspect compared to previous works, namely the adoption of real geographical data to obtain a more realistic ABM.

The model discussed in [43] was used mainly as a proof-of-concept of the proposed approach, namely it allowed to explore possible methods to merge self-reported behaviour (encapsulated in the decision algorithm of the agents in the model) and observed behavior (the real, historical data on PV installed power in the Emilia–Romagna region). The agent-based model was created and its parameters were tuned using methodologies described in the following section, but after the fine-tuning no validation was performed. In particular, the model was trained using the historical data gathered in the 2007–2013 time frame; no assessment of its predictive capacity was performed, in part due to a lack of sufficient observed data. Conversely, the current work provides a comprehensive evaluation of the overall proposed methodology and of the predictive capability of the approach. In the last years, additional observed data was collected (4 more years, 2014–2017) and used to properly validate the ABM, after the parameter tuning.

The model was composed of two types of agents: house owners and the region. The region agent provided regional incentives to house owners; at the start of the period there are some initial funds and each year the region receives a further constant budget to foster the installation of PV plants. The house owners (also referred to as households) are the main actors in the simulation: each house owner decides whether to install a PV panel or not, based on a decisional process illustrated in the following sections. Each house owner is described by a set of attributes: age class, education level, income, family size, consumption, roof area, budget, geographical coordinates and social class. These attributes cooperate to define the household behaviour and to build a social network linking all agents.

The ABM was built in two stages. Firstly, the simulation environment was set up and the virtual world was populated with agents (the households)—this was the configuration phase. The placement of houses followed the actual buildings distribution in the Emilia–Romagna region, in particular taking into account houses positions and their roofs. Then, the social network among household agents was built; the network was created depending on the reciprocal, physical distance between households and the distance in terms of attributes such as class, income, age, and so on. The social network's main contribution was defining how the information about PV system was spread across the simulated world and its agents.

After the configuration, a second stage takes place, called simulation phase in the rest of the paper. During this stage, the simulated world comes to life and the agents begin to ponder whether to install a PV panel or not. The simulation itself can be decomposed as a series of smaller steps, each of them lasting for six months. The installation decision was influenced by several factors, ranging from financial considerations such as the household income and the initial investment cost (and related payback time), to other aspects such as the environmental sensitivity and the neighbours behaviour (neighbours in the social network). The influence of these different factors was encapsulated in four expressions (also referred to as utility functions; these four expressions were then combined in order to establish the desire level of each agent—if the desire remains below a certain threshold, then the household does not install a PV panels, otherwise it proceeds with the investment.

5.1. Configuration Phase

The virtual environment initial conditions are defined during the configuration stage; moreover, the simulated world was populated with the agents, placing the buildings and assigning them a roof size according to the actual distribution observed in the Emilia–Romagna region (data made publicly available by the region itself). First, the world-area was filled with buildings and their roofs (fixed geographical coordinates). Secondly, the families-households were created (as many as specified by an input parameter) and each one was placed inside a building (buildings are not shared); households with higher income and the more numerous ones get the buildings with larger roofs.

The positions of the buildings in Emilia–Romagna were obtained by parsing the Ersi shape-files publicly available (http://dati.emilia-romagna.it), which are the results of territorial surveys conducted by the region; in these files each building was represented by a polygon encapsulating multiple information about the building. The agent-based model proposed here requires only position (spatial coordinates) and roof size, hence only these relevant information were extracted. QGIS [44] (an open source Geographic Information Systems, GIS) was the tool employed to parse the Ersi files and collect the needed information (position and roof size).

An important aspect that had a strong effect on the adoption of PV panels (and innovation in general) is the presence of incentive mechanism, aimed at fostering the diffusion of new (or less known) technology. From 2006 to 2014 the Italian government offered national incentives to private citizens willing to install PV panels, namely feed-in tariffs referred to as *Conto Energia*. There have been a few different tariff schemes during the incentives years ([45–48]), differing for the price guaranteed for the produced electricity. The national incentives are available to all house owners and the tariffs are those actually offered during the considered period. On top of the Italian feed-in tariffs, regional policy makers in Emilia–Romagna devised a number of different additional regional incentive mechanisms, such as investment grants, fiscal incentives, loans, interest funds, etc (see [42] for more details). Historically, the national incentives outweighed the regional ones by at least one order of magnitude therefore their influence has been much stronger [49,50], thus regional incentives have little or no impact on the agents behaviour.

5.1.1. Social Classes

The agent households were characterized with a set of attributes, whose values permitted to define the category of the family; these attributes were: income, size (number of components), education level, age class (average value among all components), number of earners, yearly energy consumption and social class. The attribute values assigned to the households followed the real distribution in the Emilia–Romagna region, obtained from the Survey on Household Income and Wealth (SHIW) provided by Bank of Italy (https://www.bancaditalia.it/statistiche/indcamp/bilfait/). On the basis of its attribute values, each agent possesses an associated budget for installing a PV panel. Clearly, the spending capability of a household is directly related to its income class, and this influences the price that each family is willing to pay for purchasing the PV panel. Households with higher income will accept to pay more for the initial investment, while a lower income was associated to a lower budget to invest in a PV system. Generally speaking, households that belonged to the same category (class) made similar decisions when deciding whether to install a PV panel. In practice, the proposed model assumed that this available budget of each family was determined by the attributes defining the family category. A linear regression model was then used to correlate the budget to the explanatory variables obtained from the SHIW data.

Households social classes serve to mimic the different adoption rates of innovative technologies observed in many scenarios [51]. For instance, the so called S-shape curve [52] has been widely used to describe the adoption of an innovation: initially, the adoption rate of a new technology is slow since it is not well understood and its benefits are unclear (or not fully perceived). In a second phase, the adoption rate rises together with the spread of the technology and associated knowledge (mass market phase). Beyond a certain point, the market gets saturated and the adoption rate flattens. Rogers [52] identifies five categories of different adopters: (1) innovators, (2) early adopters, (3) early majority, (4) late majority and (5) laggards. The different categories were usually reflected on characteristics defining each adopter, such as socio-economic status (i.e., high-income individuals can afford to invest more on new and not yet well-established technologies). Each house owner fell in one of the five adopters categories, depending on three of its attributes (the most important features): age class, education level and income. K-means was used to identify the five clusters and group the agents belonging to the same class.

5.1.2. The Social Network

As discussed previously, the behavior of a household-agent is significantly impacted by its social network. For this reason, during the configuration phase the social links between agents are created, namely each family has a set of friends (other households). Since previous research has shown (see [53–56]) that a small-world topology maps well the real network of relationships that exists between people, the social network adopted in this ABM has small-world properties. Small-world networks are characterised by a shortest-path distance between nodes that increases relatively slowly as a function of the number of nodes in the network [57].

The extended version of the rank-based model proposed by Liben–Nowell et al. [58] was used to get the small-world properties. The probability that a link between node u and node v existed was proportional to a ranking function which depended both on the geographical proximity of the nodes (physical neighbours) and on the attribute proximity of the nodes (how the nodes are similar w.r.t.

their attributes). After a network was built using the extended rank-based method, randomness was added through long-range links. These links drastically reduced the average path length because they connect distant parts of the network. The randomization process takes every edge and rewires it with an empirically obtained probability p.

5.2. Simulation Phase

In the simulation phase the system evolved as previously described: the decision regarding the installation of new panels took place between 2007 and 2013, then the simulator ran until 2036 to consider the lifetime of PV panels. As described at the beginning of Section 5, each agent had a particular desire level that encapsulates its willingness to invest in a PV panel. The desire level of each household is computed during the configuration phase (Section 5.1), depending on the agent's set of attributes and mathematically expressed as an utility function. The function is a weighted combination of different factors: household income, payback period (of the initial investment), perceived and expected environmental benefits, and the pressure from neighbours (as identified by the social network among agents). The weights were used to combine these factors depending to the household class, or category (see Section 5.1.1). The actual values of the weights cannot be analytically obtained and were instead tuned via model calibration, exploiting the real historical data about PV power installed in the Emilia-Romagna region in the 2007–2013 period (detail in Section 6).

The average lifetime of a PV system was 20 years; the expenses and gains cumulated during this lifespan served to estimate the return on equity (ROE) of a PV panel. The yearly cash flow was computed by subtracting the yearly total expenses from the yearly earnings—clearly considering only PV-related financial movements. The yearly expenses can be obtained by summing the cost of the system divided by its lifetime (mortgage payment), the maintenance costs and the interests on eventual loans. Potential yearly earnings comprised the surfeit electricity sold to the national electrical grid and the electricity bill savings granted by self-production. Alongside, there could be national and/or regional incentives, with a profound impact on the overall profitability of an investment. The incentives can influence yearly cash-flows in different ways: for instance, the gains are directly linked to the Italian national feed-in tariffs, while yearly expenses depend on the initial cost (affected by regional investment grants) and loan interests (target for several incentive schemes).

Each household had to find the optimal size for the PV system (the size that maximises the ROE); if the set of conditions characterizing an agent were unfavourable, the house owners can also opt not to install a PV panel. The problem was solved with an heuristic algorithm based on Simulated Annealing [59]. The proposed model assumed that households aimed at making well informed decisions, for example by getting advice from PV installers in order to properly understand the available options. Hence, agents are supposed to purchase PV panels with the goal of maximising their reward w.r.t. energy production and financial savings.

The Utility Function

One of the most important component of the proposed approach is the criterion used by agents/households to decide whether to install a PV panel. As mentioned earlier, this decision is taken by each agent based on the values of its attributes. The decision criterion for agent v is expressed by the utility function (also referred to as desire level):

$$U(v) = w_P(c_v)u_P(v) + w_B(c_v)u_B(v) + w_E(c_v)u_E(v) + w_N(c_v)u_N(v) \quad (1)$$

where c_v is the class of the agent (details in Section 5.1.1). The utility function is a weighted combination of four components:

1. the investment payback time $u_P(v)$, representing the expected payback period of the PV panel;
2. the available budget $u_B(v)$, that strongly impacts the possibility to make the initial investment (without considering external incentives);

3. the impact of the neighbours' choices $u_N(v)$;
4. the potential benefits generated by investing in a PV panel $u_E(v)$ (estimated in relation to a decrease in consumption of electrical energy from other non-renewable sources).

Each factor in Equation (1) was weighted by a class-dependent parameter; these weights are $w_P(c_v)$, $w_B(c_v)$, $w_N(c_v)$ and $w_E(c_v)$—the notation serves to express their dependency on the agent class c_v. Each agent had its set of specific weights; agents of the same social class do not share the same weights (at least not necessarily—this still could happen as a byproduct of the parameters tuning procedure). A key aspect of the proposed approach was assigning correct values to these weights: this is the crucial operation where the self-reported behaviour obtained with questionnaires and interviews (which guided the definition of the agents behaviour) was merged with the observed behaviour (historical data of installed PV power). This passage will be described in detail in Section 6.

The first factor in Equation (1) regards the payback period, pp. To obtain a balanced influence of all factors in the utility functions, all factors where normalized in a [0,1] range. In the case of the payback time, the normalization takes advantage of the bounds on the minimum payback period $min(pp)$ (assumed to be equal to one year) and on the maximum payback period $max(pp)$, assumed to be 21 years since the expected useful life for PV systems is 20 years. Hence, the payback influence for agent v is computed following [28] and expressed by this equation:

$$u_P(v) = \frac{max(pp) - pp(v)}{max(pp) - min(pp)} = \frac{21 - pp(v)}{20} \quad (2)$$

where $pp(v)$ is the payback period for the initial investment. Its value is computed using the net present value (NPV) of the PV system—the NPV typically starts with negative values (due to the initial cost of the investment) and it gradually gets closer to zero, while the initial cost is offset by yearly gains due to electrical bill savings and sale of own-produced energy. When the NPV turned from negative to positive it indicated the point when the investment became profitable. The computation of the NPV was based on the yearly cash flows—each agent measures its expenses and gains (taking into account also national and regional incentives) and computes its yearly NPV accordingly.

The household budget $u_B(v)$ is given by:

$$u_B(v) = \frac{e^{v_{budget}}}{v_{equity}} \quad (3)$$

with the initial investment v_{equity} computed as the PV panel installation cost minus any applicable incentive. v_B is the disposable budget of the household.

The third factor contributing to the agents' decision is the environmental benefit that can be gained by adopting PV technology, instead of consuming electrical energy coming from non-renewable resources. These benefits are measured in terms of oil saved, which is in turn correlated with an overall decrease in CO_2 production. The Italian Regulatory Authority for Electricity and Gas provides a factor to convert the produced energy (expressed in MWh) to the equivalent in tonnes of oil (TOE) (A TOE is defined as the amount of energy released by burning one tonne of oil, or 0.187 TOE for each MWh produced). Thanks to this conversion, the ecological benefits can be computed with the following equation:

$$u_E(v) = \frac{1}{e^{oil_{notConsumed} - oil_{consumed}}} \quad (4)$$

The final component of the desire function (Equation (1)) is the influence of the other members of the social network of the agent. This factor is identified by $u_N(v)$ and it encapsulates the importance of the neighbours' choices in shaping the household behaviour. As previously mentioned, the agent's neighbours are the nodes (other households) with a shared links; the vicinity of two nodes depends on geographical proximity and social class similarity. The neighbourhood influence contribution is computed with the following equation:

$$u_N(v) = \frac{1}{1 + e^{\frac{1}{2}L_{v,tot}L_{v,adopter}}} \quad (5)$$

with $L_{v,tot}$ being the total number of links of agent v and $L_{v,adopter}$ the number of links shared with adopters.

6. Parameter Tuning

So far, only the agent-based model has been described. The model has been built upon the insights gained by analysing the self-reported behaviour, gathered through questionnaires and interviews. By leveraging this information it was possible to identify barriers and drivers that affect the decision criterion regulating the installation of a PV panel (see Section 4). These factors were then used as a guide for the behaviour of the agents in the simulation world; however, the algorithm describing the agents' behaviour hinged on a set of parameters that cannot be easily obtained through analytical tools. At this point, the second stage of the proposed methodology came into play, namely the observed behaviour. Historical data will be used to fine-tune the model parameters, thus obtaining a model which is capable to faithfully describe real world dynamics and that can be used to make accurate predictions, as reported in Section 7.

The historical data of the PV panels installation trend in the Emilia–Romagna region was gathered looking at the data provided by the Italian government [49], in particular, the PV installation trend in Emilia-Romagna from 2007 to 2017 (Unluckily the data regarding years earlier than 2007 is very scarce due to the almost negligible consideration given by the Italian government to PV technology). The data set is divided in two chunks: training set, from 2007 to 2013, and test set, from 2014 to 2017. The training set is used to fine-tune the model parameters (trying to fit the simulated trend to the observed one). Afterwards, the trained model can be used to predict the PV installation rate during the test period; then, it is possible evaluate the quality of the prediction and thus the accuracy of the model, by comparing the historical data with the predicted one.

As a reminder, the parameters that needed to be tuned were the weights of the utility function: $w_{pp}(c_v)$, $w_B(c_v)$, $w_N(c_v)$, and $w_E(c_v)$. In practice, the scope was to find the weights values that better fit the curve representing the Emilia–Romagna PV power installation rate. The tuning problem can be seen as fitting a model to real data; there exist several methods to perform this task. After a preliminary evaluation of different methods, a genetic algorithm (GA) [60] came across as the technique that provided best results without requiring excessive computational resources. This happens because GAs are apt at finding solutions in spaces where it is hard to derive analytical models and it is not easy to mathematically find global optima. Another benefit derived from the use of a genetic algorithm is the fact that they have been proven to be very effective at dealing with problems where small changes in the weight configuration can lead to a great impact on the final outcome. This is the case of the proposed agent-based model: the four factors of the desire function are strongly intertwined and linearly combined (see Equation (1)). Moreover, the decision criterion is influenced also by the social interaction: the weights assigned to a given agent can modify its decision, which in turn has an impact on the decision process of other (possibly many) neighbours.

The genetic algorithm began with a random initial population of parameters configurations (the "individuals" in GA terminology). The initial population was then evaluated by running the agent-based model and observing the PV installed power by all households, given the currently applied parameters. Since during the training phase, the target PV power is available (the real, historical PV installation data in Emilia–Romagna), it was possible to assess the accuracy of the fit of the current population, by measuring the difference between target and simulated power. After the evaluation, the GA selected the next generation of individuals (a different set of model parameters), which were evaluated as well, with the goal of finding the best fitting population. To generate the next population the tournament selection [61] was employed: k individuals are selected from the actual population using n tournaments of j individuals. From every tournament emerges a winner (the individual with the highest fitness—the one generating the smallest distance between simulated and historical PV power) and this is the parameter configuration selected for the next generation.

The new population was not deterministically decided with the tournament mode, but a certain degree of randomness is introduced through crossover and offspring mutation. The former mechanism randomly chose two individuals for reproduction and one or multiple children were bred from them; in the proposed genetic algorithm one-point crossover has been used. A single crossover point on both parents' configuration is selected and a new child configuration is created via a swap of the values beyond the crossover point. The second random mechanism, mutation, works by randomly modifying values (parameters of the agent-based model) in randomly chosen individuals. The evolution process (creation of offspring, random mutations, evaluation) was repeated four hundred times.

7. Model Validation

After the parameter tuning via genetic algorithm described in the previous section, the resulting model accuracy needed to be measured. For this analysis, the ABM was composed by by 2000 agents; each simulation required around 10 s with a 2.40GHz Intel QuadCore (i7-5500U CPU) with a 16GB of RAM. The genetic algorithm used a population of 50 individuals and the overall time required to calibrate the model was around 30 h. As mentioned before the parameters tuning was made using observed data in the period 2007–2013; observed data in the 2014–2017 range was used only during testing. To summarize, the experimental setup was the following: (1) create an ABM and calibrate its parameters with the genetic algorithm; (2) simulate the 2007–2017 period with the fine-tuned ABM; (3) observe the simulated PV cumulative installed power and adoption rate—the difference between observed and simulated data in 2007–2013 measures the quality of the fine-tuning technique, while the difference measured in the 2014–2017 period serves to evaluate the predictive capability of the model. This scheme was repeated 30 times to obtain statistically significant values; in the rest of the paper, only mean values were reported (in both graphs and table). As metrics for the evaluation, we considered the mean absolute error (MAE), the root mean squared error (RMSE), the mean absolute percentage error (MAPE), and the coefficient of determination (R^2). These are standard metrics and are defined by the following equations (the equation for R^2 is not reported for the sake of clarity):

$$MAE = \frac{1}{N}\sum_{i=1}^{N} |o_i - s_i| \tag{6}$$

$$RMSE = \sqrt{\frac{1}{N}\sum_{i=1}^{N} (o_i - s_i)^2} \tag{7}$$

$$MAPE = \frac{100}{N}\sum_{i=1}^{N} \frac{|o_i - s_i|}{o_i}, \tag{8}$$

where N is the number of runs (30), s_i is the simulated value, for instance the PV power installed in one year in the simulated environment of the ABM, and o_i is the observed value.

We considered two outcomes to measure the fine-tuning and prediction results: (1) the yearly installed PV power and (2) the cumulative installed power, that is the total value obtained by summing previous years installed power and the current year's installed capacity. The former value better reflects the yearly changes in adoption rate while the latter could be more useful to policy makers to devise strategies for reaching long-term goals (i.e., a certain amount of total PV power by year 2020). Table 1 reports the validation results after the parameter tuning. Each row corresponds to a year; the table includes both the period used as training set (2007–2013) and the period used to evaluate the predictive capability (2014–2017), separated by a horizontal line. The last row reports the average values computed over the whole time frame 2007–2017. The first column corresponds to the year;

the following three columns report MAE, RMSE and MAPE for the yearly installed PV power; the last three columns show the same metrics computed on the cumulative installed power.

Table 1. Validation and prediction results of installed photovoltaic (PV) power; yearly rate and cumulative installations.

Year	*Yearly Installed Power*			*Cumulative Power*		
	MAE	RMSE	MAPE	MAE	RMSE	MAPE
2007	0.02	0.04	0.001	0.0003	0.0004	3.33
2008	0.53	0.62	15.41	0.006	0.008	12.93
2009	0.24	0.35	4.76	0.004	0.004	3.83
2010	0.37	0.35	3.03	0.006	0.007	0.26
2011	1.07	1.43	4.04	0.012	0.023	2.65
2012	0.3	0.35	1.39	0.011	0.018	1.58
2013	0.3	0.41	3.04	0.015	0.016	1.83
2014	0.27	0.21	4.05	0.013	0.02	1.43
2015	1.74	1.82	32.95	0.012	0.018	0.51
2016	0.44	0.64	7.55	0.0002	0.0012	0.07
2017	0.51	0.72	8.59	0.017	0.021	1.68
Average	0.52	0.69	7.71	0.008	0.009	2.73

Let us consider first the yearly adoption rate. Following the type of validation made by most other works in the literature, that was looking only at the results on the training set, it can be noted that the fit was extremely good. The average R^2 on the training set was equal to 0.997, the MAE was equal to 0.385, the RMSE was equal to 0.483, and the MAPE was equal to 4.52. These values indicate that the parameter tuning was very effective. A higher MAPE can be observed for the year 2008 (15.41); the simulated installed PV power was lower than the observed one and this could indicate that the current version of the ABM did not include some drivers that boosted the adoption of PV panels in early years. Moving on and considering also the predictive capability of the proposed approach, the results remain promising. However, it can be observed that the accuracy decreases in the test set, especially in 2015, where the simulated installed PV is significantly lower than the observed one; the proposed ABM overestimates w.r.t. the historical data. The statistical results obtained considering both the train and the test set were the following: R^2 = 0.991, MAE = 0.524, RMSE = 0.693, and MAPE = 7.71. Conversely, if we computed the average values only over the test period, these were the results: MAE = 0.896, RMSE = 1.16, and MAPE = 16.36. The precision was skewed especially by the large error made in 2015 (MAPE equal to 32.95). Probably this was due to the fact that the ABM underestimates the impact of the reduced national and regional incentives on the households decision process; as a reminder, national incentives by the Italian government ceased at the end of 2013.

In order to contextualize these values, the comparison with the validation results proposed by a recent related work could be considered, namely Lee et al. [39] (see Section 2 for details); this work was chosen because it provided easily comparable metrics and has a similar approach (tuning the parameters of a ABM using historical data). The comparison might not be completely fair since Lee et al. used different sets for parameter tuning and validation, respectively, building-based data and area-wide (summing the power of all PV panels installed in a neighbourhood) data while our ABM was trained and validated using region-wide yearly adoption data. However, the comparison is legitimate because we include the test set not used for the parameters calibration, thus increasing the difficulty of the task tackled by the proposed approach. Lee et al. proposed three different ABMs, with MAPE equal to 14.86, 13.57, and 62.52 (average value computed over all years). Our ABM instead has an average MAPE (considering both training set and test set) equal to 7.71, a significantly lower value. Moreover, it can be worth to notice that even the results on the test set alone (MAPE = 16.36) are similar to those obtained by Lee et al., although these numbers are not really comparable since they refer to entirely different tasks, pure validation (Lee et al.) and prediction (the approach proposed in this paper).

Figure 2 shows the PV installation growth rate in the considered time period (2007–2017). The solid blue line corresponds to the PV growth rate predicted by the agent-based model while the black dashed lines depicts the real installation trend in Emilia–Romagna. The x-axis displays the year and the y-axis reports the PV power growth in percentage. The figure contains the results for both the training set (period 2007–2013) and the test set (2014–2017). In this way it is possible to observe the quality of the proposed fitting mechanism, by observing the lines discrepancies in the training set. At the same time, the figure reveals that the ABM does not suffer from overfitting and the proposed methodology can be used to create models that generalize well and, consequently, that can be used for predictive purposes—by looking at the differences between the lines in the test set. The visual analysis revealed results similar to those presented with the quantitative analysis. In fact, it can be noted that the parameter tuning was very effective (in practice the two lines overlap in the training set) and that the predictive capability are good as well, albeit slightly less accurate. Focusing on the PV case study, there were also useful insights that can be gained by looking at the installation trends, both real and simulated. Both curves clearly indicated that the initial growth rate has been relatively slow (2007–2009), possibly motivated by an initial reluctance due to limited knowledge and doubts about the PV technology. As the knowledge gets more widespread and the Italian national incentives increased (higher feed-in tariffs were offered in the years 2010 and 2011), the installation growth increases steeply, with a distinct peak around 2011. After 2011, there was a steady decline in number of new PV panels installed, as the financial benefits for homeowners start to become smaller, due to a decrease in national incentives not compensated by regional incentives or by a sufficient decrease in the cost of the technology.

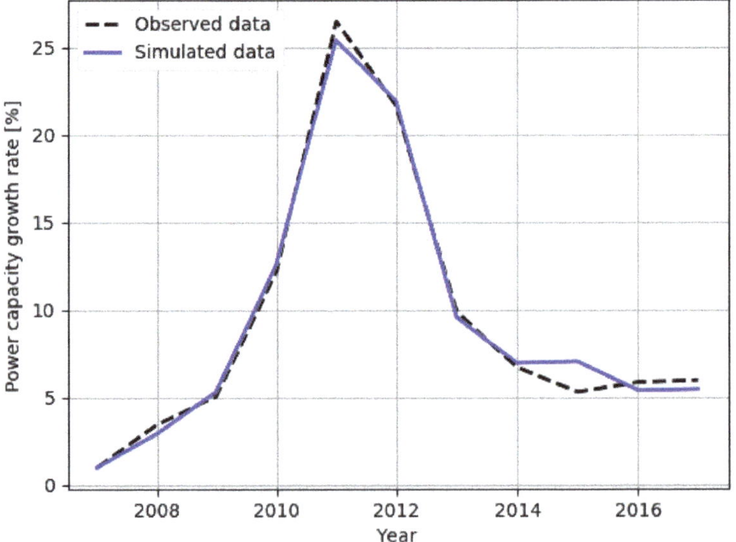

Figure 2. Model calibration results—yearly installed photovoltaic (PV) power.

This could be partially explained by the extremely complex situation that occurred in the Italian PV technology field in the last few years, with longstanding incentive mechanisms that abruptly came to an end and different regulations following one another. This situation generated a marked discontinuity in the installation trend, a discontinuity that was very hard to forecast. Clearly, there is still room for improvement in terms of ABM accuracy, but it is important to notice that the proposed approach can already emulate the observed behaviour with a precision more than sufficient to help policy makers in their decisions.

Now we look at the validation results with the cumulative installed power (last three columns of Table 1). The effect of the worse accuracy computed in the test set is amplified by the smaller magnitude of the installation rates observed in the 2014–2017 period w.r.t. the peak values observed in previous years. If we consider the total PV installed over the years the results of the parameters calibration are still very good and the "small-values" effect noticed in the test set loses its influence. Figure 3 shows the cumulative installed PV power in Emilia-Romagna in 2007–2017, both according to the historical data (black dashed line) and to the agent-based model (red continuous line). Since our simulator considers only 2000 agents against the millions of households in Emilia–Romagna, the historical and simulated absolute values differ by orders of magnitude; in order to render the comparison possible both sets (observed and simulated) were normalized dividing by the maximum value (year 2017).

As both the quantitative analysis and the graph reveal, the parameters tuning works even better for the cumulative installed power. In this case, the average values computed over the whole time frame were the following: $R^2 = 0.999$, MAE = 0.008, RMSE = 0.009, and MAPE = 2.73. The error was lower w.r.t. the case of the yearly PV installed power because the yearly changes in adoption rate are "smoothed" by the sum operation. This effect is even more pronounced when considering the test set alone. In this case the MAE is equal to 0.011, RMSE equal to 0.015 and MAPE equal to 0.92; the MAPE in particular was even lower than the training set case—this happened because in the last years (2014–2017) the PV panels installation rate has greatly slowed down, hence the errors made with the years in the test set have a relatively smaller impact on the total installed power.

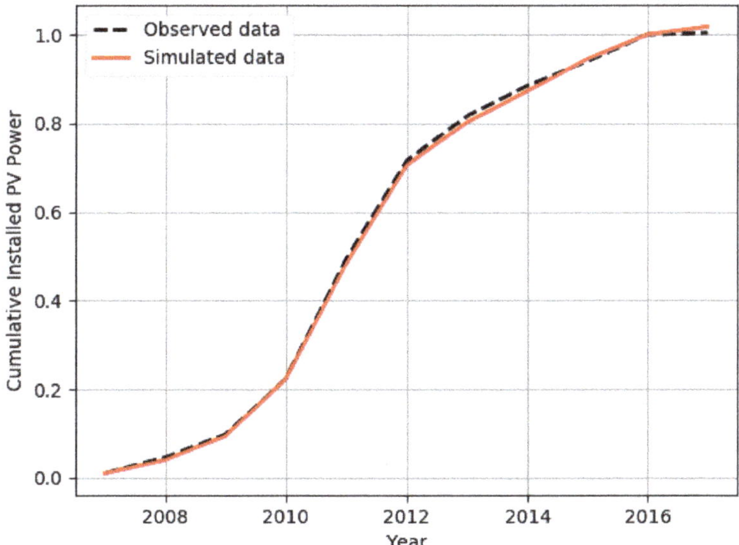

Figure 3. Model calibration results—normalized cumulative installed PV power.

8. Conclusions

In this paper we presented a novel methodology to fill the gap between self-reported behaviour and observed behaviour, by means of agent-based model and empirical parameter tuning. As a case study, we considered the diffusion of photovoltaic power in the Emilia–Romagna region of Italy. The first step of the approach consisted in a data collection phase to enable the identification of the drivers and the barriers that influence the decision making process of house owners faced with the possibility of installing a PV panel on the top of their houses. The data have been collected through online questionnaires and interviews.

These drivers and barriers were then used to model the decision process of the agents composing the simulation. Having both self-reported and observed behaviour, parameters attached to the various decision factors can be empirically tuned, thus enabling agent-based models to be used also for predictions, even in an approximate manner. The idea is that given the agent-based model based on the self-reported behaviour, its parameters can be adjusted and tuned exploiting past real data. Hence, a ABM that takes into account economic, social and geographical factors to emulate the self-reported behaviour has been proposed. The model is characterized by a set of parameters that were fine-tuned using a Genetic Algorithm. Finally, the accuracy of the model prediction has been evaluated, by analysing the difference between the historical PV installation rate and the results produced by the simulator. The results are very promising and the proposed approach can be used by policy makers to guide their decisions.

The future research directions that have yet to be explored are the following. First, the ABM can be refined in order to achieve a even greater prediction accuracy. Second, it is important to test the proposed methodology in different conditions (different region/countries, extended time period). Another possible direction to explore consists of scaling up the simulation size, up to the point of including hundreds of thousands of agents; this should lead to result even closer to the observed data. In our opinion, the most promising direction is to integrate the proposed approach and predictive model in a larger scheme aimed at helping policy makers with their task. After having bridged the gap between self-reported behaviour and observed behaviour, the following step would be to reach a target behaviour, i.e., the desired level of photovoltaic power production. For this purpose, the agent-based model could be used to extract the best guidelines for policy makers to achieve the desired strategic objectives.

Author Contributions: Conceptualization, A.B. and M.M.; data curation, A.B.; funding acquisition, M.M.; investigation, A.B.; Methodology, A.B.; Project administration, M.M.; resources, M.M.; software, A.B.; supervision, M.M.; validation, A.B.; writing—original draft, A.B.; writing—review & editing, A.B. and M.M.

Funding: EU ePolicy project (FP7/2007–2013), g.a. 288147.

Conflicts of Interest: The authors declare no conflict of interest.

References

1. Europe 2020 Strategy. Available online: https://ec.europa.eu/info/strategy/european-semester/framework/europe-2020-strategy_en (accessed on 20 May 2019).
2. Eurostat. Smarter, Greener, More Inclusive? Indicators to Support the Europe 2020 Strategy. 2016. Available online: http://ec.europa.eu/eurostat/documents/3217494/7566774/KS-EZ-16-001-EN-N.pdf/ac04885c-cfff-4f9c-9f30-c9337ba929aa (accessed on 20 May 2019).
3. Dasgupta, P.S.; Hammond, P.J.; Maskin, E.S. The Implementation of Social Choice Rules: Some General Results on Incentive Compatibility. *Rev. Econ. Stud.* **1979**, *46*, 185–216. [CrossRef]
4. Norberg-Bohm, V. Creating Incentives for Environmentally Enhancing Technological Change: Lessons From 30 Years of U.S. Energy Technology Policy. *Technol. Forecast. Soc. Chang.* **2000**, *65*, 125–148. [CrossRef]
5. Solangi, K.; Islam, M.; Saidur, R.; Rahim, N.; Fayaz, H. A review on global solar energy policy. *Renew. Sustain. Energy Rev.* **2011**, *15*, 2149–2163. [CrossRef]
6. Loulou, R.; Labriet, M. ETSAP-TIAM: The TIMES integrated assessment model Part I: Model structure. *Comput. Manag. Sci.* **2008**, *5*, 7–40. [CrossRef]
7. General Equilibrium Model for Economy–Energy–Environment. Available online: https://ec.europa.eu/clima/sites/clima/files/strategies/analysis/models/docs/gem_e3_long_en.pdf (accessed on 20 May 2019).
8. Lee, M.; Hong, T.; Jeong, K.; Kim, J. A bottom-up approach for estimating the economic potential of the rooftop solar photovoltaic system considering the spatial and temporal diversity. *Appl. Energy* **2018**, *232*, 640–656. [CrossRef]

9. Graziano, M. Spatial Patterns of Solar Photovoltaic System Adoption: The Influence of Neighbors and the Built Environment. *J. Econ. Geogr.* **2014**, *15*, 815–839 [CrossRef]
10. Korcaj, L.; Hahnel, U.J.; Spada, H. Intentions to adopt photovoltaic systems depend on homeowners' expected personal gains and behavior of peers. *Renew. Energy* **2015**, *75*, 407–415. [CrossRef]
11. Vasseur, V.; Kemp, R. The adoption of PV in the Netherlands: A statistical analysis of adoption factors. *Renew. Sustain. Energy Rev.* **2015**, *41*, 483–494. [CrossRef]
12. Sommerfeld, J.; Buys, L.; Vine, D. Residential consumers' experiences in the adoption and use of solar PV. *Energy Policy* **2017**, *105*, 10–16. [CrossRef]
13. Jager, W. Stimulating the diffusion of photovoltaic systems: A behavioural perspective. *Energy Policy* **2006**, *34*, 1935–1943. [CrossRef]
14. Kazhamiaka, F.; Jochem, P.; Keshav, S.; Rosenberg, C. On the influence of jurisdiction on the profitability of residential photovoltaic-storage systems: A multi-national case study. *Energy Policy* **2017**, *109*, 428–440. [CrossRef]
15. Macy, M.W.; Willer, R. From factors to factors: Computational sociology and agent-based modeling. *Annu. Rev. Sociol.* **2002**, *28*, 143–166. [CrossRef]
16. Gilbert, N. *Agent-Based Models*; Sage Publishing: London, UK: 2008.
17. Schwarz, N.; Ernst, A. Agent-based modeling of the diffusion of environmental innovations—An empirical approach. *Technol. Forecast. Soc. Chang.* **2009**, *76*, 497–511. [CrossRef]
18. Pegoretti, G.; Rentocchini, F.; Marzetti, G.V. An agent-based model of innovation diffusion: network structure and coexistence under different information regimes. *J. Econ. Interact. Coord.* **2012**, *7*, 145–165. [CrossRef]
19. Kiesling, E.; Günther, M.; Stummer, C.; Wakolbinger, L.M. Agent-based simulation of innovation diffusion: A review. *Central Eur. J. Oper. Res.* **2012**, *20*, 183–230. [CrossRef]
20. Zhang, H.; Vorobeychik, Y. Empirically grounded agent-based models of innovation diffusion: A critical review. *Artif. Intell. Rev.* **2017**, 1–35. [CrossRef]
21. Abrahamson, E.; Rosenkopf, L. Social network effects on the extent of innovation diffusion: A computer simulation. *Organ. Sci.* **1997**, *8*, 289–309. [CrossRef]
22. Chatterjee, R.A.; Eliashberg, J. The innovation diffusion process in a heterogeneous population: A micromodeling approach. *Manag. Sci.* **1990**, *36*, 1057–1079. [CrossRef]
23. Jacobsson, S.; Johnson, A. The diffusion of renewable energy technology: an analytical framework and key issues for research. *Energy Policy* **2000**, *28*, 625–640. [CrossRef]
24. Zhang, H.; Vorobeychik, Y.; Letchford, J.; Lakkaraju, K. Predicting rooftop solar adoption using agent-based modeling. In Proceedings of the 2014 AAAI Fall Symposium Series, Arlington, VA, USA, 13–15 November 2014.
25. Zhang, H.; Vorobeychik, Y.; Letchford, J.; Lakkaraju, K. Data-driven agent-based modeling, with application to rooftop solar adoption. *Auton. Agents Multi-Agent Syst.* **2016**, *30*, 1023–1049. [CrossRef]
26. Alyousef, A.; Adepetu, A.; Meer, H. Analysis and Model-based Predictions of Solar PV and Battery Adoption in Germany: An Agent-based Approach. *Comput. Sci.* **2017**, *32*, 211–223. [CrossRef]
27. Zhao, J.; Mazhari, E.; Celik, N.; Son, Y.J. Hybrid agent-based simulation for policy evaluation of solar power generation systems. *Simul. Model. Pract. Theory* **2011**, *19*, 2189–2205. [CrossRef]
28. Palmer, J.; Sorda, G.; Madlener, R. Modeling the diffusion of residential photovoltaic systems in Italy: An agent-based simulation. *Technol. Forecast. Soc. Chang.* **2013**, *106*, 106–131. [CrossRef]
29. Borghesi, A.; Milano, M. Multi-agent simulator of incentive influence on PV adoption. In Proceedings of the 2014 International Conference on Renewable Energy Research and Application (ICRERA), Milwaukee, WI, USA, 19–22 October 2014; pp. 556–560. [CrossRef]
30. Robinson, S.A.; Stringer, M.; Rai, V.; Tondon, A. GIS-integrated agent-based model of residential solar PV diffusion. In Proceedings of the 32nd USAEE/IAEE North American Conference, Anchorage, AK, USA, 28–31 July 2013; pp. 28–31.
31. Davidson, C.; Drury, E.; Lopez, A.; Elmore, R.; Margolis, R. Modeling photovoltaic diffusion: An analysis of geospatial datasets. *Environ. Res. Lett.* **2014**, *9*, 074009. [CrossRef]
32. Zhang, H.; Vorobeychik, Y.; Letchford, J.; Lakkaraju, K. Data-driven agent-based modeling, with application to rooftop solar adoption. In Proceedings of the 2015 International Conference on Autonomous Agents and Multiagent Systems, Istanbul, Turkey, 4–8 May 2015; International Foundation for Autonomous Agents and Multiagent Systems: Richland, SC, USA, 2015; pp. 513–521.

33. Macal, C.; Graziano, D.; Ozik, J. *Modeling Solar PV Adoption: A Social-Behavioral Agent-Based Framework*; AAAI Press: Menlo Park, CA, USA, 2014.
34. Rai, V.; Robinson, S.A. Agent-based Modeling of Energy Technology Adoption. *Environ. Model. Softw.* **2015**, *70*, 163–177. [CrossRef]
35. Lee, T.; Yao, R.; Coker, P. An analysis of UK policies for domestic energy reduction using an agent based tool. *Energy Policy* **2014**, *66*, 267–279. [CrossRef]
36. Johnson, P.; Balke, T.; Gilbert, N. Report on the Policy Instruments Considered and Their Likely Effectiveness. 2014. Available online: http://cordis.europa.eu/docs/projects/cnect/7/288147/080/deliverables/001-D43.pdf (accessed on 20 May 2019).
37. Adepetu, A.; Alyousef, A.; Keshav, S.; de Meer, H. Comparing solar photovoltaic and battery adoption in Ontario and Germany: An agent-based approach. *Energy Inform.* **2018**, *1*, 6. [CrossRef]
38. Sinitskaya, E.; Gomez, K.J.; Bao, Q.; Yang, M.C.; MacDonald, E.F. Examining the Influence of Solar Panel Installers on Design Innovation and Market Penetration. *J. Mech. Des.* **2019**, *141*, 041702. [CrossRef]
39. Lee, M.; Hong, T. Hybrid agent-based modeling of rooftop solar photovoltaic adoption by integrating the geographic information system and data mining technique. *Energy Convers. Manag.* **2019**, *183*, 266–279. [CrossRef]
40. ePolicy: Engineering the POLicy-Making LIfe CYcle. Available online: http://www.epolicy-project.eu/node (accessed on 20 May 2019).
41. Balke, T.; Gilbert, N. Description of the Actor Calibration. 2014. Available online: http://cordis.europa.eu/docs/projects/cnect/7/288147/080/deliverables/001-deliverable42.pdf (accessed on 20 May 2019).
42. Borghesi, A.; Milano, M.; Gavanelli, M.; Woods, T. Simulation of incentive mechanisms for renewable energy policies. In Proceedings of the European Conference on Modeling and Simulation (ECMS2013), Ålesund, Norway, 27–30 May 2013.
43. Iachini, V.; Borghesi, A.; Milano, M. Agent Based Simulation of Incentive Mechanisms on Photovoltaic Adoption. In Proceedings of the Congress of the Italian Association for Artificial Intelligence, Ferrara, Italy, 23–25 September 2015; Springer: New York, NY, USA, 2015; pp. 136–148.
44. QGIS Development Team. *QGIS Geographic Information System*; Open Source Geospatial Foundation: Beaverton, OR, USA, 2009.
45. Criteri e Modalità pper Incentivare la Produzione di Energia Elettrica Mediante Conversione Fotovoltaica Dell'energia Solare, 19 February 2007. Available online: https://www.gse.it/dati-e-scenari/rapporti (accessed on 20 May 2019).
46. Incentivazione della Produzione di Energia Elettrica Mediante Conversione Fotovoltaica della Fonte Solare, 6 August 2010. Available online: http://www.gse.it/it/ContoEnergia/Fotovoltaico/EvoluzionedelContoEnergia/ (accessed on 20 May 2019).
47. Incentivazione della Produzione di Energia Elettrica da Impianti Solari Fotovoltaici, 5 May 2011. Available online: http://www.gse.it/en/feedintariff/Photovoltaic/Fourthfeed-intariff/ (accessed on 20 May 2019).
48. Incentivazione della Produzione di Energia Elettrica da Impianti Solari Fotovoltaici (c.d. Quinto Conto Energia), 5 July 2012. Available online: http://www.gse.it/en/feedintariff/Photovoltaic/FifthFeed-inScheme/ (accessed on 20 May 2019).
49. GSE Statistical Reports. 2017. Available online: http://www.gse.it/it/Statistiche/Pages/default.aspx (accessed on 20 May 2019).
50. Feed-In Scheme Results. 2017. Available online: http://www.gse.it/en/feedintariff/Supportmechanismsoutcomes/ (accessed on 20 May 2019).
51. Schilling, M.; Izzo, F. *Gestione dell'innovazione*; Collana di istruzione scientifica; Serie di discipline aziendali; McGraw-Hill Education: New York, NY, USA, 2013.
52. Rogers, E.M. Diffusion of preventive innovations. *Addict. Behav.* **2002**, *27*, 989–993. [CrossRef]
53. Aggarwal, C.C. An introduction to social network data analytics. In *Social Network Data Analytics*; Springer: Berlin/Heidelberg, Germany, 2011; pp. 1–15.
54. Telesford, Q.K.; Joyce, K.E.; Hayasaka, S.; Burdette, J.H.; Laurienti, P.J. The ubiquity of small-world networks. *Brain Connect.* **2011**, *1*, 367–375. [CrossRef]

55. Leskovec, J.; Horvitz, E. Planetary-scale views on a large instant-messaging network. In Proceedings of the 17th International Conference on World Wide Web, Beijing, China, 21–25 April 2008; ACM: New York, NY, USA, 2008; pp. 915–924.
56. Amaral, L.A.N.; Scala, A.; Barthelemy, M.; Stanley, H.E. Classes of small-world networks. *Proc. Natl. Acad. Sci. USA* **2000**, *97*, 11149–11152. [CrossRef]
57. Watts, D.J.; Strogatz, S.H. Collective dynamics of 'small-world' networks. *Nature* **1998**, *393*, 440–442. [CrossRef]
58. Liben-Nowell, D.; Novak, J.; Kumar, R.; Raghavan, P.; Tomkins, A. Geographic routing in social networks. *Proc. Natl. Acad. Sci. USA* **2005**, *102*, 11623–11628. [CrossRef]
59. Kirkpatrick, S.; Gelatt, C.D.; Vecchi, M.P. Optimization by simulated annealing. *Science* **1983**, *220*, 671–680. [CrossRef]
60. Goldberg, D.E. *Genetic Algorithms in Search, Optimization and Machine Learning*, 1st ed.; Addison-Wesley Longman Publishing Co., Inc.: Boston, MA, USA, 1989.
61. Goldberg, D.E.; Deb, K. A comparative analysis of selection schemes used in genetic algorithms. *Found. Genet. Algorithms* **1991**, *1*, 69–93.

Article

Improving Computational Efficiency in Crowded Task Allocation Games with Coupled Constraints

Ming Chong Lim and Han-Lim Choi *

Department of Aerospace Engineering & KI for Robotics, Korea Advanced Institute of Science and Technology, Daejeon 34141, Korea; mclim@kaist.ac.kr

* Correspondence: hanlimc@kaist.ac.kr

Received: 17 April 2019; Accepted: 17 May 2019; Published: 24 May 2019

Abstract: Multi-agent task allocation is a well-studied field with many proven algorithms. In real-world applications, many tasks have complicated coupled relationships that affect the feasibility of some algorithms. In this paper, we leverage on the properties of potential games and introduce a scheduling algorithm to provide feasible solutions in allocation scenarios with complicated spatial and temporal dependence. Additionally, we propose the use of random sampling in a Distributed Stochastic Algorithm to enhance speed of convergence. We demonstrate the feasibility of such an approach in a simulated disaster relief operation and show that feasibly good results can be obtained when the confirmation and sample size requirements are properly selected.

Keywords: multi-agent systems; multi-agent planning and scheduling; potential game; equilibrium selection

---

1. Introduction

Application of multi-agent systems have often been considered for large-scale problems that are otherwise difficult or impossible to solve with only a single agent. Despite the difficulties in coordinating multiple agents, systems with multiple heterogeneous agents have often been preferred over a single omnipotent agent and the belief that the collective effort of multiple agents is superior to that of an individual can be attributed to concepts of robustness, parallelism, and cost issues.

Therefore, challenges in managing a multi-agent system have been extensively studied with the proposal of a variety of approaches and algorithms [1–8] which are applicable to a multiplicity of systems ranging from military operations to resource distribution [2,6,8–10]. One notable approach is the introduction of game-theoretic framework in multi-agent systems. With the continuous development of intelligence in autonomous systems, concerns of social awareness and acceptability [11] have been frequently raised. In traditional systems which are designed to achieve near-optimal solutions, agents are often programmed to be greedy and to seek maximum rewards for their actions leading to a competitive environment within the system, such that an agent may act in a manner that jeopardize the overall objective of the group for their own good. This concern leads to the adaptation of a game-theoretic framework which provide each individual agent freedom over its actions while at the same time, ensure that the solutions proposed by the collective action are socially acceptable and stable despite the players being self-regarding. As such, many game-theoretic models have been proposed to meet the requirements of different application scenarios and equilibrium selection algorithms ranging from constraint optimization [2,12] to learning [13–15] have been designed and shown to be feasible.

In this paper, we consider a game-theoretic framework for the coupled-constraint task allocation problem which is highly applicable to situations concerning military and disaster relief operations where the number of tasks is, generally, significantly large and are frequently spatially and temporally correlated. In such emergency situations, higher levels of emphasis are placed on the speed of decision-making rather than the optimality of the decisions, and as such, the main objective of this paper is to propose a game-theoretic model that enables autonomous multi-agent systems to make quick decisions that are feasible despite possibly having lower levels of optimality. We build on our previous work [16] which considers a game-theoretic framework for spatially constrained task allocation environments, taking into account additional temporal constraints and, at the same time, propose some modifications to the Distributed Stochastic Algorithm to allow quicker convergence in congested problems.

2. Preliminaries

2.1. Coupled-Constraint Consensus-Based Bundle Algorithm (CCBBA)

CCBBA [8] is an extension to a market-based task allocation model called Consensus-Based Bundle Algorithm (CBBA) [3]. CCBBA was designed to address the inadequacies of CBBA, namely the inability to feasibly resolve problems with complex spatial and temporal relationships. As far as we know, CCBBA is the first algorithm to propose such relationships which mimic the requirements of emergency operations in the real world. Coupled constraints, which affect both space and time, can be succinctly described as follows:

Definition 1. *(Spatial coupled constraints)*

1. *Unilateral dependency: Task A can only be assigned if task B is assigned;*
2. *Mutual dependency: Tasks A and B must either both be assigned or not at all;*
3. *Mutual exclusivity: Only either task A or B can be assigned at each time.*

Definition 2. *(Temporal coupled constraints)*

1. *Simultaneous: Tasks A and B must begin at the same time;*
2. *Before: Task A must end before task B begins;*
3. *After: Task A must begin after task B ends;*
4. *During: Task A must begin when task B is in progress;*
5. *Not during: Task A must either end before task B begins, or begin after task B ends;*
6. *Between: Task A must begin after task B ends and end before task C begins.*

These constraints may be non-exhaustive but are more than sufficient to meet the requirements of real-world applications. Mathematically, the constraint relationships between the tasks can be denoted as two matrices—the dependency matrix, $\mathcal{D}$, and temporal matrix, $\mathcal{T}$. The entry (q, p) in $\mathcal{D}_{q,p}$ describes the spatial constraints between the qth and p^{th} elements in the form of a coded variable shown in Table 1 while in $\mathcal{T}_{q,p}$, the value (q, p) in the temporal matrix specifies the maximum amount of time q can begin after p begins.

Table 1. Code for Dependency matrix entry $\mathcal{D}_{q,p}$.

Code	Relationship
0	p is independent of q
1	p is unilaterally dependent on q
−1	p and q are mutually exclusive

Although CCBBA can handle most coupled-constraint task allocation, it faces convergence issues in certain problem scenarios due to latency. As the problematic scenarios cannot be accurately predicted beforehand, it is infeasible to depend solely on CCBBA in emergency operations.

2.2. Potential Games

The concept of a potential game was first conceived by Monderer [17] and lays the necessary foundations in most of today's work in game theory.

Definition 3. *(Ordinal potential game) [18] A game is an ordinal potential game if and only if a potential function $\phi(\mathcal{A}) : \mathcal{A} \mapsto \mathbb{R}$ exists such that:*

$$U_i(a_i'', a_{-i}) - U_i(a_i', a_{-i}) > 0 \Leftrightarrow \phi(a_i'', a_{-i}) - \phi(a_i', a_{-i}) > 0 \qquad \forall a_i, a_i' \in \mathcal{A}_i \quad \forall i \in \mathcal{N} \tag{1}$$

where $\mathcal{A}_i$ is player i's action set, $\mathcal{N}$ is the set of all players, U_i is the local utility of player i, a_i' and a_i'' are the actions of player i, and a_{-i} is the action of all other players except i.

Effectively, a game is an ordinal potential game if and only if any unilateral change in actions by an individual that leads to a change in its local utility affects the potential function in the same manner. The suitability of a game-theoretic framework for task allocation problems lies in the characteristics of a potential game. Foremost, every finite potential game possesses the finite improvement property and has at least one pure-strategy equilibrium. Additionally, for every game with a finite improvement property, it will always converge to a Nash equilibrium [17,19–21]. Therefore, by modeling a task allocation problem as a finite potential game, we can guarantee the existence of a feasible pure-strategy allocation that is socially acceptable.

2.3. Game-Theoretic Models

Multiple game-theoretic models for task allocation have been introduced [1,2,5,6,22] and our previous work introduced a game-theoretic model for task allocation problems with spatial coupled constraints [16]. Consider a set of agents $\mathcal{N} = \{1, \cdots, n\}$ and a set of tasks $\mathcal{M} = \{k_1, \cdots, k_m\}$. The action set of agent i, $\mathcal{A}_i$, is then an ordered combination of all compatible tasks available inclusive of the null action $\mathbf{0}$, and the action profile $a_i \in \mathcal{A}_i$ is the (ordered) path that agent i will take. The reward of each task, u^{k_j}, for a generic game-theoretic task allocation model is then defined as

$$u^{k_j}(a) = \begin{cases} v_{k_j}(a) & \text{if } t^c_{k_j}(a) \leq t^d_{k_j}, \\ 0 & \text{otherwise,} \end{cases} \tag{2}$$

where $v_{k_j}(a)$ is the reward of the task, $t^c_{k_j}(a)$ is the completion time, and $t^d_{k_j}$ is the deadline. The amount of reward provided and the time which the task will be completed are both dependent on the collective action of all agents. To consider the spatial relationships between the tasks, the reward structure in Equation (2) can be adapted [16] to give

$$u^{k_j}(a) = \begin{cases} -C_{k_j}(a) & \text{if spatial constraints are violated,} \\ v_{k_j}(a) & \text{else if } t^c_{k_j}(a) \leq t^d_{k_j}, \\ 0 & \text{otherwise,} \end{cases} \tag{3}$$

where $C_{k_j}(a)$ is a function dependent on the collective action of all agents that returns a real non-zero positive value. With either reward structure, the marginal utility of each agent for each task is then

$$\mu_i^{k_j}(a_i, a_{-i}) = u^{k_j}(a_i, a_{-i}) - u^{k_j}(0, a_{-i}), \tag{4}$$

and the local utility of any agent is given as the sum of its marginal utilities,

$$u_i(a_i, a_{-i}) = \sum_{k_j \in \mathcal{M}} \mu_i^{k_j}(a_i, a_{-i}). \tag{5}$$

The agents in the model act in a greedy manner at each iteration, selecting an action profile a_i^* that maximizes their local utility,

$$a_i^* = \arg\max_{a_i \in \mathcal{A}_i} u_i(a_i, a_{-i}^*), \tag{6}$$

and the global utility is the sum of all task utilities,

$$u_g(a) = \sum_{k_j \in \mathcal{M}} u^{k_j}(a). \tag{7}$$

For a spatial coupled-constraint task allocation problem, the above game is an exact potential game which always converge to a feasible solution when function $C_{k_j}(a)$ is well-designed. The proofs for the game model can be found in [23].

Recall the definition of an agent's action set in the model above. The size of an agent's action set is approximately $\mathcal{O}(\bar{m}!)$, where $\bar{m}$ is the number of tasks that the agent can service (i.e., tasks that are compatible with the agent), since the action profile is an ordered sequence of task to service (i.e., an ordered path). This essentially means that the size of the action set blows up when the number of compatible tasks is large, making the model highly impractical in crowded problems which require quick decision-making due to the large number of action combinations that needs to be evaluated. In fact, this is a common problem that plagues most game-theoretic task allocation models as the computational load in equilibrium selection is generally highly dependent on the size of the task set. Chapman [2] attempted to overcome this issue by approximating a single large game as a series of smaller static potential games within a limited time interval where the agent's action is a vector of tasks to attend to during the interval $[t, t+w]$, given as $a_i = \{k^t, k^{t+1}, \cdots, k^{t+w}\}$. While this approach may help to alleviate the issue in most scenarios, it alone cannot guarantee improvements in extreme situations when many tasks are present in any of the time intervals.

2.4. Equilibrium Selection

An equilibrium selection algorithm is a negotiation mechanism that allows players to determine a Nash equilibrium, which is the stable state in the game where all players reach an agreement on their collective action. A variety of equilibrium selection algorithms have been proposed, each considering wildly different approaches. Chapman [24] methodically categorized these approaches into three main categories—a learning process [13–15], a traditional constraint optimization [2,12], or a heuristic search [22,25–27].

As seen previously, game-theoretic task allocation models tend to be impractical when considering large games due to high computational load. As such, the study on application of game-theoretic task allocation in crowded problems have largely been avoided. While some have attempted to propose algorithms that can accelerate this equilibrium selection, none seem to have convincingly tackled the problem at hand. Borowski [27] proposed a fast convergence algorithm whose convergence time is roughly linear, instead of exponential, with the number of agents but did not provide any answers to the more demanding issue—the dependency on the number of tasks.

Distributed Stochastic Algorithm (DSA)

DSA is a myopic, greedy, local search algorithm that employs a random parallel schedule, in which each agent will, with some probability, called the degree of parallel executions, change its action. The preferred action will be determined by the arg max decision rule. The motivation to implement such a schedule is to minimize the phenomenon known as "thrashing", where having all agents change their actions at the same time unintentionally leads to a suboptimal Nash equilibrium. However, a random parallel schedule cannot completely eliminate thrashing although it may be minimized. Furthermore, DSA almost surely converges to a Nash equilibrium in potential games due to the finite improvement property as there is a probability of moving towards the Nash equilibrium, which is an absorbing state, at every time step [24].

3. Game Design

We extend the model in our previous work [16]. To begin, it is necessary to determine a variable in the model that relates the agents' actions to time and the most obvious variable for consideration will be the task completion time, $t^c(a)$. If it is possible to influence the variable in such a way that the temporal constraints are reflected in the task completion time, then explicit alteration of the reward structure for the game model will not be required. (i.e., If the task completion time is determined in such a way that the temporal constraints are always satisfied, then the agent's reward will be reduced if the chosen action cannot provide a feasible solution since the task will be incomplete.) In essence, a task allocation problem with both spatial and temporal coupled constraints can be considered to be two sub-problems—an allocation and a scheduling problem. The game model for allocation penalizes agents for actions that violate spatial constraints, while the scheduling algorithm eliminates possible rewards for actions that violate temporal constraints.

To reflect the temporal relationships between the tasks, a temporal matrix $\mathcal{T}$ is introduced. This temporal matrix differ from the matrix in CCBBA [8] in that the entries in the matrix do not represent time value restrictions but instead describes the explicit relationships between tasks using a coded variable in Table 2 for simplicity purposes. This difference is trivial as the temporal matrix and the proposed algorithm can be adapted to consider time relationships similar to CCBBA if required.

Table 2. Code for Temporal matrix entry $\mathcal{T}_{q,p}$.

Code	Relationship
0	p is independent of q
1	(*Simultaneous*) p must begin at the same time with q
2	(*Before*) p must end before q begins
3	(*After*) p must begin after q ends
4	(*During*) p must begin while q is in progress
5	(*Not During*) p must either end before q begins, or after q ends

While CCBBA describes six types of temporal coupled constraints, only five types of constraints need to be considered, less the *between* coupled constraint. Recall that the *between* coupled-constraint states that task A must begin after task B ends and end before task C begins. Therefore, it is possible to decompose a *between* coupled constraint into a *before* and an *after* constraint as seen in Figure 1.

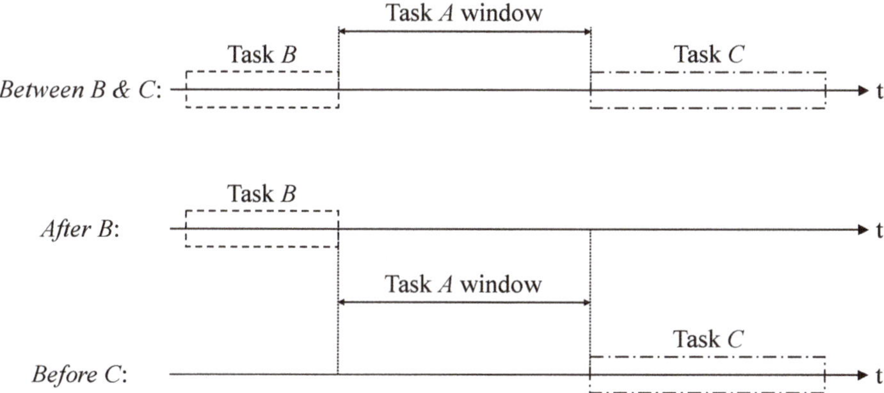

Figure 1. Decomposition of *between* coupled constraint.

3.1. Learnability of $C_{k_j}(a)$

When considering only spatial coupled constraints, the function $C_{k_j}(a)$ in Equation (3) is well-designed if it is learnable [1,28,29]. However, when considering games with both spatial and temporal constraints, additional restrictions on function $C_{k_j}(a)$ need to be imposed.

Example 1. *Consider that agent i selects a path* $a_i = \{k_1, k_2\}$ *such that some spatial constraints are violated by* k_1*. Agent i's local utility can be easily computed as*

$$u_i(\{k_1, k_2\}, a^*_{-i}) = -C_{k_1}(a) + v_{k_2} + c \tag{8}$$

where v_{k_2} *and c are some positive rewards that agent i gained from attending* k_2*. Now, assume that there exists a task* $k_3 \in a^*_{-i}$ *that* k_2 *is spatially dependent on. Furthermore,* k_3 *is also temporally dependent on* k_2 *such that if agent i considers* $a_i = k_2$*, the temporal constraints for* k_3 *cannot be satisfied and* k_3 *will be unassigned. In this way, agent i's local utility for* $a_i = k_2$ *is*

$$u_i(k_2, a^*_{-i}) = -C_{k_2}(a). \tag{9}$$

The (possibly) only feasible solution is $a_i = \mathbf{0}$ *which leads to zero local utility for i. However, for some game settings (e.g.,* $C_{k_1}(a)$ *is the number of spatial constraint violations for task* k_1 *as a result of collective action, a, and reward* v_{k_2} *is significantly large), it is possible that*

$$u_i(\{k_1, k_2\}, a^*_{-i}) \geq u_i(\mathbf{0}, a^*_{-i}) \geq u_i(k_2, a^*_{-i}) \tag{10}$$

and thus, path $a_i = \{k_1, k_2\}$*, which is infeasible, is the preferred action. For the game to converge to a feasible solution, it is imperative that*

$$u_i(\{k_1, k_2\}, a^*_{-i}) < u_i(\mathbf{0}, a^*_{-i}) \tag{11}$$

$$-C_{k_1}(a) + v_{k_2} + c < 0 \tag{12}$$

or rather,

$$-\sum_{k_j \in \mathcal{K}_{constrained}} C_{k_j}(a) + \sum_{k_i \in \mathcal{K}_{unconstrained}} v_{k_i} + c < 0 \quad \forall k_i, k_j \in a_i, \tag{13}$$

where $\mathcal{K}_{unconstrained}$ *are tasks assigned to agent i that do not violate any spatial constraints, and* $\mathcal{K}_{constrained}$ *are tasks assigned to agent i that violate some spatial constraints.*

In other words, the magnitude of the penalty needs to be sufficiently large to ensure that agents will never prefer a path that is infeasible. A possible design for $C_{k_j}(a)$ which satisfies Equation (13) and is learnable can then be

$$C_{k_j}(a) = G \times N_{k_j}(a) \tag{14}$$

where G is a gain, and $N_{k_j}(a)$ is the number of agents attending to task k_j based on the collective action a. Any sufficiently large gain should satisfy the constraint.

3.2. Scheduling Algorithm

Before describing the scheduling algorithm, some terminologies that will be used in the discussions are first defined along with the assumptions that are made.

Definition 4. *(Allocation) A task is allocated if the task is in the path of any agent (i.e.,* k_j *is allocated if* $k_j \in a$*).*

Definition 5. *(Assignment) A task is assigned if there exists a task completion time determined by the scheduling algorithm.*

The definitions essentially differentiate allocation and assignment such that

$$\mathcal{M}_{\text{assigned}} \subseteq \mathcal{M}_{\text{allocated}} \subseteq \mathcal{M}. \tag{15}$$

In other words, not all allocated tasks must be assigned. Allocated tasks that are incompatible with the temporal constraints will not be assigned. In this way, the scheduling algorithm can be thought of as an oracle which advises the agents on the suitability of their actions. All constraint violations are then determined based on assignments rather than allocations.

Assumption 1. *(Complete information game) Every agent has complete information with regards to the environment. They know the locations and details concerning all agents and tasks. This assumption, though demanding, is not overly restrictive. Firstly, in multi-agent systems, it is not unusual for an agent to understand the basic capabilities of other agents within the system, especially in cases where the agents are expected to cooperate. Secondly, the main research objective of this paper covers that of task allocation rather than path planning or mapping. Therefore, it is fair to assume that there is sufficient knowledge regarding the environment when assigning agents to tasks.*

Assumption 2. *(Optimistic agents) If an agent arrive at a task that it cannot complete due to insufficient information on the schedule resulting in the inability to determine a feasible service time window, then the agent will hold at this task for a user-specified time period before leaving. If sufficient information can be gathered during this hold period, then the agent will attempt to complete the task (i.e., determine a task completion time). In other words, agents are optimistic. In our application, this hold period is trivially considered in the form of number of iterations rather than true time units.*

The scheduling algorithm (Algorithm 1) is designed using a greedy approach—assignment is based on the earliest-first principle. Additionally, cooperation between agents to complete a single task is permitted. The duration of service for each task, and thus completion time, is then determined based on some function f_s that is dependent on the number of agents and starting times.

To begin, the availability of each agent, which can be computed from the expected time of arrival at the targeted task in its path a_i, needs to be determined. (i.e., The initial availability of an agent is first determined based on the expected time of arrival at the first task in its path a_i, which is also the initial target. After leaving the first task, the agent targets the next task in the path and determine its availability based on the targeted task. Target transition and availability computation continue in such a manner until the agent reaches the final task in its path whereby having no target after leaving the final task, the availability is set to be infinity.) At each iteration, the agent with the earliest availability is selected (line 4) and the serviceability of the allocated task is determined by the temporal constraints that are enforced on the task. For example, if task A is to end before task B begins, then task A will only be serviceable if an expected start time for task B exists. In fact, it is for this particular reason that the second assumption is necessary since most tasks can be unserviceable in complicated scenarios if the agents are pessimistic. Additionally, it should be noted that for a task to enforce some temporal constraints, it has to be allocated (e.g., If task C must begin after task D, but task D is not allocated, then task C can be considered to be having no temporal constraints). Temporal constraints do not actively enforce allocations, but rather, consider that if dependent tasks are allocated, then their assignments must be constrained.

If the allocated task is unserviceable, then the agent is placed on hold and the next available agent is considered. Otherwise, a service time window can be computed, and the agent determines the expected completion time for the task and updates its availability based on the next task in its path. At any iteration, if any agent has been placed on hold at a task for longer than the specified hold period, then it leaves the current task and transits to the next task in its path (line 11 to 16) and will most probably never return to the skipped task. If an agent completes all its allocated task, then it will no longer be considered even if it has the earliest availability. The algorithm converges when all agents can no longer proceed for any reason (e.g., all allocations have been assigned, all agents' allocated tasks are temporally constrained, etc.).

Algorithm 1 Scheduling Algorithm: *schedule*

Input: $a \leftarrow (a_i, a_{-i})$
Initialize: $\mathbf{L} = \mathbf{1}^n, \mathbf{t}^s = -\mathbf{1}^n, \mathbf{t}^c = -\mathbf{1}^n, \text{hold} = \mathbf{0}^n, Con(i, \mathbf{L}, a, \mathbf{t}^c, \mathbf{t}^s) = \{0, 1\}$
Sort:
$i \leftarrow \arg\min_{i \in \mathcal{N}}(\text{availability})$
Constraint:
if $Con(i, \mathbf{L}, a, \mathbf{t}^s, \mathbf{t}^c) = 1$ **then**
    if $\text{hold}_i < \text{maxHold}$ **then**
        $\text{hold}_i = \text{hold}_i + 1$
        $i \leftarrow \arg\min_{i \in \mathcal{N} \setminus \{i\}}(\text{availability})$
        goto *Constraint*
    else
        $\text{hold}_i = 0$
        $\mathbf{L}_i = \mathbf{L}_i + 1$
        $update(\text{availability})$
        goto *Sort*
    end if
Schedule:
else if $Con(i, \mathbf{L}, a, \mathbf{t}^s, \mathbf{t}^c) = 0$ **then**
    $\mathbf{L}, \mathbf{t}^s, \mathbf{t}^c \leftarrow compute\_schedule(i, \mathbf{L}, a, \mathbf{t}^s, \mathbf{t}^s, f_s)$
    $\mathbf{L}_i = \mathbf{L}_i + 1$
    if not first **then**
        $\mathbf{L}_0 = \mathbf{L}$
        $\mathbf{L}, \mathbf{t}^s, \mathbf{t}^c, \text{hold}, \text{availability} \leftarrow align(\mathbf{L}, a, \mathbf{t}^s, \mathbf{t}^c)$
        if $\mathbf{L} \neq \mathbf{L}_0$ **then**
            $\mathbf{L}, \mathbf{t}^s, \mathbf{t}^c, \text{hold}, \text{availability} \leftarrow constraint(\mathbf{L}, a, \mathbf{t}^s, \mathbf{t}^c)$
            if $ismember(\mathbf{L}, \mathbf{L}_{\text{hist}})$ **then**
                return
            else
                $append(\mathbf{L}, \mathbf{L}_{\text{hist}})$
            end if
        end if
    end if
    goto *Sort*
end if

In the event that multiple agents have been allocated to the same task, then agents will only participate if the task has yet to be completed at their time of arrival. If an agent were to participate in servicing a task, then it will update the expected completion time of the task, its own availability and also the availability of all other (previous and current) participating agents to reflect the change based on f_s. Immediately after updating the schedule, the algorithm will invoke an alignment procedure (Algorithm 2) and possibly a temporal constraint check procedure (Algorithm 3) before moving on to the next iteration.

Algorithm 2 Alignment procedure: *align*

```
for all agents i participating in task k do
    find k in a_i
    if index(k ∈ a_i) + 1 ≠ L_i then
        t^s, t^c, hold_i ← drop(tasks after k)
        L_i = index(k ∈ a_i) + 1
    end if
end for
update(availability)
```

Algorithm 3 Constraint check procedure: *constraint*

```
if Con(i, L, a, t^s, t^c) = 1 then
    loop:
    for k which violate constraints do
        for i ∈ N do
            t^c, t^s, hold ← drop(k and tasks after)
            affected tasks ← tasks after
            update(availability)
        end for
        while len(affected tasks) ≠ 0 do
            for i ∈ N do
                t^c, t^s, hold ← drop(affected tasks and tasks after)
                affected tasks ← tasks after
                update(availability)
            end for
        end while
    end for
    if Con(i, L, a, t^s, t^c) = 1 then
        goto loop
    end if
end if
```

The alignment procedure (Algorithm 1, Line 23) is crucial in this algorithm when multiple agents' availability are updated simultaneously to ensure that the agents' availability are in-line with their targeted tasks. The importance of this procedure is shown in the following example.

Example 2. *Consider an agent i with ordered path $a_i = \{k_1, k_2, k_3\}$. Agent i was previously expected to complete task k_1 and k_2 at time t_1 and t_2, respectively and its availability to begin work at k_3 is expected to be t_3. However, agent j now decides to participate in task k_1 and the expected completion time for task k_1 is brought forward to $t_1 - w$. Agent j then updates agent i's expected availability to be some time $t_2 - w'$ as j expects i to be at the next step in the path, k_2, because it cannot easily predict i's availability any further than the next step. This leads to a loss of alignment between path step and expected availability and any further scheduling will only be incorrect.*

Therefore, an alignment invoked at task k is effectively a procedure to bring agents back to k such that the agents must reschedule for all tasks in their path after k. When an alignment procedure is invoked, temporal constraint violation checks (Algorithm 1, Line 25) are necessary as a result of dropping some tasks when aligning the agents. If no agents are realigned, then a temporal constraint violation check is not necessary. However, if a temporal constraint violation is found during the checking procedure, then tasks with violations (and all future tasks in the path) will be dropped. Further alignment for all agents (Algorithm 3, Line 9) is required before checking for temporal constraint violations again.

One major issue resulting from the alignment and temporal constraint check procedure is that cycling may occur. Cycling refers to the phenomenon when, as a result of removing some conflicting task schedules, the overall schedule reverts to a historic state. Since the scheduling process is deterministic, then the process is stuck in an infinite loop. Maintenance of a history on the scheduling outcome, such as the path step of every agent, **L**, whenever a constraint check procedure is invoked will help to identify cycling and if a cyclic state transition is observed (Algorithm 1, line 26 to 27), then the scheduling is considered to have converged as defined previously since it is infeasible to continue with the scheduling process. Please note that the scheduling algorithm is anytime (with respect to temporal constraints) as temporal constraint checks are considered at every iteration when necessary. Therefore, any schedule proposed by the algorithm at the end of an iteration is feasible, even when a cyclic state transition is present. The convergence of the algorithm when faced with cycling can be thought of as an early termination of the scheduling process which provides a feasible but incomplete schedule. Hence, the impact of cycling can be considered trivial as such phenomenon leads to low global and local utility due to "incompleteness" in the scheduling process where agents are unlikely to prefer such collective action. Generally, cycling tends to occur when the number of agents considered for the problem is too low when compared to the hold period and therefore, proper selection of hold period will minimize the occurrence of cycling.

Theorem 1. *Using the proposed game model and scheduling algorithm for a game with spatial and temporal coupled constraints, the game will always converge to a feasible solution where all the coupled constraints are satisfied.*

Proof. We prove by contradiction. Consider a game with both spatial and temporal coupled constraint which converged to an infeasible solution where agent i selected a path a_i' which contains k_j that violates y spatial coupled constraints. We also know that there exists a null action which is always feasible. If a_{-i}^* violates $x \in \mathbb{R}^+$ number of spatial coupled constraints, then

$$u^{k_j}(a_i', a_{-i}^*) = -G(x+y), \quad u^{k_j}(0, a_{-i}^*) = -Gx \tag{16}$$

$$\mu_i^{k_j}(a_i', a_{-i}^*) = -Gy < \mu_i^{k_j}(0, a_{-i}^*) = 0 \tag{17}$$

$$u_i(a_i', a_{-i}^*) < u_i(0, a_{-i}^*) \tag{18}$$

$$a_i' \neq \arg\max_{a_i \in \{a_i', 0\}} u_i(a_i, a_{-i}^*) \tag{19}$$

a_i' is not the argument which maximizes the local utility of agent i and thus, a_i' cannot be the preferred action at equilibrium. □

Explicit consideration on temporal constraint violations in the proof is not necessary as the scheduling algorithm will always propose a temporally feasible schedule. Since violations in the solution are determined by the assignments rather than allocations, any task in any path that do not meet the temporal requirements will not be assigned. Therefore, it is only necessary to show that agents will never choose an action that violates spatial constraints given current observations to prove the feasibility of the solution.

4. DSA with Sampling

For equilibrium selection, the DSA algorithm was considered to be the basis for improvement. The DSA is preferred over other algorithms such as log-linear learning for its relatively low overhead, good solutions [2,24] and ease of decentralization. In theory, DSA uses a best-reply dynamic given a complete action set and therefore, when considering a very large action set, DSA requires significant computational power and time to determine the best reply, making implementation impractical. Hence, by somehow placing a limit on the number of evaluations required, it is possible to implement DSA in a game with significant number of tasks.

One obvious way to do so will be to constrain the action set of every agent to reduce the computational load, where at each iteration, the action set is randomly sampled with a pre-defined sample size s (Algorithm 4, Line 4). It is necessary to further ensure that the null action, which is always feasible, is included in the constrained action set so that agents will always have a feasible action to consider. An agent will then choose to either maintain its action, or select an action in the constrained set using the best-reply dynamics. Marden [26] took a similar approach in his payoff-based implementation of log-linear learning albeit for different reasons, and consider only one randomly chosen action at every iteration, without the restriction of having a null action. In our implementation, the rate of convergence is expected to improve due to the reduction in number of evaluations required at every iteration. Recall that the motivation for using a parallel degree of execution in DSA is to introduce stochasticity into the model in hopes of escaping from a local optimum. This also means that the search path taken to select the Nash equilibrium is stochastic and in a problem with multiple Nash equilibria, the game can terminate at either equilibrium when played multiple times despite having similar levels of parallel degree of execution. Instead of using parallel executions, stochasticity is introduced into the equilibrium selection through having constrained action sets. Intuitively, DSA with sampling will still arrive at a Nash equilibrium as $t \to \infty$. A basic implementation of the game-theoretic coupled-constraint task allocation is shown in Algorithm 4.

Algorithm 4 Game-Theoretic Implementation

```
1: Input: 𝒩, 𝒜, s, t_con, t = 1, T = 1, a_{T=0} = 0
2: while t ≤ t_con do
3:     for i ∈ 𝒩 do
4:         𝒜_{i,T} = datasample(𝒜_i, s)
5:         assignment_{i,T−1} = schedule(a_{i,T−1}, a_{−i,T−1})
6:         u_{i,T−1} = utility(assignment_{i,T−1})
7:         mμ = 0, bestAction_{i,T} = a_{i,T−1}
8:         for a_{i,T} ∈ 𝒜_{i,T} do
9:             assignment_{i,T} = schedule(a_{i,T}, a_{−i,T−1})
10:            u_{i,T} = utility(assignment_{i,T})
11:            if u_{i,T} − u_{i,T−1} > mμ then
12:                mμ = u_{i,T} − u_{i,T−1}
13:                bestAction_{i,T} = a_{i,T}
14:            end if
15:        end for
16:        a_{i,T} = bestAction_{i,T}
17:    end for
18:    if a_{i,T} = a_{i,T−1}  ∀i ∈ 𝒩 then
19:        t = t + 1
20:    else
21:        t = 0
22:    end if
23:    T = T + 1
24: end while
```

The proposed equilibrium selection algorithm should have a lower dependency on the number of tasks and the rate of convergence is affect by two main parameters—the number of confirmations t_{con} (Algorithm 4, Line 2) and the sample size s (Algorithm 4, Line 4). Therefore, by carefully varying the number of confirmations and sample size, the equilibrium selection will be capable of satisfying the time requirements of the game.

5. Results and Discussion

We assess the game design and equilibrium selection algorithm in a simulated disaster relief operation in a 10-by-10 grid world. In this operation, there are three types of mission-specific autonomous vehicles, each with varying response capabilities, to be allocated to different types of disaster situations. The engineering vehicles are capable of clearing wreckage from collapsed structures while the rescue vehicles are required to extract casualties to a safe location and the firefighting vehicles are equipped to deal with fire outbreaks in the region. The agent-task compatibilities are depicted in Figure 2. Furthermore, some sites may be struck with more than one type of disaster which will then require agents to work together albeit with some constraints.

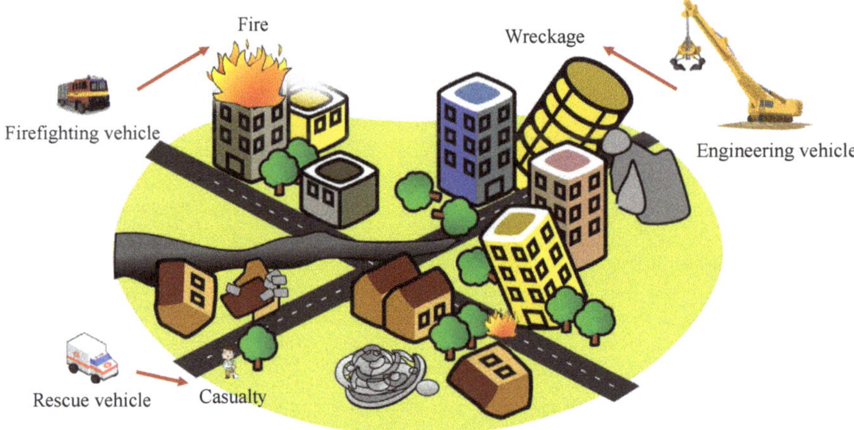

Figure 2. Agent-Task compatibilities in a disaster relief operation [30–34]. The figure is a composite image constructed from the various sources cited.

5.1. Mission Coupled Constraints

1. In an area where a fire broke out due to collapse of structures, cooperation of both engineering vehicles and firefighting vehicles will be required. To allow the engineering vehicles to begin clearing the wreckage efficiently, the fire needs to be first extinguished. *Wreckage clearance task is dependent on the response to fire and must begin after the fire has been extinguished.*
2. In a fire outbreak, some casualties may be trapped in the fire. Assuming that the rescue vehicles are well-equipped to withstand some levels of heat such that casualty extraction is possible when firefighting vehicles are on site to provide assistance in controlling the fire. Furthermore, casualty extraction will not be possible after the fire has been extinguished as the casualties will have likely suffocated during the process of firefighting. *Casualty rescue is dependent on firefighting operation and must begin during firefighting.*
3. The rescue vehicles do not possess the heavy lifting capabilities required to rescue casualties trapped in a wreckage. Assistance from the engineering units is required. *Casualty rescue is dependent on wreckage clearance task and must begin after the wreckage have been cleared.*

These constraints are only applicable when the tasks are in the same position (e.g., casualties trapped in a burning house.) The mission coupled constraints for these sites with multiple types of disaster can be reflected in the dependency matrix (Table 3) and temporal matrix (Table 4) as follows:

Table 3. Mission Dependency matrix.

	Fire	Casualty	Wreckage
Fire	0	1	1
Casualty	0	0	0
Wreckage	0	1	0

Table 4. Mission Temporal matrix.

	Fire	Casualty	Wreckage
Fire	0	4	3
Casualty	0	0	0
Wreckage	0	3	0

5.2. Feasibility of Game Model

Several agents and disaster sites at random locations of the grid world were considered. The summary of the disaster relief operations and agent parameters are provided in Table 5 with 20 different relief operations, each simulated as a game to be played 20 times each. The reward for the successful completion of each task was given as a time-decaying function

$$v_{k_j}(a) = \nu_{k_j} \exp(-\lambda_{k_j} t^c_{k_j}(a)) \tag{20}$$

where ν_{k_j} is the intrinsic value of the task, and λ_{k_j} is the discount factor for task k_j, to reflect the urgency of the tasks and motivate the agents to work in an efficient manner.

Table 5. Disaster relief operation details and simulation parameters.

Agents	6	Tasks	27	Parameters	
Engineer	2	Fire	3	Path length	4
Rescue	2	Casualty	3	Sample size	20
Firefighter	2	Wreckage	3	Confirmations	100
		Casualty $\xrightarrow{\text{dep}}$ Fire	3	Hold period	4
		Casualty $\xrightarrow{\text{dep}}$ Wreckage	3		
		Wreckage $\xrightarrow{\text{dep}}$ Fire	3		

$\xrightarrow{\text{dep}}$ depicts a dependency relationship (i.e., a task of each type in the same location leading to a coupled constraint).

The feasibility of the game model using DSA with sampling can be inferred from the satisfaction of all coupled constraints in all the solutions obtained and the stochasticity of DSA is evident from the variation in global score values and computational times across the games for every operation scenario. For a centralized approach, the computational time for each game is below 1 min and when considering a synchronous decentralized approach, the estimated computational time falls to generally below 10 s. In a decentralized approach, each agent will run the scheduling algorithm, based on other agents' previous actions, independently. After evaluating the possible schedules, the agent will decide on its choice of action and communicate its decision to all other agents. Without explicitly considering the means of communication but assuming synchronous, the simulation times of a decentralized approach is estimated based on the slowest agent for each iteration and the overheads accumulated in the communication process. For a synchronous decentralized approach, the model can provide a feasible solution quickly.

Figure 3 provides the graphical interpretation of a proposed schedule for one of the simulated games and information on the labels for the schedule are provided in Table 6.

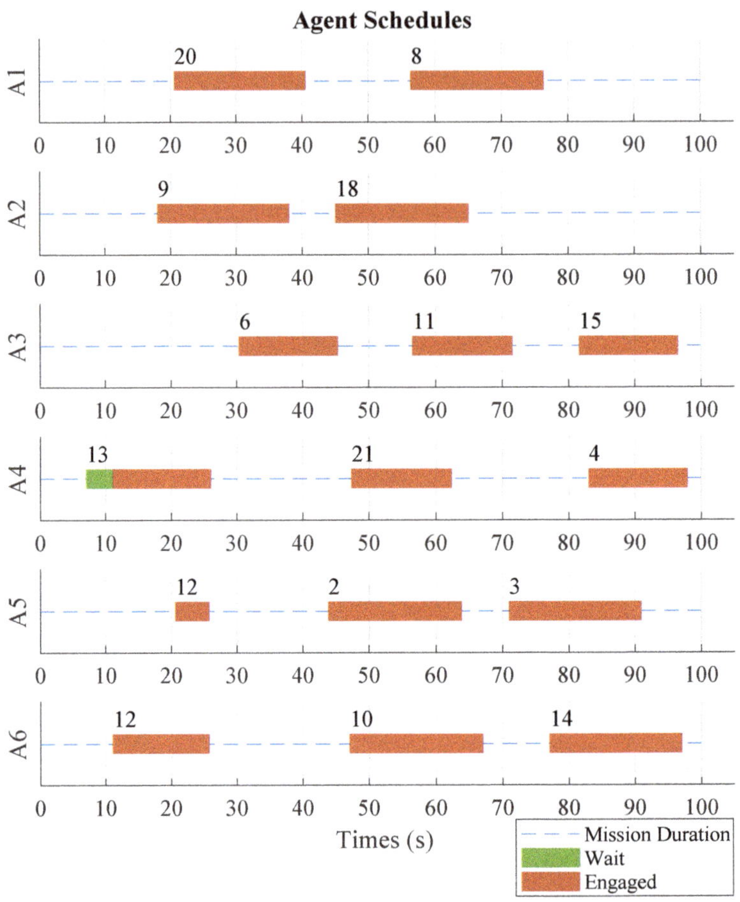

Figure 3. Example of a feasible schedule.

Table 6. Labeling information for agents and tasks.

Label	Type
A1, A2	Engineer
A3, A4	Rescue
A5, A6	Firefighter
1, 2, 3	Fire
4, 5, 6	Casualty
7, 8, 9	Wreckage
$11 \xrightarrow{dep} 10$, $13 \xrightarrow{dep} 12$, $15 \xrightarrow{dep} 14$	Casualty $\xrightarrow{dep}$ Fire
$17 \xrightarrow{dep} 16$, $19 \xrightarrow{dep} 18$, $21 \xrightarrow{dep} 20$	Casualty $\xrightarrow{dep}$ Wreckage
$23 \xrightarrow{dep} 22$, $25 \xrightarrow{dep} 24$, $27 \xrightarrow{dep} 26$	Wreckage $\xrightarrow{dep}$ Fire

The schedule in Figure 3 represents but one of many possible solutions to the simulated scenario. For a complex multi-agent multi-task allocation problem coupled with a flexible hold period, multiple Nash equilibria are likely to exist and the stochasticity of the agents' actions due to random sampling leads to variation in the route of progression for each game, and thereby terminating at various different solutions. Regardless of the progression, the feasibility of the scheduling algorithm to satisfy the temporal constraints are evident from the complementary behaviors portrayed in the solution (e.g., A *rescue*-type agent (*A4*) waits at a *casualty*-type task (*13*) for a short period so that the temporal constraint which require the *rescue* (*13*) to occur during *firefighting* (*12*) can be satisfied, *rescue*-type agent (*A4*) attempts a *casualty* rescue (*21*) task after the *wreckage* (*20*) have been cleared by an *engineer*-type agent (*A1*), etc.)

Moving beyond providing a feasible solution, the game should converge within a reasonably short time period to ensure the practically of a game-theoretic implementation in real-world operations and hence, further studies on the number of confirmations t_{con} and sample size s, which are the main parameters influencing the rate of convergence, are presented in the subsequent sections.

5.3. Number of Confirmations

In this section, the effects of variation in number of confirmations on the global score and computation times are examined. Simulation of an operation setting similar to the example in Table 5 but with varying number of confirmations at 10, 50, 100, 500, and 1000, respectively, were considered and the results are shown in Figure 4 with the means and standard deviations for the various aspects of the simulation presented in Tables 7–9. The observed global scores are generally similar at the selected levels of confirmations and the standard deviation generally decreases with increasing number of confirmations, providing a tighter bound to the range of scores. Intuitively, the decrease in standard deviation is in relation to the increasing probabilities of the solution being a Nash equilibrium (refer to Appendix A).

With only 10 confirmations, the mean global score is recognizably lower, and the range is also obviously greater as compared to the global scores for games with higher levels of confirmation. This lower levels of performance can be attributed to the extremity of the constraints placed on the agents' action set. With a sample size of 20 and 10 confirmations, the agents are expected to be exposed to a maximum of 191 unique actions which barely covers approximately 5% of all possible actions. The exposure of an agent refers to the number of actions that it has seen from the moment when it last changed its action. As such, it is plausible that the agents do not have sufficient understanding of the full range of possible actions, leading to agreements that are less efficient. However, it should be noted that such inefficiencies lead to a global score which is merely 3% lower as seen in Figure 5. Indeed, it can be argued that agents do not need to understand the entirety of their action sets to arrive at reasonable solutions. With 50 confirmations, the maximum exposure is at 951 unique actions, or approximately 26% of all possible actions. Yet, mean global scores comparable to those at higher levels of exposure are observed. With 1000 confirmations, it is highly likely that every agent has seen its entire possible action set since there are only 3610 possible actions while it is exposed to 19,001 non-unique actions. Interestingly, the maximum scores achieved for each category are similar with a value of 259.862 except for simulations with only 10 confirmations having a slightly lower maximum of 259.791. This phenomenon will be amplified in dense task allocation settings since most tasks will be closely grouped together and most action sequences will provide similar levels of reward. As such, most Nash equilibria will provide near-optimal solutions and it is, therefore, unnecessary for agents to have a complete understanding of their possible actions since emphasis is not placed on achieving the solution with an optimal global score or the Nash equilibrium with the maximum global score. When constraints are provided at every iteration, agents have lesser numbers of choices to make at each iteration, allowing them to arrive at a general agreement more quickly and easily as evident in the short computation times for games with only 10 confirmations. Improvements to the agreement are then made as the agents continue to explore the possible action space for each iteration.

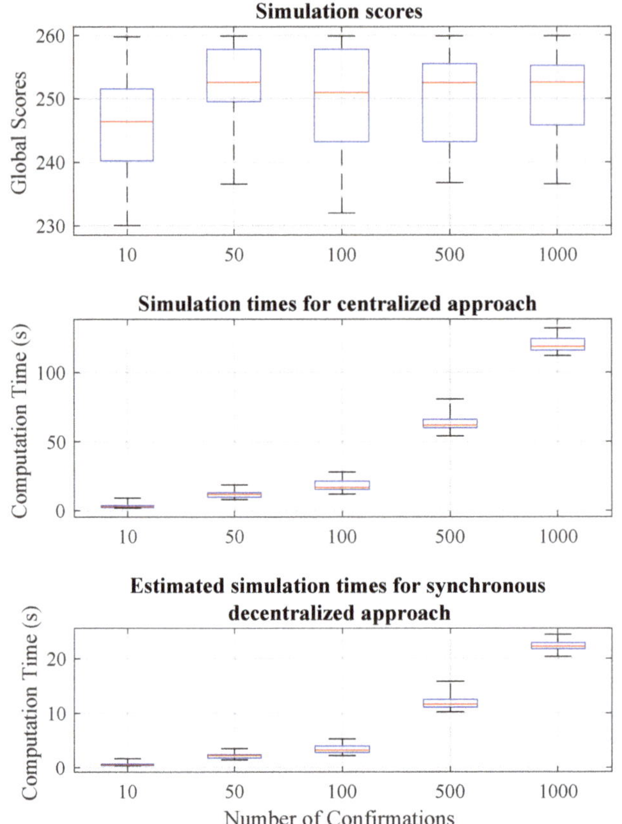

Figure 4. Variations in global scores and computation times due to number of confirmations.

However, it is necessary to note that increasing the confirmations do not necessarily lead to increase in global scores for every play. Recall that in DSA, progression is stochastic and hence the Nash equilibrium obtained for every play may differ and therefore, there is a non-zero probability of reaching a Nash equilibrium that has the lowest possible score despite considering a high number of confirmations in one play while having a Nash solution with a higher score when considering lower number of confirmations.

Another observation made, from Figure 6, is that the computation times increase with the number of confirmations as expected. In essence, a confirmation is defined as an iteration where all agents do not change their actions. This also means that in the new constrained action set of any agent at the current iteration, there are no actions that can improve any agent's local utility. For t_{con} confirmations, all agents do not change their actions for t_{con} consecutive iterations. When the agreement between the agents is not a Nash equilibrium, then the possibility of such an event occurring decreases with increasing values of t_{con}. Therefore, as the number of confirmations increases, confidence on the solution being Nash increases, but at the expense of increasing computation time due to the additional evaluations made by the agents.

Table 7. Mean and standard deviation for global scores.

Confirmations	Mean	Standard Deviation
10	245.302	8.088
50	251.692	8.114
100	249.176	8.420
500	250.786	6.581
1000	251.295	6.556

Table 8. Mean and standard deviation for centralized computation times.

Confirmations	Mean	Standard Deviation
10	3.772	1.901
50	12.083	2.749
100	18.086	4.173
500	64.859	7.391
1000	120.583	5.699

Table 9. Mean and standard deviation for estimated decentralized computation times.

Confirmations	Mean	Standard Deviation
10	0.720	0.383
50	2.268	0.512
100	3.400	0.818
500	12.152	1.508
1000	22.410	1.063

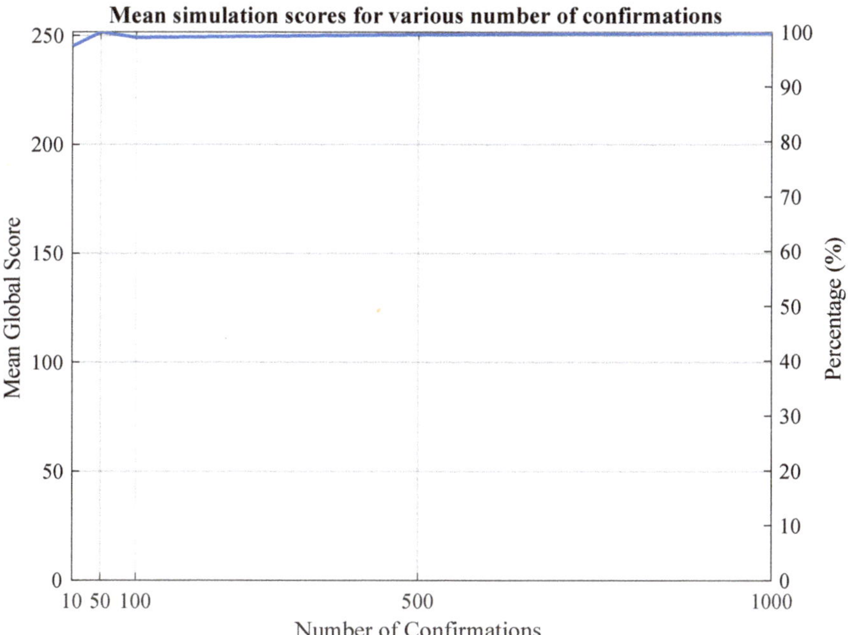

Figure 5. Mean global score values for various number of confirmations.

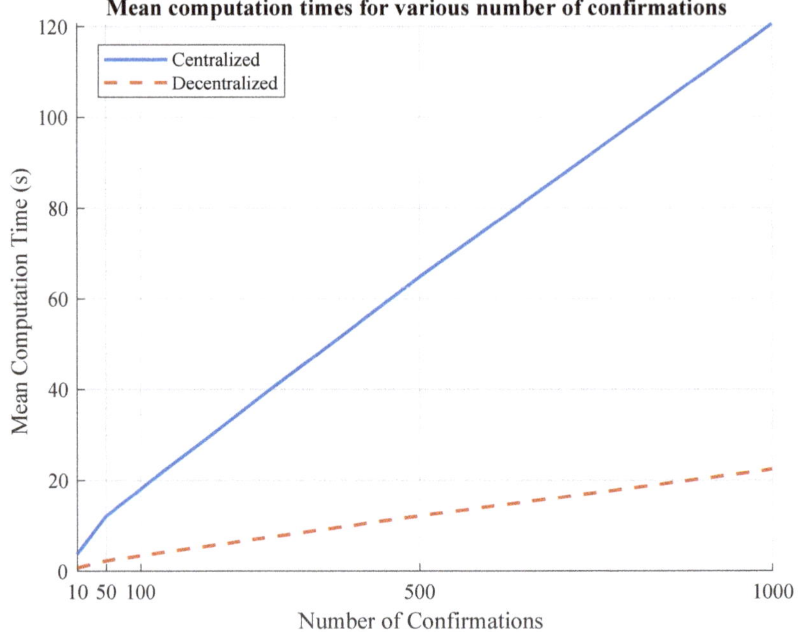

Figure 6. Mean computation times for various number of confirmations.

5.4. Size of Samples

To investigate the impact of variation in sample size, a dense allocation setting given in Table 10 is considered. The world remains as a 10-by-10 grid but the number of agents and tasks increased significantly and the sample sizes were varied to be 20, 40, 60, 80, and 100. The simulation results for the dense task allocation game are displayed in Figure 7, Tables 11–13.

Table 10. Crowded disaster relief operation details and simulation parameters.

Agents	15	Tasks	90	Parameters	
Engineer	5	Fire	10	Path length	4
Rescue	5	Casualty	10	Confirmations	100
Firefighter	5	Wreckage	10	Hold period	4
		Casualty $\xrightarrow{\text{dep}}$ Fire	10		
		Casualty $\xrightarrow{\text{dep}}$ Wreckage	10		
		Wreckage $\xrightarrow{\text{dep}}$ Fire	10		

Predictably, the mean global score values increase with increasing size of sample. The sample sizes of 20, 40, 60, 80, and 100 provided each agent with exposure to approximately 0.3%, 0.6%, 0.9%, 1.2% and 1.4% of its entire action set, respectively. Such low levels of understanding of the action set usually means that the agreements are suboptimal since the probability of making improvements to the existing agreement is rather high. Despite the suboptimality of the solutions, it should be noted that a 5-fold increase in sample size from 20 to 100 led to improvements in score by a mere 5% as seen in Figure 8. The logarithmic increment trend for the global scores further supports our belief that agents can make reasonably good scheduling decisions despite having an incomplete, and possibly small, understanding of their action set in a dense task allocation setting.

More importantly, from Figure 9, the computation times were observed to increase exponentially with the sample size. Without the use of sampling in the equilibrium selection, a game-theoretic approach for task allocation in a dense emergency operation will simply be impractical. Using random sampling, the computation times were suppressed to provide reasonably good solutions, enabling game-theoretic implementations in emergency operations which tend to be crowded, complex, and time-sensitive. In the game simulated, the mean computation times for both the centralized and decentralized approach were kept well below 60 min and 10 min, respectively, when the sample sizes are kept smaller than 60. Feasibly good solutions were also obtained quickly when considering a sample size of only 20 in the simulations.

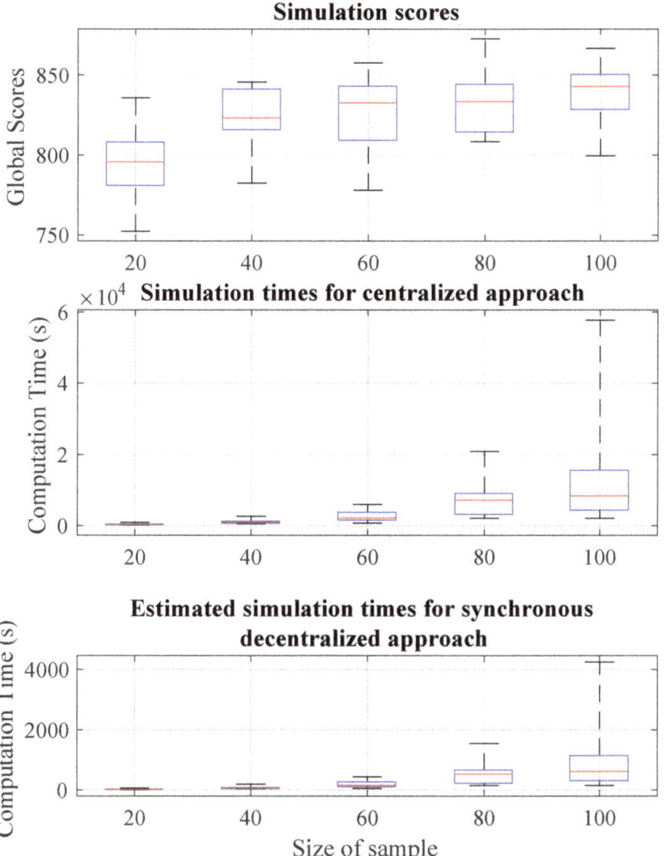

Figure 7. Variations in global scores and computation times due to size of sample.

Table 11. Mean and standard deviation for global scores.

Sample Size	Mean	Standard Deviation
20	795.294	21.166
40	823.923	16.742
60	825.086	21.365
80	831.913	17.918
100	839.281	18.486

Table 12. Mean and standard deviation for centralized computation times.

Sample Size	Mean	Standard Deviation
20	404.854	165.431
40	1093.975	476.654
60	2804.496	1593.538
80	7360.438	5086.329
100	12,632.142	12,861.328

Table 13. Mean and standard deviation for estimated decentralized computation times.

Sample Size	Mean	Standard Deviation
20	30.295	12.386
40	80.517	35.030
60	205.754	116.449
80	542.142	376.058
100	928.077	944.542

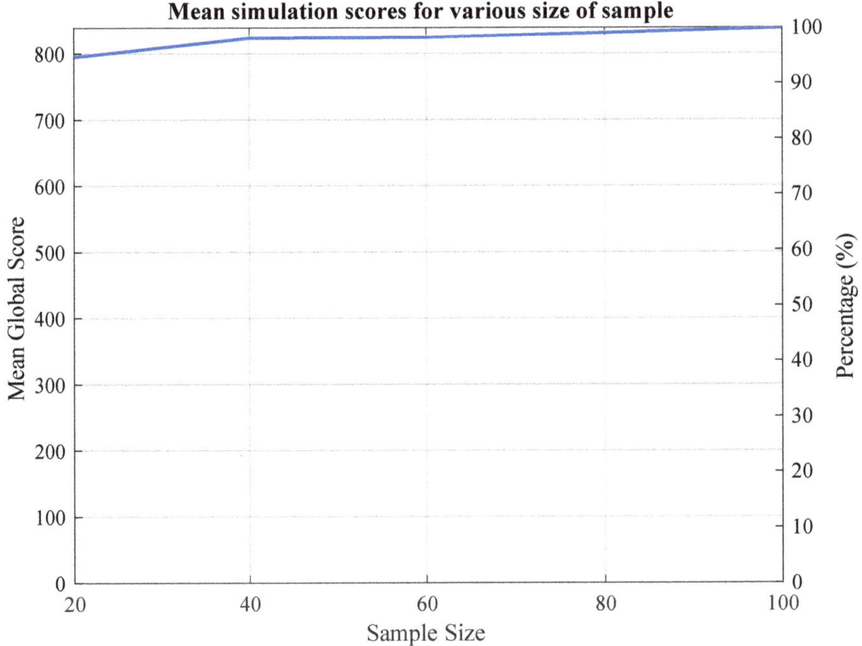

Figure 8. Mean global score values for various sample sizes.

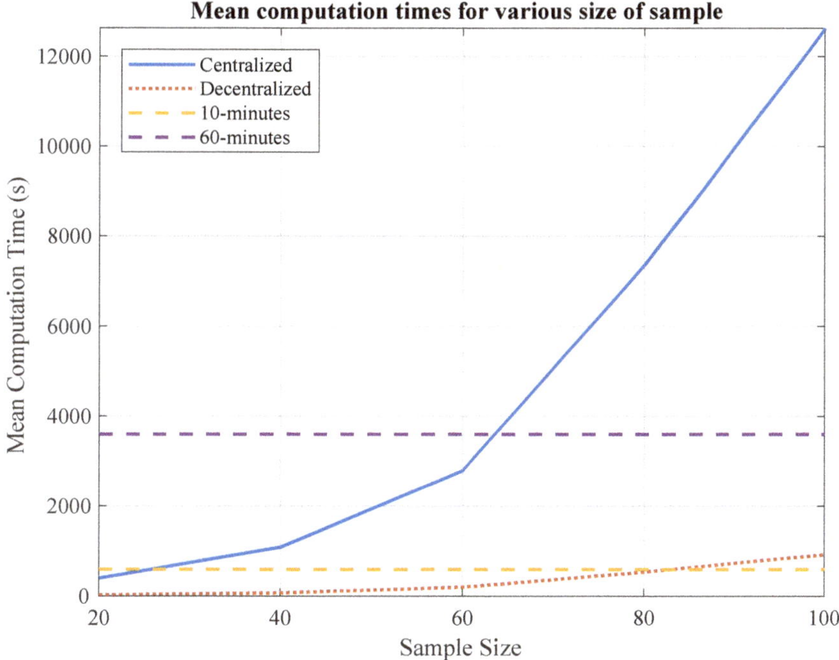

Figure 9. Mean computation times for various sample sizes.

6. Conclusions

In this paper, we introduced a game-theoretic framework for allocation of tasks with spatial and temporal coupled constraints first seen in CCBBA. A game-theoretic modeling of such problems helps to overcome potential convergence issues faced in a market-based allocation model through leveraging on the properties of potential games. A well-designed game model alone can effectively overcome spatial relationships among tasks and when coupled with the scheduling algorithm, allows feasible assignment of tasks with both spatial and temporal dependencies. Additionally, existing equilibrium selection algorithms have commonly been restricted by the size of the problem, making implementation of game-theoretic theoretic task allocation impractical for large-scale problems. However, by using random sampling in DSA, it is possible to obtain a feasibly good solution within a short time, allowing game-theoretic implementations in emergency operations which frequently consist of large numbers of tasks with complex relationships and yet require quick allocation at the same time.

The greatest limitation for the proposed methodology lies in the inability to quantify or guarantee the optimality of any given solution, in terms of global allocation, despite its feasibility to provide an allocation solution that considers the spatial and temporal relationships between the different types of task. A plausible explanation may be due to the existence of multiple Nash equilibria and the probabilistic approach in DSA. Additional studies to limit the game model to have only a single Nash equilibrium or to consider deterministic equilibrium selection algorithms that are a capable of selecting the Nash equilibrium with the maximum global allocation scores will help to overcome this deficiency.

Despite the current flaws in the proposed methodology, reasonably good solutions can still be obtained quickly to tackle coupled-constraint task allocation problem in a crowded space.

Author Contributions: Conceptualization, M.C.L. and H.-L.C.; methodology, M.C.L.; software, M.C.L.; validation, M.C.L. and H.-L.C.; formal analysis, M.C.L.; investigation, M.C.L.; resources, H.-L.C.; data curation, M.C.L.; writing—original draft preparation, M.C.L.; writing—review and editing, M.C.L. and H.-L.C.; visualization, M.C.L.; supervision, H.-L.C.; project administration, H.-L.C.; funding acquisition, H.-L.C.

Funding: This work was supported in part by Information Communication Technology Research and Development program of Institute for Information & Communications Technology Planning and Evaluation (IITP) grant funded by the Korea government Ministry of Science and ICT (NO.R1711055271, Development of High Reliable Communications and Security Software for Various Unmanned Vehicles).

Conflicts of Interest: The authors declare no conflict of interest. The funders had no role in the design of the study; in the collection, analyses, or interpretation of data; in the writing of the manuscript, or in the decision to publish the results.

Abbreviations

The following abbreviations are used in this manuscript:

CCBBA	Coupled-Constraint Consensus-Based Bundle Algorithm
CBBA	Consensus-Based Bundle Algorithm
DSA	Distributed Stochastic Algorithm

Appendix A

A Nash equilibrium is defined as a state where all players cannot improve its own position through unilateral change in strategy and can be identified when all players stop changing their actions for t iterations. In theory, Nash can only be guaranteed as $t \to \infty$. For some equilibrium selection algorithms, this may be the only means of identification and the solution obtained cannot be guaranteed to be a Nash equilibrium since implementation of $t \to \infty$ is impossible. That said, it is still possible to qualify the Nash properties of a solution through probability.

As the game model proposed in this paper is an exact potential game, there exists at least one collective action a^* that is the Nash equilibrium. Assuming that all players except i is at the Nash equilibrium, then the only possible reason that $a_i(t) \neq a_i^*$ at iteration t is that $a_i(t-1) \neq a_i^*$ and a_i^* is not in the constrained action set for the current iteration since we know that the proposed equilibrium selection algorithm will either maintain its previous action, or select an action in the constrained action set that maximizes its local utility and a_i^* is the maximizer. Therefore, the probability that $a_i \neq a_i^*$ is upper bounded by

$$\delta_i \leq \frac{(|\mathcal{A}_i| - s)(|\mathcal{A}_i| - 1)}{|\mathcal{A}_i|^2} \tag{A1}$$

where $\mathcal{A}_i$ is the complete action set for player i and s is the sample size. Given that sampling is independent for each player at each iteration, then if all players do not change their actions for t_{con} iterations, the probability that the solution is a Nash equilibrium is at least

$$1 - \delta \geq \prod_{i \in \mathcal{N}} 1 - \left(\frac{(|\mathcal{A}_i| - s)(|\mathcal{A}_i| - 1)}{|\mathcal{A}|^2} \right)^{t_{\text{con}}}. \tag{A2}$$

Since $\frac{(|\mathcal{A}_i|-s)(|\mathcal{A}_i|-1)}{|\mathcal{A}_i|^2} < 1$, then $1-\delta$ approaches to 1 as $t_{\text{con}} \to \infty$ and it is obvious that DSA with sampling will converge to a Nash equilibrium with probability 1 regardless of the size of the sample. Equation (A2) allows the qualification of the solution with regards to simulation parameters s and t_{con} by providing a lower bound on the probability of a solution being Nash. While such a qualification does not reflect the optimality of the solution in terms of global score, it can be used as a guide for the selection of simulation parameters to balance between the confidence and speed of convergence as and when required. Increase in either the number of confirmations t_{con} or sample size s will not only increase the confidence on the solution being a Nash equilibrium but also the computational times as seen in the results and discussion. Therefore, when emphasis is placed on either of the aspects—being socially efficient or fast convergence, the simulations parameters can be modified accordingly to meet the objectives.

References

1. Arslan, G.; Marden, J.R.; Shamma, J.S. Autonomous vehicle-target assignment: A game-theoretical formulation. *J. Dyn. Syst. Meas. Control* **2007**, *129*, 584–596. [CrossRef]
2. Chapman, A.C.; Micillo, R.A.; Kota, R.; Jennings, N.R. Decentralised dynamic task allocation: A practical game: Theoretic approach. In Proceedings of the 8th International Conference on Autonomous Agents and Multiagent Systems, Budapest, Hungary, 10–15 May 2009; Volume 2, pp. 915–922.
3. Choi, H.L.; Brunet, L.; How, J.P. Consensus-based decentralized auctions for robust task allocation. *IEEE Trans. Robot.* **2009**, *25*, 912–926. [CrossRef]
4. Gerkey, B.P.; Mataric, M.J. Sold!: Auction methods for multirobot coordination. *IEEE Trans. Robot. Autom.* **2002**, *18*, 758–768. [CrossRef]
5. Li, N.; Marden, J.R. Designing games to handle coupled constraints. In Proceedings of the 49th IEEE Conference on Decision and Control (CDC), Atlanta, GA, USA, 15–17 December 2010; pp. 250–255.
6. Li, N.; Marden, J.R. Decoupling coupled constraints through utility design. *IEEE Trans. Autom. Control* **2014**, *59*, 2289–2294. [CrossRef]
7. Parker, L.E.; Tang, F. Building multirobot coalitions through automated task solution synthesis. *Proc. IEEE* **2006**, *94*, 1289–1305. [CrossRef]
8. Whitten, A.K.; Choi, H.L.; Johnson, L.B.; How, J.P. Decentralized task allocation with coupled constraints in complex missions. In Proceedings of the 2011 American Control Conference, San Francisco, CA, USA, 29 June–1 July 2011.
9. Netto, R.; Ramalho, G.; Bonatto, B.; Carpinteiro, O.; Zambroni de Souza, A.; Oliveira, D.; Braga, R. Real-Time Framework for Energy Management System of a Smart Microgrid Using Multiagent Systems. *Energies* **2018**, *11*, 656. [CrossRef]
10. Han, Q.; Tan, G.; Fu, X.; Mei, Y.; Yang, Z. Water resource optimal allocation based on multi-agent game theory of HanJiang river basin. *Water* **2018**, *10*, 1184. [CrossRef]
11. Baldoni, M.; Baroglio, C.; May, K.M.; Micalizio, R.; Tedeschi, S. Computational Accountability in MAS Organizations with ADOPT. *Appl. Sci.* **2018**, *8*, 489. [CrossRef]
12. Marden, J.R.; Arslan, G.; Shamma, J.S. Joint strategy fictitious play with inertia for potential games. *IEEE Trans. Autom. Control* **2009**, *54*, 208–220. [CrossRef]
13. Bowling, M.; Veloso, M. *An Analysis of Stochastic Game Theory for Multiagent Reinforcement Learning*; Technical Report; Carnegie-Mellon Univ Pittsburgh Pa School of Computer Science: Pittsburgh, PA, USA, 2000.
14. Heinrich, J.; Silver, D. Deep reinforcement learning from self-play in imperfect-information games. *arXiv* **2016**, arXiv:1603.01121.
15. Wang, Y.; Pavel, L. A Modified Q-Learning Algorithm for Potential Games. *IFAC Proc. Volumes* **2014**, *47*, 8710–8718. [CrossRef]
16. Lim, M.C.; Choi, H.L. A Game-Theoretic Approach for Multi-Robot Task Allocation with Dependency Constraints. In Proceedings of the 5th International Conference of Robot Intelligence Technology and Applications, Daejeon, Korea, 13–15 December 2017.
17. Monderer, D.; Shapley, L.S. Potential games. *Games Econ. Behav.* **1996**, *14*, 124–143. [CrossRef]
18. Chapman, A.; Rogers, A.; Jennings, N.R. A Parameterisation of Algorithms for Distributed Constraint Optimisation via Potential Games. 2008. Available online: https://eprints.soton.ac.uk/265208/ (accessed on 5 December 2016).
19. Lã, Q.D.; Chew, Y.H.; Soong, B.H. *Potential Game Theory: Applications in Radio Resource Allocation*; Springer: Berlin, Germany, 2016.
20. Marden, J.R.; Arslan, G.; Shamma, J.S. Cooperative control and potential games. *IEEE Trans. Syst. Man Cybern. Part B Cybern.* **2009**, *39*, 1393–1407. [CrossRef] [PubMed]
21. Choi, H.L.; Lee, S.J. A potential game approach for information-maximizing cooperative planning of sensor networks. *IEEE Trans. Control Syst. Technol.* **2015**, *23*, 2326–2335. [CrossRef]

22. Macarthur, K.S.; Stranders, R.; Ramchurn, S.D.; Jennings, N.R. A Distributed Anytime Algorithm for Dynamic Task Allocation in Multi-Agent Systems. In Proceedings of the Twenty-Fifth AAAI Conference on Artificial Intelligence, San Francisco, CA, USA, 7–11 August 2011; pp. 701–706.
23. Lim, M.C. A Game-Theoretic Approach for Coupled-Constraint Task Allocation. Master's Thesis, Korea Advanced Institute of Science and Technology, Daejeon, Korea, 2018.
24. Chapman, A.C.; Rogers, A.; Jennings, N.R. Benchmarking hybrid algorithms for distributed constraint optimisation games. *Auton. Agents Multi-Agent Syst.* **2011**, *22*, 385–414. [CrossRef]
25. Garivier, A.; Kaufmann, E.; Koolen, W.M. Maximin action identification: A new bandit framework for games. In Proceedings of the Conference on Learning Theory, New York, NY, USA, 23–26 June 2016; pp. 1028–1050.
26. Marden, J.R.; Shamma, J.S. Revisiting log-linear learning: Asynchrony, completeness and payoff-based implementation. *Games Econ. Behav.* **2012**, *75*, 788–808. [CrossRef]
27. Borowski, H.; Marden, J.R.; Frew, E.W. Fast convergence in semi-anonymous potential games. In Proceedings of the IEEE 52nd Annual Conference on Decision and Control, Florence, Italy, 10–13 December 2013; pp. 2418–2423.
28. Tumer, K.; Agogino, A.K.; Wolpert, D.H. Learning sequences of actions in collectives of autonomous agents. In Proceedings of the First International Joint Conference on Autonomous Agents and Multiagent Systems: Part 1, Bologna, Italy, 15–19 July 2002; pp. 378–385.
29. Tumer, K.; Wolpert, D. A survey of collectives. In *Collectives and the Design of Complex Systems*; Springer: Berlin, Germany, 2004; pp. 1–42.
30. Juhele. City after Earthquake [PNG File]. 2016. Available online: https://openclipart.org/detail/250253/city-after-earthquake (accessed on 23 February 2019).
31. Markacio. Construction Crane [PNG File]. 2012. Available online: https://openclipart.org/detail/168285/construction-crane (accessed on 23 February 2019).
32. Rdevries. Fire Truck [PNG File]. 2014. Available online: https://openclipart.org/detail/190874/fire-truck (accessed on 23 February 2019).
33. Ginkgo. Isometric Ambulance [PNG File]. 2016. Available online: https://openclipart.org/detail/252628/isometric-ambulance (accessed on 23 February 2019).
34. Oksmith. Injured [PNG File]. 2017. Available online: https://openclipart.org/detail/285043/injured (accessed on 23 February 2019).

Article

Woc-Bots: An Agent-Based Approach to Decision-Making

Sean Grimes and David E. Breen *

Department of Computer Science, Drexel University, Philadelphia, PA 19104, USA; spg63@drexel.edu
* Correspondence: david@cs.drexel.edu

Received: 30 September 2019; Accepted: 29 October 2019; Published: 1 November 2019

Abstract: We present a flexible, robust approach to predictive decision-making using simple, modular agents (WoC-Bots) that interact with each other socially and share information about the features they are trained on. Our agents form a knowledge-diverse crowd, allowing us to use Wisdom of the Crowd (WoC) theories to aggregate their opinions and come to a collective conclusion. Compared to traditional multi-layer perceptron (MLP) networks, WoC-Bots can be trained more quickly, more easily incorporate new features, and make it easier to determine why the network gives the prediction that it does. We compare our predictive accuracy with MLP networks to show that WoC-Bots can attain similar results when predicting the box office success of Hollywood movies, while requiring significantly less training time.

Keywords: classification; prediction; multi-agent; wisdom-of-crowds; Hollywood; feature-extension; collective-intelligence; swarm

---

1. Introduction

We, humans, want to predict the future; disease outbreak and risk factors, business success, economics, and many more applications can benefit from better forecasting. Researchers have developed many tools to help us make predictions, with artificial neural networks (ANNs) being a current popular choice. ANNs can be used for classification, allowing us to take, for example, a series of features about an upcoming movie and determine, with fairly high accuracy, if the movie will be successful. ANNs, however, typically require a large amount of training data and compute time, and they do not generalize well to other topics. We cannot use an ANN trained on Hollywood movies to help us determine if some sports team will win an upcoming game or where the next 'hot spot' in an epidemic will be; they are inherently inflexible. Recent efforts are improving their flexibility by adding to their basic design, as seen in transfer learning [1], however this increases complexity and compute/data requirements while further obfuscating the internal workings of an ANN, making it *even more* difficult to answer the "why" about some outputted classification [2].

Prediction markets (PM) are designed to determine the probability of a future event taking place. Well-designed PMs encourage agents, human or computer-based, to contribute information to the market through trading shares and incentivizing correct, truthful information sharing, and then aggregate the information from individual agents into a collective knowledge [3]. PMs work because the aggregate knowledge of the group will generally be more precise and complete than the knowledge that any individual within the group holds. However, participants are expected to be well-informed in the topic being predicted, which as of current technology, requires human participants [4]. Additionally, computer agent-based PMs are difficult and programmer-intensive to create. Othman said on computer-based agents, "agent-based modeling of the real world is necessarily dubious. Attempting to model the rich tapestry of human behavior within economic structures—both the outstandingly bad and the terrifically complex—is a futile task [5]." Even if it were possible to model the complexity of human knowledge and decision-making within some narrow topic it would be extremely difficult to generalize across topics. We can consider simpler agents that don't attempt to model human interaction. These agents have been studied in simple, academic, scenarios with some success when their behavior is limited in possible actions and their opponents are not adversarial [6,7]. However, work done by Othman and Sandholm [8] has shown that simply changing the order in which the agents participate in the market can drastically impact the outcome of the market, indicating that "markets may fail to do any meaningful belief aggregation."

An alternative to PMs is Wisdom of the Crowd (WoC). WoC takes the approach that the opinion of a large, diverse group will be more accurate than any individual opinion within the group given a sufficiently competent aggregation mechanism [9]. The classic example that demonstrates this is guessing how many jelly beans are in a jar at a county fair. Typically no individual is consistently able to get close to the correct amount, but the aggregate opinion of the group is generally very close to the correct number of jelly beans. WoC doesn't expect or require expert knowledge, Scott Page said "the squared error of the collective prediction equals the average squared error minus the predictive diversity" [10]. This means the more diverse the crowd, the smaller the predictive error.

In this paper we present a robust, computer-agent-based approach to making predictions about the success of Hollywood movies that can be easily distributed across multiple computational nodes [11]. We take a WoC approach, using simple agents (WoC-Bots) without expert knowledge that are trained with different, small, subsets of features that describe the movies. This initially gives us a group of agents with a diverse and independent set of knowledge. The agents interact with one another socially, sharing some knowledge, determining the trust they have in other agents and the confidence they have in their own opinion, and changing their opinion given enough evidence. Following this interaction an overall conclusion is drawn from the crowd using a trust and performance-based aggregation mechanism. Our system was compared with traditional multilayer perceptron (MLP) networks trained with the full set of features available to the agents, as well as a subset of the most highly correlated features. We show that WoC-Bots are able to achieve more accurate classification results, with reduced training time and resistance to feature drop-out.

2. Methods & Design

The test scenario for this research involves predicting if a movie will be a success. Success was defined as the reported revenue being greater than 2× the reported budget for the movie. It is difficult to determine exactly what revenue is considered a success, and it differs on a movie-by-movie basis. However, advertising and promotion budgets are generally less than the production budget, which indicates studios should start to see some positive cash flow if a movie makes 2× the production budget [12]. Additionally, defining success as we did split our data, discussed more in Section 2.1, roughly equally between success (47.5%) and failure (52.5%).

2.1. Datasets & Libraries

We primarily used two datasets for this work:

1. The Movie Database (TMDb) (https://www.kaggle.com/tmdb/tmdb-movie-metadata/)
2. MovieLens (ML) https://www.kaggle.com/grouplens/movielens-20m-dataset/) [13]

The MovieLens dataset provides information for more than 27,000 movies, while the TMDb dataset includes 5000 movies. Only movies found in both datasets, with complete information for all features used for classification, were considered. The features used for classification are listed in Table 1. A note about the genre feature; only the first two listed genres were considered for each movie in classifiers that used the genre feature(s). Each genre was assigned a unique numeric ID. We were left with 4722 possible movies for testing and training, however 1023 movies appeared to contain incorrect information, e.g., negative values for movie budget or movie revenue; these movies were removed from our testing and training subsets. We used both datasets to help reduce sparse areas in the data for less popular and older movies. The data was split into two subsets, testing and training, with 2959 randomly selected examples used for training and 740 examples used for testing. The training subset was randomly selected from the full dataset at the start of each simulation, with the remaining examples being used for testing.

Table 1. Features available for classification.

Features	Description
`budget`	given to all agents, reported budget for movie
`tmdb_popularity`	dynamic variable from TMDb API attempting to represent interest in movie
`revenue`	used for sanity checks, reported revenue
`runtime`	unreliable metric for success without including genre information
`tmdb_vote_average`	average score from TMDb, can be combined with ML average
`tmdb_vote_count`	total votes for a movie from TMDb, can be combined with ML count
`ml_vote_average`	average score from ML, can be combined with TMDb average
`ml_vote_count`	total votes for a movie from ML, can be combined with TMDb count
`ml_tmdb_genres`	combined genre information from TMDb & ML; first 2 listed genres used
`vote_average`	combined `tmdb_vote_average` and `ml_vote_average`
`vote_count`	combined `tmdb_vote_count` and `ml_vote_count`

The data was transformed in the following ways:

- Movies were matched between the two datasets based on ID, using "movieId" and "tmdbId" values provided in the ML dataset.
- The ML dataset used a 0–5 rating system while the TMDb dataset used a 0–10 rating system, the ML ratings were multiplied by 2.
- Only overlapping genres from each dataset were considered; e.g., if, for Toy Story, the ML dataset lists it as "action, animation, family" and the TMDb dataset lists Toy Story as "family, adventure, animation", the movie was considered to fall into only the "animation" and "family" genres.

Sample training and testing CSV files encompassing the transformed data can be found at the following url https://data.mendeley.com/datasets/gj66mt4s4j/2, while the code required to reproduce the results presented in this article can be found at https://github.com/spg63/MDPIApplSciCodeRepo. The code will be made available upon request to Sean Grimes or David E. Breen. Eclipse Deeplearning4j (DL4J) (https://deeplearning4j.org/) [14] is "an open-source, distributed deep-learning project in Java and Scala spearheaded by the people at Skymind (https://skymind.ai/), a San Francisco-based business intelligence and enterprise software firm." DL4J (versions 0.9.1 and 1.0.0-beta3) was used as a neural network library, providing the core multilayer perceptron classifier used by each agent (discussed in more detail in Section 2.2.1). Additionally, DL4J was used to build and test the larger MLP classifiers that were compared with our agent-based approach. All non-DL4J library code was written in Kotlin (https://kotlinlang.org/) (versions 1.3.20 - 1.3.41), running on the Java Virtual Machine (JVM) (https://www.java.com/en/download/) (versions 1.8.0_151 - 1.8.0_211). All feature, agent history data, and trained agents were stored in various databases, using SQLite3 (https://www.sqlite.org/index.html) as the database engine.

2.2. Agent Design

Agents are designed to be modular, presenting an interface that includes different algorithms for all aspects of their behavior. Agents are responsible for coordinating and managing the following functions:

- Classifier: MLP classifier with a single hidden layer
- Classifier configuration: shape, depth, activation and optimization algorithms
- Initialization algorithm: How the agents are initialized within the interaction space
- Movement algorithm: How the agents move within the interaction space
- Interaction algorithm: How (if) an agent interacts with other agents
- Scoring algorithm: How an agent reaches the conclusion it does; a combination of the internal classifier and information learned while interacting with other agents

2.2.1. Classifier

Each agent contains a very small, very simple MLP classifier. All agents were configured with similar classifiers, each containing 2–4 input nodes (one for each feature), a single hidden layer containing `numInputNodes * 2` number of nodes, and an output layer with two output nodes, one for each of the two output classes, "success" and "failure". All classifiers used the DL4J implementation of the Adam updater [15] (learning rate), softmax activation function [16], and traditional stochastic gradient descent for optimization.

Each agent's classifier was given a single feature in common with all other agents, the movie `budget`. Other features were spread across multiple agents, occasionally in pairs (e.g., `budget` & `vote_count` & `vote_average`), but more frequently a single feature in addition to `budget`. Classification performance was the determining characteristic in assigning features to agents, with very low (<50% accuracy) performing combinations dropped during early testing in favor of decreased computational complexity.

2.2.2. Agent Initialization & Movement

All agents currently participate and interact within a centralized 'interaction arena' (`Arena`) managed by a central controller responsible for registering agents, confirming that their initial location is valid, and validating each movement. All agents must be initialized within the bounds of the arena and into an empty space. All movements must be within the bounds of the arena and there can be at most two agents occupying any space within the arena. Interactions between three or more agents at the same location are not currently supported. Centralized control is currently being used to ease implementation, however it is not a requirement. Agents are capable of validating location and movement on their own, or if required, having some number n other agents confirm all positioning as valid in a decentralized manner. The `Arena` can take any 3D shape comprised of rectangles allowing for arrangements from a simple 4×4 square to something more complex, with multiple rooms, floors, and restricted movement between each, e.g., a simulated building.

Agents are currently responsible for maintaining a history of their movements within the arena, a history of interactions, and a history of how each interaction affected their internal belief. Historical information for each agent is stored in memory during each iteration and dumped to individual SQLite tables for long-term storage.

Agents are initialized with an `InitializationAlgorithm` and a `MovementAlgorithm` which implement simple interfaces, `init()` and `move()`, respectively. `init()` has a single goal, initialize the agent in the arena in an empty space. Initialization can be random, or account for complexities like location of other agents, placing agents with similar features close together, localizing similar information, or spreading them out to facilitate transmission of information between dissimilar agents. Localizing similar information may allow a group of similar agents to come to an optimal conclusion, whereas spreading similar agents out may allow the best performing agents to convince others of their opinions [17]. The `InitializationAlgorithm` used in this work randomly initialized agents within a rectangular arena.

`move()` also has a simple goal, move the agent within the arena. `move()` can be as simple or complex as necessary; randomly selecting a space within the arena and 'teleport' the agent to the new space, or it can require the agent to move towards some target location or another agent. Agents interact when two agents move onto the same space. The `MovementAlgorithm` used for this work randomly moved agents in a "Manhattan-like" fashion, allowing each agent to move one step north, south, east, or west, within the bounds of the arena.

2.2.3. Interaction & Scoring

The `InteractionAlgorithm` and `ScoringAlgorithm` are both designed to be modular, with the `InteractionAlgorithm` being responsible for deciding with whom an agent should interact, truthfulness, and trust updating. The `InteractionAlgorithm` is required to implement three functions, `shouldInteract()` which determines if the agent is interested in interacting with another agent, `truth()` which determines if the agent should be truthful with another agent, and `updateTrust()` which tries to update the other agent's trust score. `updateTrust()` is allowed to be a NO-OP function when it is not desirable to update other agents' trust scores. This work assumes all interactions are acceptable, doesn't limit repeat interactions, and requires all interactions to be truthful.

The `ScoringAlgorithm` determines how interactions update an agent's internal belief state. Agents are initialized with specific internal values, referenced in Table 2, that are (in part) updated during each interaction. Many of these values are made available to other agents during interaction, allowing each agent to determine how certain another agent is, what that agent's initial classification values were, and how much influence it will allow the agent to have over its current belief.

Table 2. Internal scoring variables.

Variable	Description
`current_prediction`	true if prediction for movie is success
`trust_score`	initialized to classifier precision, updated by other agents
`features`	a list of features used by the agent's classifier
`prior_performance`	long-term history of agent performance, varied between 0.7 and 1.3 where 1.0 is average performance
`certainty`	an average of classifier accuracy and precision, multiplied by `prior_performance`, bounded by 0.5 and 1.5
`eval_accuracy`	initial classification accuracy
`eval_precision`	initial classification precision
`eval_recall`	initial classification recall
`confidence`	biased value based on an average of accuracy, precision, and recall favoring whichever is deemed most important

Similar to the previous algorithm interfaces, the `ScoringAlgorithm` used by each agent allows the scoring to be implemented in a way most appropriate to the given problem. The algorithm is required to implement a single function, `updatePrediction` which updates the current binary prediction based on information from the most recent interaction. The `ScoringAlgorithm` used in this study works as follows: initially the agent (agent a) determines how willing it is to accept information from another agent (agent b), this is a function of a's current `certainty`, where $a_{certainty}$ represents a's current `certainty` and $a_{acceptance}$ represents a's willingness to accept information from b

$$a_{acceptance} = 1.0 - a_{certainty} \quad (1)$$

Agent a then determines how much influence b should have ($b_{influence}$).

$$b_{influence} = b_{confidence} * a_{acceptance} * b_{trustCertainty}, \tag{2}$$

where $b_{confidence}$ represents b's `confidence` and $b_{trustCertainty}$ represents b's `trust_score * certainty`. b's influence is modified based on its prior performance where $b_{priorPerf}$ represents b's prior performance, a value between 0.7 and 1.3, as noted in Table 2, and $b_{correctedInfluence}$ represents this modified value,

$$b_{correctedInfluence} = b_{priorPerf} * b_{influence}. \tag{3}$$

$b_{correctedInfluence}$ is multiplied by -1 if b's opinion (success or failure) differs from a's opinion. a's new certainty, $a_{certainty}$ is now calculated by Equation (4), where a's certainty is increased if both a and b have the same belief and is diminished if they disagree,

$$a_{certainty} = a_{certainty} + b_{correctedInfluence}. \tag{4}$$

a now checks if it should flip its opinion, which it does if $a_{certainty}$ is less than 0.50. Finally, a updates its certainty if its opinion changed,

$$a_{certainty} = 1.0 - a_{certainty}. \tag{5}$$

2.3. Opinion Aggregation

Effective opinion aggregation is an open question with many different possible approaches [18]. This research hopes to contribute more to this area in the future. We implement a voting system, where each agent receives a maximum of 100 possible votes for their preferred outcome, success or failure. We considered three methods of vote aggregation. The first and simplest method gives equal weight to each agent regardless of performance, the Unweighted Mean Model (UWM) [19]. The second method gives each agent votes based on prior accuracy, where an 80% accuracy rate would result in 80 votes, similar to the Weighted Voter Model presented in [20]. Agents are initially allowed 50 votes each until an accuracy for prior performance can be determined. The third method we used is similar to the second, however it also takes into account the trust score that other agents are allowed to modify, giving more granular control over how much influence an agent has on the aggregate opinion. Total votes for agent a is represented by $a_{totalVotes}$, where $a_{priorAccuracy}$ represents a's prior accuracy and a_{trust} represents a's trust score,

$$a_{totalVotes} = \left(\left(a_{priorAccuracy} + a_{trust}\right)/2\right) * 100. \tag{6}$$

During an interaction the agent, a, is allowed to modify another agent's, b, trust score (b_{trust}) based on how agent b has performed in the past, if agent a and b are in agreement ($doAgree$), and if prior information that a has received from b was correct. Agent a will check its interaction history, look for any interactions with b to determine what percent of interactions gave advice that was correct ($b_{percCorrect}$). If there are no prior interactions the trust score will not be modified. The trust score can be modified a maximum of 5% during each interaction.

$$b_{trust} = b_{trust} + 0.05\left(b_{percCorrect} * b_{priorPerf}\right) * doAgree \tag{7}$$

A high-level overview of the training, interaction, and voting process can be seen in Figure 1. The internal MLP for each agent is trained using available training data, the agents are presented with a binary question, they are initialized in an arena where they move and interact for some number of steps (based on time or total interactions). Agents are then assigned some number of votes based on the system described in Section 2.3 and they vote at the end of the interaction period.

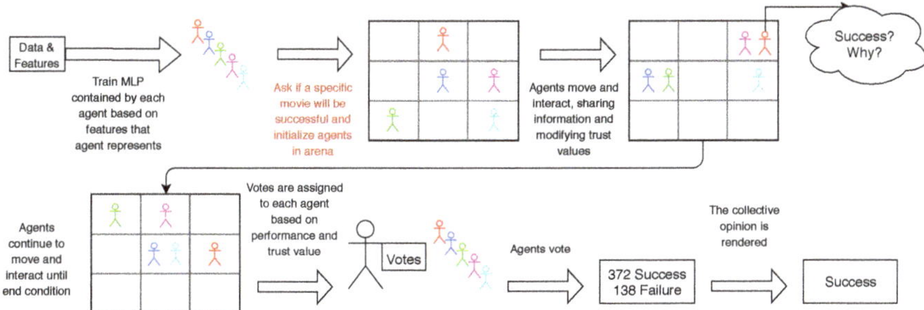

Figure 1. Agent training, initialization, movement, interaction, and voting.

3. Results

We compared the results from our social, agent-based approach to the results produced by multiple configurations of a traditional, monolithic MLP developed in DL4J. All agent classifiers were trained on 4× Nvidia GTX 1070 GPUs using Cuda (versions 10.0–10.1 update 2) through the DL4J library. Table 3 shows the configuration and accuracy under different training conditions for each of the 10 agents. All agent classifiers were trained in parallel, taking an average of 2.8 s for 5 epochs and 22 s for 50 epochs; there was no accuracy improvement beyond 50 epochs. Once trained, the agents can participate in any decision-making configuration without re-training their classifiers. The best performing agent classifier was the `budget`, `vote_count`, `vote_average` agent, with the most important feature being `budget`, followed by `vote_count`. The worst performing agent was the `budget`, `runtime` agent.

Table 3. Agent Classifier Accuracy for 5 and 50 epochs.

Features	5 Epochs	50 Epochs
`budget`, `revenue`	98%	100%
`budget`, `vote_average`, `vote_count`	77.2%	77.6%
`budget`, `tmdb_popularity`, `vote_average`, `vote_count`	75.4%	75.7%
`budget`, `vote_count`	75.7%	75.5%
`budget`, `tmdb_popularity`, `tmdb_vote_average`, `tmdb_vote_count`	72.8%	74.9%
`budget`, `tmdb_vote_count`, `ml_vote_count`	73%	73.4%
`budget`, `ml_vote_average`, `ml_vote_count`	62.2%	64.1%
`budget`, `ml_vote_count`	60.3%	61.9%
`budget`, `tmdb_vote_average`	60.9%	61.4%
`budget`, `runtime`	53.9%	56.4%
Average (`budget`, `revenue` agent removed)	67.93%	68.99%

We tested 10 MLP networks, with a variety of feature sets, and with one containing the final `revenue`. Figure 2a shows the accuracy for five classifier configurations when trained for 5 and 50 epochs. The figure shows the change in classifier accuracy when various features have been removed as inputs into the network.

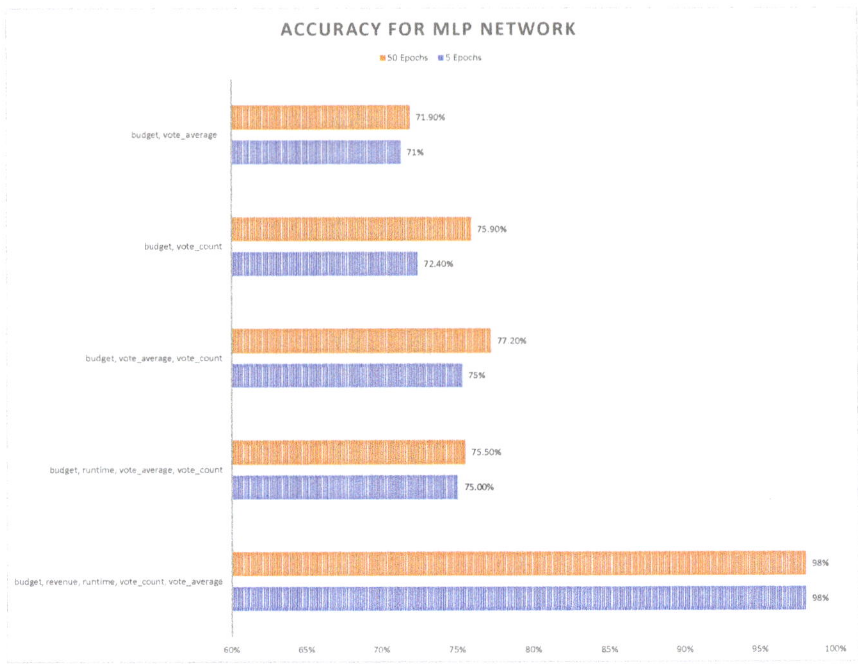

(**a**) Accuracy for single hidden layer MLP classifier. The y-axis shows which feature were included.

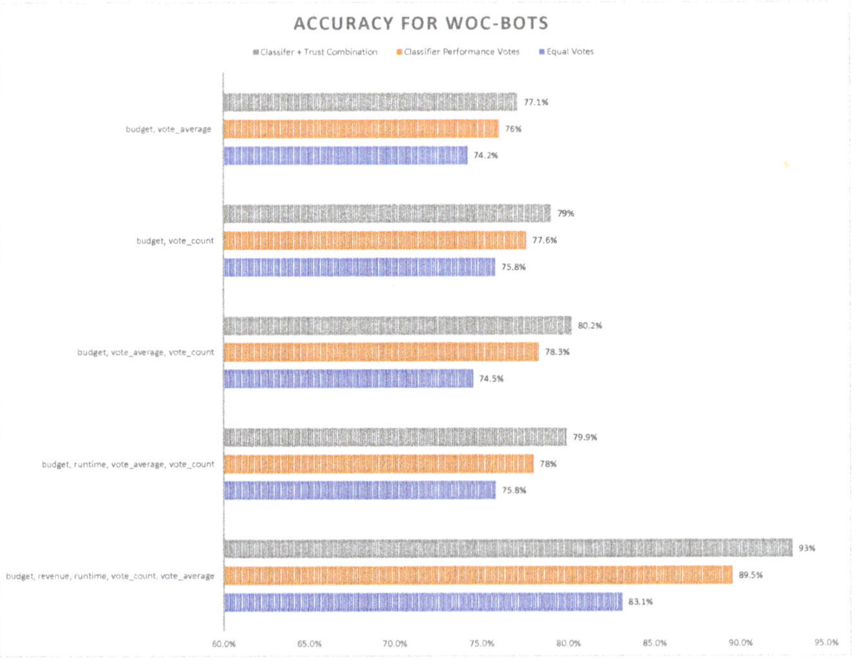

(**b**) Accuracy for agent interaction. The y-axis shows features included.

Figure 2. Comparison of MLP and Woc-Bots performance.

All MLP classifiers were trained individually with an average training time of 2.6 s for 5 epochs and 21.2 s for 50 epochs. Comparing training times with the social agents, training 10 MLP classifiers in parallel took 1 min and 1 s (3 min and 32 s if computed sequentially) vs. 22 s to train 10 agents for 50 epochs. It should be noted that inference is slower using WoC-Bots; it takes an average of 260 milliseconds to test 740 examples using an MLP classifier incorporating all the features listed in Table 1, while it takes the WoC-Bots, encompassing the same feature set, an average of 13.4 s to test the same 740 examples. But, once trained, the agents can be reconfigured to compute new prediction results for different feature sets, without requiring retraining, unlike a monolithic MLP.

Accuracy results from five configurations of our social, agent-based prediction system can be found in Figure 2b. We show results for three aggregation mechanisms after 50 epochs of training: (1) unweighted equal voting, (2) votes assigned based on initial classifier performance, and (3) votes assigned based on classifier performance and agent trust (described in Section 2.3), with method (3) consistently out-performing methods (1) and (2). Method (1) is represented by the blue bar, method (2) by the orange bar, and method (3) the grey bar. We tested similar configurations across our agents and MLP networks. Data from Movielens and TMDb was combined, as described in Section 2.1, with no agent receiving information from only one source.

Similar to Figure 2a, Figure 2b's labels show which features were included in the interaction. Feature distribution across agents was optimized for accuracy, within the limits of available features. Five agents participated in each interaction, with the `budget, vote_average` and `budget, vote_count` interactions being comprised of five copies of the same agent.

WoC-Bots out-performed the MLP classifier in all cases except where final revenue was included as a feature, indicating that our aggregation method does not give enough weight to an agent with exceptionally good performance. We tested removing a highly correlated (http://ibomalkoc.com/movies-dataset/) feature, `vote_count`, which caused a performance decline in both the MLP and social agents, with the MLP network accuracy declining 4% compared to a decline of 1.9% in WoC-Bots, indicating our agents are more resistant to feature drop-out. We also tested removing an unimportant feature, `runtime` which showed a 1.7% performance increase in the MLP network and only a 0.3% increase for WoC-Bots, indicating poorly performing agents have little impact on other agents during the interaction period and receive few votes during opinion aggregation. Statistical analysis confirms that `runtime` is not highly correlated in both the TMDb dataset and an ensemble dataset combining the Movielens and TMDb data, as used in this article [21,22].

Figure 3 shows the performance of MLP networks and WoC-Bots as features are systematically added. The results presented in this figure are produced via the classifier performance & trust aggregation mechanism. The agents are configured to allow for maximum agent participation without duplicating agents in any simulation testing more than two features. Four copies of an agent, representing `budget` and `vote_count`, participated in the first simulation. Four agents participated in the `budget, vote_count, popularity` simulation, eleven agents participated in each of the following four-feature simulation. Twenty-six agents participated in the `budget, vote_count, vote_average, runtime, popularity` simulation with one agent receiving five features, five agents receiving four features, 10 agents receiving three features, and 10 agents receiving two features.

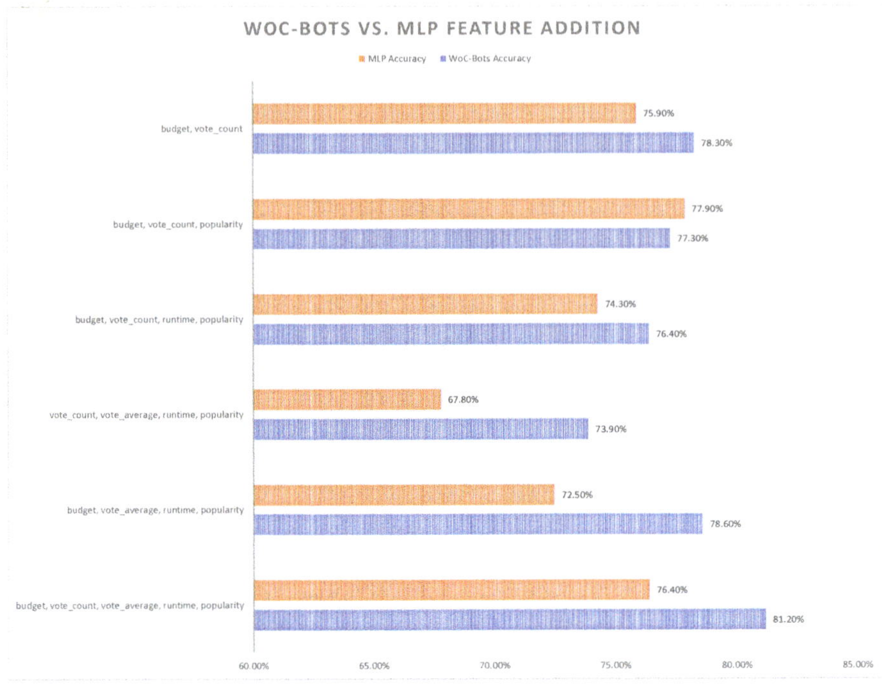

Figure 3. Accuracy of MLP vs WoC-Bots w/Max agent configuration while adding features.

In five out of six simulations WoC-Bots out-performed the MLP network, and significantly out-performed the MLP network when the most important feature, `budget`, was removed. WoC-Bots performed best when all features were available, and when the maximum number of unique agents were participating in the simulation. The MLP network performed best when the three most highly correlated features were the only features being considered. This performance difference indicates WoC-Bots are able to gather additional information from features that are less correlated with `revenue` without a net negative impact to their accuracy from the additional feature noise.

Figure 4 shows the accuracy for training epochs 1–50 for an MLP network and WoC-Bots. The network was configured with five features, `budget`, `vote_count`, `vote_average`, `runtime`, and `popularity`. Five agents participated in the simulation with four agents receiving two features and one agent receiving five features. Each two-featured agent received `budget` as a feature and one other feature from the list of features. No agent was duplicated. We choose this agent and feature configuration to make as fair and direct comparison with an MLP network as possible despite WoC-Bots performing better when more agents participate in the simulation, as seen in the `budget`, `vote_count`, `vote_average`, `runtime`, `popularity` simulation in Figure 3 where 26 agents were allowed to participate. WoC-Bots are out-performed in this configuration, slightly, by the MLP network when trained for more than 40 epochs, however they are able to more quickly integrate information compared to the MLP network, reaching an optimum (76.3%) at 20 epochs vs. the MLP optimum (76.8%) at 40 epochs.

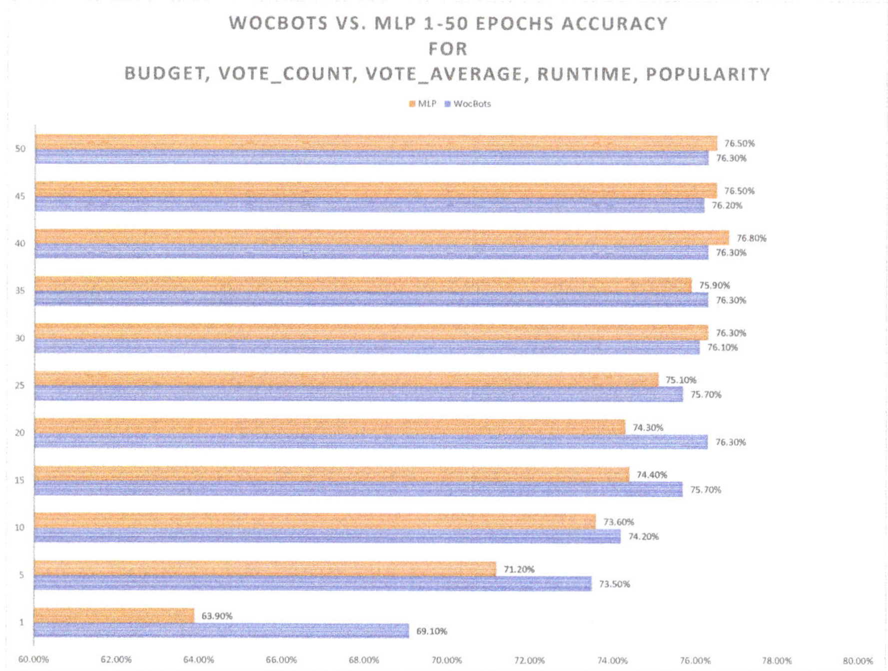

Figure 4. Accuracy of MLP and WoC-Bots for five features over 1–50 epochs.

4. Discussion

Our system can easily include new features by creating a new agent to represent those features, allowing the agent to be added to the next "interaction" without the need to re-train a network or employ complex, dynamically expandable networks found in [2] when incorporating new data. Additionally, this design allows us to quickly test the impact of removing features or testing various combinations of features without the time-consuming re-training step required when changing features in a full MLP network. This allows us to easily removing features, like `runtime`, to test how they impact the final prediction.

WoC performance, or a computer-agent-based version of it, depends on two attributes, a diverse and independent crowd and an aggregation mechanism that assigns appropriate weights to individuals within the crowd to reach the correct collective decision [23]. We also know from Othman [5] that it is not reasonable to develop a computer agent that accurately represents the intricate and diverse knowledge that humans have. Therefore, we need to find the right balance between independently thinking agents within a crowd and information sharing to better represent the diverse knowledge of human agents.

5. Conclusions & Future Work

We have demonstrated a robust, flexible alternative to traditional ANN methods for making predictions about specific future events. Our implementation takes ideas from prediction markets, wisdom of crowds, and multi-agent systems to use simple, modular agents in a social setting to answer binary questions. Our results show that we can attain similar results to that of a multilayer perceptron when classifying Hollywood movies, while requiring less training time and offering more flexibility and prediction options. Further, our system is robust, demonstrating only a 1.9% loss in accuracy when losing the `vote_count` feature versus a 4% loss in accuracy when the same feature was removed from the MLP network.

We have three main areas to focus on for improvement in the future. Recent work on deep neural networks is starting to explain "why" we get certain output. However, there is still a long way to go before we have the ability to easily answer this question [24]. Our system offers a framework, using a multi-agent approach, that should allow us to answer "why" more easily; at the core of each agent is a *very* simple, single hidden layer MLP. Agents track all interactions, how those interactions affect their internal belief, and how they change the trust value of other agents. Given the state of the system during a simulation, the internal belief and trust scores of each agent, as well as each agent's interaction history, we can follow the history of each agent, starting with its initial belief post-classification, through each interaction, allowing us to see when and why an agent's belief changed (or stayed the same).

The two other areas we will address in future work are (1) the interaction, movement, and initialization algorithms, allowing us to change and optimize the distribution and flow of information and (2) the aggregation mechanism. We will use theories from swarm intelligence [25] to better aggregate the information that each agent possesses in a manner that better extracts information from the **correct** agents while limiting the impact that **incorrect** agents have on the collective opinion. Unanimous A.I. (https://unanimous.ai/) has a unique, swarm-based aggregation method that is capable of arriving at a collective answer. Unanimous A.I. maintains a "human-in-the-loop" approach [26], where their 'swarm' is comprised of humans, answering binary and non-binary questions by working together to move a virtual puck to the collective answer [27]. We prefer a computer-agent-based approach that allows for new agents to be created as needed to answer questions as they come up. Our future work will focus on implementing a swarm-based algorithm to produce an "emergent prediction" from a group of relatively simple, modular agents.

Author Contributions: Conceptualization, S.G.; methodology, S.G. and D.B.; software, S.G.; investigation, S.G. and D.B.; resources, S.G. and D.B.; data curation, S.G.; writing–original draft preparation, S.G.; writing–review and editing, S.G. and D.B.; supervision, D.B.

Funding: This research received no external funding.

Conflicts of Interest: The authors declare no conflict of interest.

Abbreviations

The following abbreviations are used in this manuscript:

MDPI	Multidisciplinary Digital Publishing Institute
DOAJ	Directory of open access journals
WoC	Wisdom of the Crowd
MLP	Multi-layer Perceptron
ANN	Artificial Neural Network
PM	Prediction Market
TMDb	The Movie Database
API	Application Programming Interface
ML	MovieLens
DL4J	Deeplearning4j
JVM	Java Virtual Machine
UWM	Unweighted Mean Model

References

1. Olivas, E.S. *Handbook of Research on Machine Learning Applications and Trends: Algorithms, Methods, and Techniques: Algorithms, Methods, and Techniques*; IGI Global: Hershey, PA, USA, 2009.
2. Yoon, J.; Yang, E.; Lee, J.; Hwang, S.J. Lifelong learning with dynamically expandable networks. *arXiv* **2017**, arXiv:1708.01547.

3. Parkes, D.C.; Seuken, S. Prediction Markets. In *Introduction to Economics and Computation: A Design Approach*; Cambridge University Press: Cambridge, UK, 2016; Chapter 18.
4. Yi, S.K.; Steyvers, M.; Lee, M.D.; Dry, M.J. The wisdom of the crowd in combinatorial problems. *Cogn. Sci.* **2012**, *36*, 452–470. [CrossRef] [PubMed]
5. Othman, A. Zero-intelligence agents in prediction markets. In Proceedings of the 7th International Joint Conference on Autonomous Agents and Multiagent Systems, Honolulu, HI, USA, 14–18 May 2007; Volume 2, pp. 879–886.
6. Chen, Y.; Reeves, D.M.; Pennock, D.M.; Hanson, R.D.; Fortnow, L.; Gonen, R. Bluffing and strategic reticence in prediction markets. In Proceedings of the International Workshop on Web and Internet Economics, San Diego, CA, USA, 12–14 December 2007; pp. 70–81.
7. Dimitrov, S.; Sami, R. Non-myopic strategies in prediction markets. In Proceedings of the 9th ACM Conference on Electronic Commerce, Chicago, IL, USA, 8–12 June 2008; pp. 200–209.
8. Othman, A.; Sandholm, T. When do markets with simple agents fail? In Proceedings of the 9th International Conference on Autonomous Agents and Multiagent Systems, Toronto, ON, Canada, 9–14 May 2010; Volume 1, pp. 865–872.
9. Ertekin, Ş.; Rudin, C.; Hirsh, H. Approximating the crowd. *Data Min. Knowl. Discov.* **2014**, *28*, 1189–1221. [CrossRef]
10. Page, S.E. *The Difference: How the Power of Diversity Creates Better Groups, Firms, Schools, and Societies-New Edition*; Princeton University Press: Princeton, NJ, USA, 2008.
11. Helsinger, A.; Thome, M.; Wright, T. Cougaar: A scalable, distributed multi-agent architecture. In Proceedings of the IEEE International Conference on Systems, Man and Cybernetics, The Hague, The Netherlands, 10–13 October 2004; Volume 2, pp. 1910–1917.
12. De Vany, A. *Hollywood Economics: How Extreme Uncertainty Shapes the Film Industry*; Routledge: Abingdon, UK, 2003.
13. Harper, F.M.; Konstan, J.A. The movielens datasets: History and context. *ACM Trans. Interact. Intell. Syst.* **2016**, *5*, 19. [CrossRef]
14. Team, D. Deeplearning4j: Open-source distributed deep learning for the JVM. *Apache Softw. Found. Licens.* **2018**, *2*.
15. Kingma, D.P.; Ba, J. Adam: A method for stochastic optimization. *arXiv* **2014**, arXiv:1412.6980.
16. Goodfellow, I.; Bengio, Y.; Courville, A. *Deep Learning*; MIT Press: Cambridge, MA, USA, 2016.
17. Garnett, P.; Bissell, J. Modelling Social Networks Reveals How Information Spreads. 2013. Available online: http://theconversation.com/modelling-social-networks-reveals-how-information-spreads-18776 (accessed on 10 June 2019).
18. Du, Q.; Hong, H.; Wang, G.A.; Wang, P.; Fan, W. CrowdIQ: A New Opinion Aggregation Model. In Proceedings of the 50th Hawaii International Conference on System Sciences, Hilton Waikoloa Village, HI, USA, 4–7 January 2017.
19. Hastie, R.; Kameda, T. The robust beauty of majority rules in group decisions. *Psychol. Rev.* **2005**, *112*, 494. [CrossRef] [PubMed]
20. Valentini, G.; Hamann, H.; Dorigo, M. Self-organized collective decision making: The weighted voter model. In Proceedings of the International Conference on Autonomous Agents and Multi-Agent Systems, Paris, France, 5–9 May 2014; pp. 45–52.
21. Awasthi, D. Exploratory Analysis of Movies on TMDB. Available online: https://rstudio-pubs-static.s3.amazonaws.com/369891_b123051c3cb64da5a6d22a8d0b6e0d84.html (accessed on 19 October 2019).
22. Malkoc, I. The Movies Dataset. Available online: http://ibomalkoc.com/movies-dataset/ (accessed on 23 August 2019).
23. Surowiecki, J. *The Wisdom of the Crowds*; Anchor Books, a division of Random House: New York, NY, USA, 2005.
24. Shwartz-Ziv, R.; Tishby, N. Opening the black box of deep neural networks via information. *arXiv* **2017**, arXiv:1703.00810.
25. Zhu, Y.F.; Tang, X.M. Overview of swarm intelligence. In Proceedings of the IEEE International Conference on Computer Application and System Modeling, Taiyuan, China, 22–24 October 2010; Volume 9, pp. 9–400.

26. Rosenberg, L. Artificial Swarm Intelligence, a Human-in-the-loop approach to AI. In Proceedings of the 30th AAAI Conference on Artificial Intelligence, Phoenix, AZ, USA, 12–17 February 2016.
27. Rosenberg, L.; Pescetelli, N.; Willcox, G. Artificial Swarm Intelligence amplifies accuracy when predicting financial markets. In Proceedings of the IEEE 8th Annual Conference on Ubiquitous Computing, Electronics and Mobile Communication, New York, NY, USA, 19–21 October 2017; pp. 58–62.

Article

An Agent-Based Model Driven Decision Support System for Reactive Aggregate Production Scheduling in the Green Coffee Supply Chain

María del Rosario Pérez-Salazar [1], Alberto Alfonso Aguilar-Lasserre [1,*], Miguel Gastón Cedillo-Campos [2], Rubén Posada-Gómez [1], Marco Julio del Moral-Argumedo [1] and José Carlos Hernández-González [1]

1 Division of Research and Postgraduate Studies, Tecnologico Nacional de Mexico, Instituto Tecnologico de Orizaba, Orizaba 94320, Veracruz, Mexico; rosario.perez.salazar@gmail.com (M.d.R.P.-S.); pgruben@yahoo.com (R.P.-G.); marcojulioarg@gmail.com (M.J.d.M.-A.); jos_car.01@live.com.mx (J.C.H.-G.)

2 Transportation Systems and Logistics National Laboratory, Mexican Institute of Transportation, San Fandila 76703, Querétaro, Mexico; gaston.cedillo@imt.mx

* Correspondence: albertoaal@hotmail.com; Tel.: +52-(272)-725-7056

Received: 14 October 2019; Accepted: 6 November 2019; Published: 15 November 2019

Featured Application: The agent-based model driven decision support system can handle the delay in the arrival of raw materials while considering planning scenarios reflecting the context of the green coffee production, scenarios where the demand is skewed towards the end of the planning horizon, where the demand is skewed towards the beginning of the planning, and where demand peaks in the middle of the planning horizon and falls under the available capacity on the first and last days of the horizon. The exhibition of the management process within the green coffee supply chain context may help practitioners and managers interested in implementing the agent-based modeling and simulation approach to increase the possibilities of successful adopting of the reactive aggregate production scheduling.

Abstract: The aim of this paper is to contribute to the thread of research regarding the need for logistic systems for planning and scheduling/rescheduling within the agro-industry. To this end, an agent-based model driven decision support system for the agri-food supply chain is presented. Inputs in this research are taken from a case example of a Mexican green coffee supply chain. In this context, the decision support agent serves the purposes of deriving useful knowledge to accomplish (i) the decision regarding the estimation of Cherry coffee yield obtained at the coffee plantation, and the Parchment coffee sample verification decision, using fuzzy logic involving an inference engine with IF-THEN type rules; (ii) the production plan establishment decision, using a decision-making rule approach based upon the coupling of IF-THEN fuzzy inference rules and equation-based representation by means of mixed integer programming with the aim to maximize customer service level; and (iii) the production plan update decision using mathematical equations once the customer service level falls below the expected level. Three scenarios of demand patterns were considered to conduct the experiments: increasing, unimodal and decreasing. We found that the input inventory and output inventory vary similar over time for the unimodal demand pattern, not the case for both the increasing and decreasing demand patterns. For the decreasing demand pattern, ten tardy orders for the initial production schedule, an 88% service level, and nineteen tardy orders from the estimated production results, a 77% service level. This value falls below the expected level. Consequently, the updated aggregate production schedule resulted in ten tardy orders and an 88% service level.

Keywords: decision support system; agent based modeling and simulation; production scheduling; green coffee supply chain

1. Introduction

A main challenge for the agri-food supply chain (ASC) relates to the need for logistic systems for planning and scheduling/rescheduling due to unpredictable variations in quality, moment and quantity in primary production; the need for high efficiency of technical equipment despite long food industry production times; and an intricate network structure where many farms and food processors trade with multinationals in the wholesaler/retail sector. This challenge is derived from the following ASC characteristics [1]: (i) for the primary producer echelon, the seasonal growth is often limited to a specific period in a geographic region, dependent on weather and agricultural practices management, (ii) for the food processor echelon, high volume and low variety production systems combined with high volume frequent deliveries, and (iii) for the wholesaler/retailer echelon, the variability of quality and quantity of supply of farm-based inputs coupled with high demands from consumers, including food safety legislation and quality standards. This challenge calls for joint decision making in order to leverage the knowledge resources in the ASC.

From an agent-based modeling and simulation approach, a system is modeled as a set of autonomous agents that interact with each other and the environment; the agents have behaviors that are influenced by agent's interactions [2]. According to an agent's behavior [3] (i) agents can respond in an event-action-mode (reactive agent), (ii) agents can have domain knowledge to undertake a sequence of actions in order to achieve a goal (deliberative agent), and (iii) agents can encompass both of this features. These behaviors have important implications in the use of the agent-based modeling approach as a valid methodology to model the supply chain (SC). Moreover, for SC researches and practitioners, Hilletofth and Lättilä [4] stressed the benefits of agent-based decision support systems including the ability to convert manager experience into agents, the ability to conduct experiments and what-if analysis through simulation-based decision support systems and the increased predictability of operations in the real system. The core functionality component of a simulation model-driven decision support system is a quantitative model and is used by decision-makers to help in analyzing a real system by means of modeling and data collection, model validation, system parameter setting, and system evaluation [5].

Agent-based modeling use and simulation for decision-making within the agri-food industry has been applied considering the integration of logistics, quality decay and sustainability modeling [6,7]. According to findings from the review conducted by Utomo et al. [8], most agent-based modeling and simulation applications in agri-food supply chains focus on the simulation of production planning and investment decisions. However, although there is a volume of literature about applications of agent-based modeling and simulation in the agri-food supply chains, there is a lack of studies that consider important actors, such as food processors and retailers in the scope of the model, since most agent-based modeling and simulation applications focus on one echelon [8].

A make-to-order ASC is considered for this study. For these types of chains, Chatfield et al. [9] classified the improvement opportunities for the SC modeling approaches into three categories: model building, model quality, and model execution. Regarding the agent-based modeling approach, the authors highlighted the issue with model quality as a measure of how well a model represents the aspects of interest in a real system and how a completely agent-based approach is not the best way to represent the entire supply chain. Consequently, as the agent-based approach focus on the behavior and decision processes of individual participants, often at the expense of event-oriented aspects of the supply chain, hybrid configurations are often necessary.

The aim of this paper is to contribute to these strands of research. To this end, an agent-based model driven decision support system for the agri-food supply chain is presented. The agent-based modeling and simulation and discrete-event simulation (DES) combination is the agent-related hybrid configuration used for the development decision support system. The agent-based modeling and simulation and DES combination are one of the agent-related hybrid configurations that have received the most attention. According to Macal [10], a hybrid modeling challenge exists and refers to the understanding of how agent-based modeling can be effectively used with other simulation and modeling techniques operating together in the same hybrid model in such a way that each technique addresses the part of the problem that it does best.

Inputs in this research are taken from a case example of a Mexican green coffee SC. The green coffee SC comprises four stakeholders from Cherry coffee cultivation to processed green coffee: (i) Cherry coffee growers; (ii) Parchment coffee producers; (iii) green coffee producers; and (iv) green coffee toasters. In this context, the decision support agent serves the purposes of deriving useful knowledge to accomplish (i) the decision regarding the estimation of Cherry coffee yield obtained at the coffee plantation, and the Parchment coffee sample verification decision, using fuzzy logic involving an inference engine with IF-THEN type rules; (ii) the plan production establishment decision, using a decision-making rule approach based upon the coupling of IF-THEN fuzzy inference rules and equation-based representation by means of mixed integer programming with the aim to maximize customer service level; and (iii) the production plan update decision using mathematical equations once the customer service level falls below the expected level.

This paper is structured as follows: Section 2 presents an overview of the strands of theory used to underpin the proposed system. Section 3 describes the methodology, while Section 4 presents the results of the system assessment in the case study. Finally, Section 5 summarizes our conclusions.

2. Literature Overview and Work Position

The present work will try to integrate the following strands of theory to underpin the decision support system for the ASC. The first strand related to the use of agents in SC modeling and simulation. Based on the concept of SC uncertainty revised by van der Vorst and Beulens [11] as a decision-making situation in which the decision-maker lacks understanding; information processing capacities and effective control actions; the authors assert that SC uncertainty could be reduced through the implementation of specific-scenario redesign strategies regarding configuration, control structure, information systems, and governance structures. Supply chain value stream mapping is a technique to leverage the knowledge of a company's supply actors [12].

van der Zee and van der Vorst [13] proposed a modeling framework for decision-making improvement based on agents modeling the SC actors as autonomous objects assigned with decision making intelligence, jobs representing the SC activities, and types of flows (goods, information, resources, and job definitions). Broadly, an agent possesses skills and knowledge to interact with the environment including applications for cooperation, communication, command and control [14]. A multi-agent system is defined by Turban et al. [15] as: "a computer-based environment that contains multiple software agents to perform certain tasks"; therefore, its scope refers to the breakdown of a complex solution into sub-problems then assigned to agents supported by a knowledge base. Furthermore, decision support systems agents could be classified into five types [15]: data monitoring, data gathering, modeling, domain managing, and preference learning. Indeed, agent-based decision support systems enable decision-making activities such as knowledge representation, knowledge reuse, reasoning, and inference techniques [16]. These properties have implications on industrial environment agent-based solution adoption, feasibility, breakdown robustness, ready reorganizability, effective response to external disruptions, and reconfigurability. Nevertheless, the adoption barriers comprise cost, guarantees for operational performance, scalability and standards definition [17]. Other issues related to barriers of agent-based decision support systems comprise [4] the difficulty to access data from partners in the SC, long development and validation time, long learning time, and the difficulty to develop agent rules that generates the wanted behavior. Consequently, the next generation of decision agents in SC management must consider these barriers in order to develop agents embedded in systems that will be distributed, dynamic, intelligent, integrated, responsive, reactive, cooperative, interactive, reconfigurable, and adaptable [18].

In the review conducted by Méndez et al. [19] regarding optimization methods for short-term scheduling, the authors stated that within the artificial intelligence field, scheduling problems have been solved by a set of individual agents which can work parallel and their coordination may bring a more effective way to find an optimal solution; the agents are expected to interact together to achieve the goals of the overall system. The second strand of theory comprises to agent-based decision making for scheduling in supply chains. According to Phanden et al. [20], the agent-based modeling technique is the most promising distributed approach to tackle the integration of process planning and scheduling adaptiveness. For their part, Barbati et al. [21] asserted that a relevant number of applications of agent-based models are devoted to SC planning problems. In that sense, industrial agent-based solution adoption has focused in following areas [17]: (i) distributed solutions for real-time manufacturing control problems; (ii) distributed solutions for complex operations management problems such as planning, scheduling, initiating execution, and monitoring; and (iii) distributed solutions for coordinating supply chains that will integrate manufacturing, sales networks, suppliers, customers, and third-party coordinators. The application areas of multi-agent systems designs have recently evolved from intra-business processes such as job scheduling and production planning and coordination to complicated decision procedures involving the management of independent companies but interacting supply chain management partners [22]. Within the manufacturing context, an agent is an intelligent entity is capable of acting and decision making to accomplish tasks, such as distributed production planning, scheduling, and execution control [3]. In multi-agent scheduling, agents manipulate both resource and order variables under their own authority [23].

The third strand of theory encompasses the issues regarding production planning and quality sorting in the ASC with focus on the coffee SC. An ASC is defined by van der Vorst et al. [24] as chains where: "agricultural products are used as raw materials for producing consumer products with higher added value." For the primary producer echelon of the ASC, the decision-making processes can broadly be divided into three stages [25]: production planning, cultivation practices, and post-harvest management and marketing; production planning relates to crop production planning based on market forces, soil testing and crop rotation practices, whilst cultivation practices encompasses decisions regarding crop nutrition and irrigation management for maximization of the total production of each crop. Indeed, a challenge for the agricultural sector relates to the need for a reactive and flexible crop production supply chains with high yield at low cost [26]. From an agricultural value chain point of view, Higgins et al. [27] argue: "multi-agent models provide a capacity to accommodate the complexity of relationships between and within value chain segments by representing these segments (or their activities) as agents." For their part, Tsolakis et al. [28] enlisted the decisions for tactically and operational planning in an ASC, including the planning of harvesting operations and logistics operations and the adoption of quality management policies. With regards to quality sorting and grading, van der Vorst et al. [29] stated that quality controlled logistics in the ASC entails an adaptive control based upon customer requirements and current agri-product quality.

From an agent-based simulation perspective, Handayati et al. [30] identified the value co-creation in a sustainable ASC, understanding sustainability as a the integration of the moral, ecological, technical, economic, and social dimensions of human activity [31]: (1) Planning: agro-input selection, cultivating and harvesting scheduling, more certain demand and price, more certain supply; (2) Cultivating and Harvesting: exporters requirement fulfillment, and good agricultural practice; (3) Post-harvest and Distribution: good post-harvest handling, cold storage system for maintaining the freshness of agri-product; and (4) Consumption: customer's requirement fulfillment.

From a sustainability perspective and with regard to the Central American coffee supply, Killian et al. [32] found that organic and fair trade certification production schemes seem to be a viable strategy for Central American farmers to receive better pricing and to improve productivity to maintain or increase farm income. Moreover, Killian et al. [33] determined the carbon footprint of the SC of Costa Rican coffee, sources of the most intense emission and mitigation possibilities.

Regarding coffee yield analysis, Espinosa-Solares et al. [34] found that Cherry coffee yield is affected more by cultivar characteristics than by harvest date in a two year study in Mexico. For their part, the study of Bosselmann et al. [35] demonstrated that shade trees not be planted with the purpose of improving beverage quality in small holder coffee agroforestry systems in Southern Colombia.

For quality sorting in the Green coffee SC, Green coffee assessment focus on Acidity, Body, and absence of Defects [36]. Livio and Hodhod [37,38] developed a fuzzy expert system for sensorial evaluation of coffee bean attributes to derive quality scoring comprising 11 attributes: fragrance, flavor, aftertaste, acidity, body, uniformity, balance, clean cup, sweetness, overall and defects. Testing results of the system shown 95% of matching results compared to the experts' evaluations. Flores and Pineda [39] also presented a fuzzy logic expert system with the aim to train Honduran coffee cuppers; testing results of the system shown 97% of matching results evaluating the attributes brew, aroma, taste, aftertaste, and body.

3. Methodology

3.1. Case Study

Green coffee production is generally characterized by both the necessity of management agricultural practices improvement and production technology implementation, which is related to production yield and quality in coffee in coffee plantations with an average yield of 2408 hectograms per hectare (hg/ha), against 5333 to 25,487 hg/ha reported by the eight countries with highest productivity [40]. The case study is an order-driven Mexican green coffee SC.

The green coffee SC actors are described below and depicted in Figure 1. The second tier suppliers are the Cherry coffee growers. In this echelon, the cultivation and harvest of the Cherry coffee take place in the coffee plantations. In the region where the case study chain is located, there is an altitudinal gradient from sea level to above 3000 m above sea level. The average annual temperature ranges from 12 °C to 24 °C, the coffee soils of the region can be classified as suitable, medium and unfit and the annual precipitation oscillates between 1000 and 3000 mm [41]. These geo-agro climatic characteristics provide a very varied mosaic wherein each of the regions you can find sites of high, medium, and low potential both production and quality of coffee.

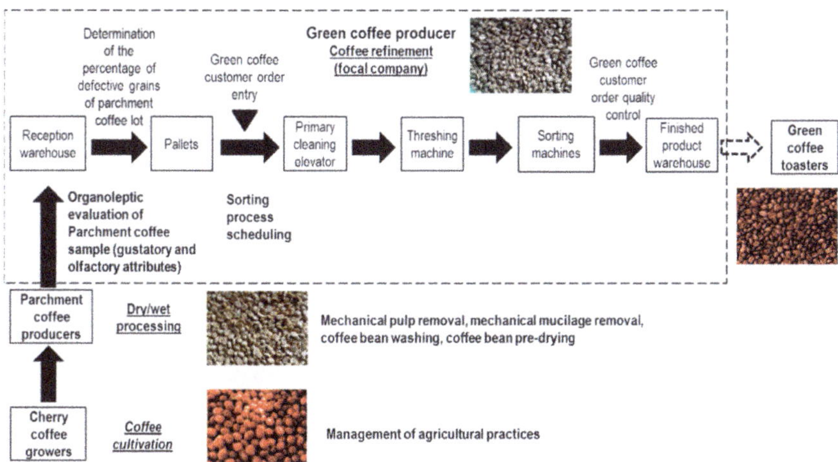

Figure 1. The green coffee SC under investigation.

The harvest refers to the cut of ripe Cherry coffee, with the cut of the fruits in a mature state, weight is gained in the scale in the sale process, the benefit process is facilitated, the production loss is reduced when the green coffee is prepared, and organoleptic quality is gained. Cherry coffee is the most frequent form of sale with a local or industrial intermediary-collector, where the process of wet profit is carried out. Cherry coffee is transported in plastic sacks or tarpaulins usually used in livestock feed or various grains.

The first tier supplier is the Parchment coffee producer, local and industrial intermediary that carry out the wet benefit process. The wet benefit process consists of mechanically removing the pulp (exocarp) of the coffee Cherry in the presence of water or without it, and followed either by (i) the removal of the mucilage (mesocarp) by fermentation or mechanical method, followed by the washing to obtain Parchment coffee, or (ii) direct drying of the pulpless grains inside the mucilaginous [42]. Parchment followed by threshing to produce semi-washed green coffee. Next, drying of the coffee beans either through the combined effect of sunlight and aeration or drying machines, to obtain coffee in the shell. The drying time is variable in each of the methods, the goal being in any case that the humidity in the coffee bean reaches within the parameter of 11.5–12.5%. The final product of the wet processing process is called Parchment coffee; the name comes from the fibrous parchment-like husk which covers the grain at the end of the process. The storage is carried out in warehouses using mainly jute bags. The process of commercialization of Parchment coffee is carried out mainly between individual producers, or through their organizations, towards industrial exporters, who will sell it in the national and international markets.

The focal company of this study is an industrial green coffee producer that stores and distributes both regional and nationally. In this echelon, the Parchment coffee refinement takes place, the dry benefit process. In the dry benefit process, the Parchment coffee received and graded is threshed to release the green coffee bean, to later be classified by its size, density, and color, as well as being cleaned of foreign objects. In the threshing process, the dry endocarp is removed from the Parchment coffee to produce green coffee. The dry benefit process starts with the Parchment coffee reception where the coffee is sampled to perform an organoleptic evaluation and physical revision. The organoleptic evaluation of coffee sample is the process of sensorial evaluation of coffee beans. In this process, the toasting and grinding of a sample of Parchment coffee are carried out, an infusion of the roasted and ground coffee sample is prepared in freshly boiled water, from which the gustatory and olfactory characteristics of the grain are evaluated, such as flavor, body, aroma, and acidity. In the physical analysis, defects are visually identified; the defects refer to irregularly shaped coffee beans and coffee beans of irregular appearance [43].

The purchase decision depends on the results of these evaluations. If the coffee is purchased, the coffee entry quality grading takes place, namely, the defective beans percentage in the coffee lot is determined, and then, the lot is stacked on pallets in the warehouse area designated for each coffee type. Next, in the pre-cleaning process, the coffee enters the pre-cleaning machine, where foreign materials of a different origin than coffee are eliminated. After the pre-cleaning process, the threshing process takes place. The pre-cleaning is the technological operation used to reduce the percentage of humidity of the Parchment coffee to a level of 10 to 12.5%, which allows threshing under satisfactory technical conditions. In the threshing process, the dried endocarp is removed from the Parchment coffee of natural coffee to produce green coffee. At the entrance to the thresher, a quality control point is present in which the metals that Parchment coffee could carry are eliminated. Finally, the sorting process is the technological operation used to eliminate foreign matter, fragments of coffee and defective grains of green coffee, and to separate healthy coffee beans according to their shape, size, and weight. The machinery, the manual labor or the combination of both, can be variable but in general the methods of this process are classified into sorting by sieve, sorting by vibration-gravity, pneumatic sorting, and optical sorting. The result of the process is the production of green coffee, with a defective beans percentage in a coffee lot called percentage of stain, and composition of defined grain size. In the sorters by sieve, the husk and the stain are separated through a fan, where the coffee that does not have the appropriate weight is discriminated. In the sorters by vibration-gravity, the coffee is classified by size and shape, separating it into first, second, third, shell, pellet (amount of broken coffee beans) and dry Cherry. In pneumatic sorters, coffee is classified by weight in first, second and third, and stones and sticks of smaller size are also eliminated. In electronic sorters, coffee is classified by color, eliminating mainly the black and yellow grain, by a computerized optical system that eliminates undesirable color grains, according to the required preparation and quality standards.

The typical demand for green coffee declines in autumn (September–December) and peaks in spring (March–June) as depicted in Figure 2. Green coffee orders from wholesale customers (industrial coffee toasters) require processing only in some sorting processes or in a certain sequence of these (pneumatic sorting, optical sorting, and sift sorting). Sorting process scheduling decision is based on (i) the size of the green coffee order and its requirement of stain percentage, and (ii) the percentage of defective beans resulted from the physical analysis of the Parchment coffee entries necessary to complete the wholesale customer's order. If the requirements of the client's order are not met, two consequences arise. The first consists of the re-entry of the coffee lot to another sequence of sorting processes, which generates reprocessing. The second consequence refers to an over-processing of the coffee bean when the quality that the customer is willing to pay is exceeded, which results in the storage of the coffee lot or its sale at a lower price.

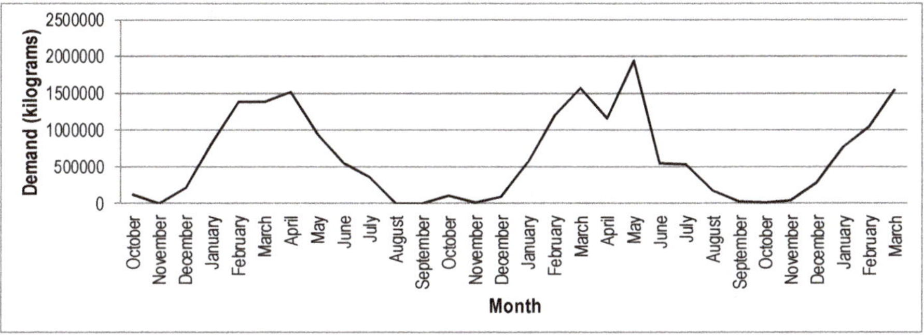

Figure 2. The typical demand for green coffee declines in autumn (September–December) and peaks in spring (March–June).

3.2. General Methodology

The proposed methodology encompasses data collection and model definition, model validation, system configurations definition, and output data analysis. The relationships between them are shown in Figure 3.

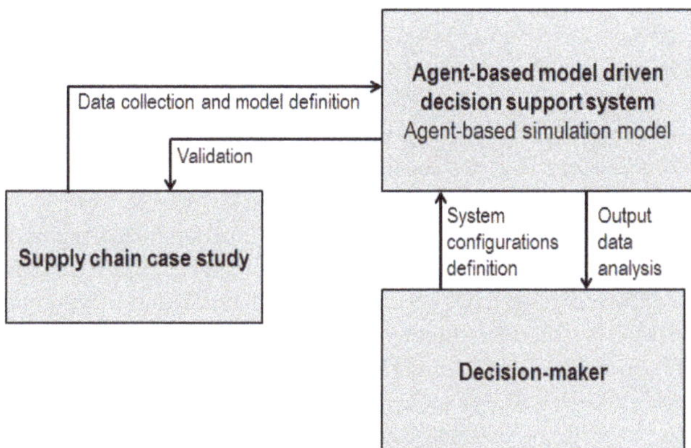

Figure 3. General methodology.

The data collections refer to collect and analyze information on case study operating procedures and control logic which is used to formulate the decision-making rules in the simulation. Once the model is built and verified, the validation process compares simulation output with the real data. The agent-based simulation model is used to support the decision-making of the case study through repeated simulations. The decision support system allows the decision-maker to iteratively set parameters and define system configurations, run simulations, and analyze the output data in order to generate strategies to implement the decisions.

3.3. Modeling the Green Coffee Supply Chain

From a process-oriented supply chain management approach, cross-functional business processes are used to structure the activities between members of a SC [44]. The SC can be represented, analyzed and configured through the application of the reference model Supply Chain Operations Reference (SCOR) developed by the Supply Chain Council [45]. The description of supply chains is made using these building blocks of processes, from simple to complex networks using a common set of definitions of performance metrics, processes, best practices, and the necessary skills to carry out the processes of the SC. The SCOR model has a sweeping overview of the SC, viewing it as something ranging from suppliers' suppliers to customers' customers and incorporating the financial, organizational and societal aspects of performance [46]. Through an exploratory study, Lockamy III and McCormack [47] investigated the relationship between supply chain management planning practices and SC performance based on four decision areas provided in SCOR: plan, source, make and deliver. The authors stated that planning processes are important in all SCOR SC planning decision areas.

The SCOR comprises three levels of process detail. Level one defines both the scope and the content for the supply network. Additionally, the competition performance targets are set. At the second level, companies implement their specific SC operations strategy through three core business models, namely, process categories (i) make-to-stock, (ii) make-to-order, and (iii) engineer-to-order. At level three, companies "fine tune" their operations strategy through (i) process element definitions, (ii) process element information inputs, and outputs, (iii) process performance metrics and iv) best practices and system capabilities required to support best practices. For the decision support system, the SCOR-process oriented approach is used to represent the ASC activities from an "as-is" state to a "to-be" state [48]. To this end, a diagram of level three SCOR process is constructed.

In this section, we attempt to fit the green coffee SC activities in the frame of the SCOR model in order to construct a diagram of level three process elements of the desired "to-be" future state, from the analysis of the "as-is" state of the chain processes. For the green coffee SC actors, the "to-be" future state of the current study scope covers the following processes.

For Cherry coffee producers, M1.3 Produce and test includes the activities of adding value for the products by having the raw material pass through several activities, in this case meaning the Cherry coffee growth yield in a coffee plantation; M1.6 Release Finished Product to Deliver relates to the harvest process and D1 Deliver Stocked Product relates to the market demand satisfaction. For the Parchment coffee producer, S1.2 Receive Product refers to the process and associated activities of receiving Cherry coffee lots from the producers; in this case, M1.3 Produce and test represents the wet benefit process in an industrial intermediary and evaluates Cherry coffee volume to determine Parchment coffee lots according to type and percentage of defective coffee beans. The different stages of the wet benefit process involve Cherry coffee receiving and feeding, mechanical removal of the pulp, mechanical removal of the mucilage, coffee bean washing, coffee bean drying, and Parchment coffee grading; and M1.6 Release Finished Product to Deliver refers to the Parchment coffee lot quality and D1 Deliver Stocked Product relates to the market demand satisfaction.

For the green coffee producer, S1.2 Receive Product refers to the process and associated activities of receiving Parchment coffee from the producer; the amount of received coffee is defined by S1.1 Schedule Product Deliveries. S1.3 Verify Product relates to the process and actions required determining product conformance to requirements and criteria, in this case, both an organoleptic evaluation and physical revision of a coffee sample. After the coffee entry quality grading, in S1.4 Transfer Product, the accepted coffee lot is stacked on pallets in the warehouse area designated for each coffee type.

P3 Plan Make comprises the establishment of courses of action over specified time periods that represent a projected appropriation of production resources to meet production requirements, while M2.1 Schedule Production Activities has the purpose of scheduling the activities, in this case, the processes of M2.3 Produce and Test: coffee bean pre-cleaning, coffee bean threshing, coffee bean sorting, and green coffee customer order quality control. M2.2 Issue Material relates to the selection of Parchment coffee entries from the warehouse. The inventory availability record will determine the coffee lots to be issued to support the production operations. Finally, and M1.6 Release Finished Product to Deliver refers to the gGreen coffee lot quality and D2 Deliver Make-to-order Product relates to the market demand satisfaction.

For the industrial coffee roasters, S2.1 Schedule Product Deliveries places the green coffee order. Figure 4 depicts the level three process elements comprising the make-to-order process at the green coffee producer. The organizational units involved in the Parchment coffee refinement are the Parchment coffee supplier, the internal departments of the green coffee producer and green coffee toaster. The Parchment coffee Delivery and Parchment Procurement level one processes are related by means of a customer-supplier connection that reflects the temporal relation between them and that the former process has a container-element link with one instance of the D1 level process. Furthermore, the instance of the M2 level two process, which is planned by an instance of the P3 level two process, is comprised of four instances of level three process elements. Regarding process elements, Figure 3 shows some of the resources created and used by them; for instance, Production Plan and Customer Reception Schedule, which are instances of the Production Schedule class. The former is created by the instance of Establish Production Plan process element, an element of a P3 level two process, and used by the instance of Schedule Production Activities. Each of these two processes perceives the Production Plan resource from a different perspective; two instances of the Resource Perspective class, named Production Orders and Production Plan. In the same way, the Delivery Orders resource perspective is presented for the Customer Reception Schedule resource, which is created by the instance of Schedule Production Activities, and it is used by the instance of Reserve Inventory Capacity and Determine Delivery Date, a level three process element corresponding to the D2 level two process.

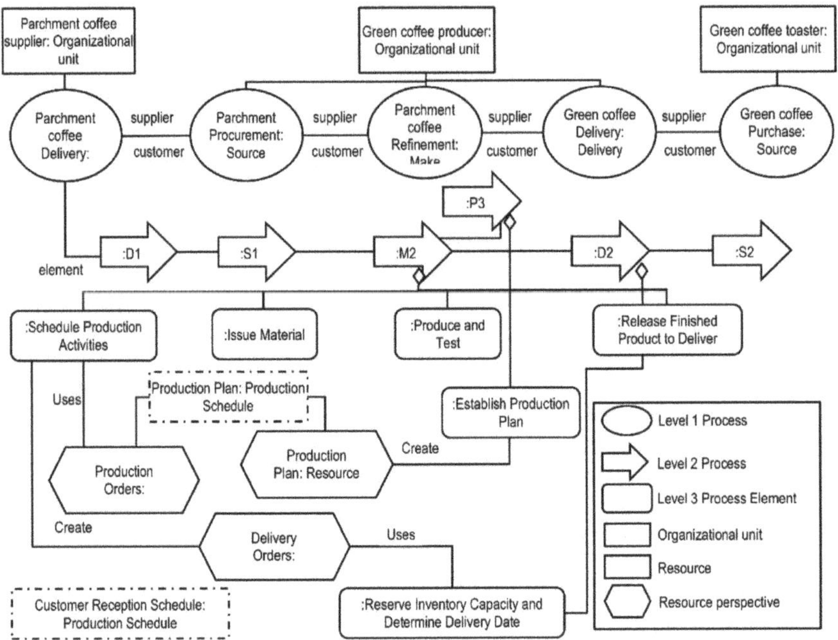

Figure 4. Level three process elements comprising the make-to-order process at the focal company.

3.4. Agent Description

Derived from this mapping in the framework of the SCOR model, the identified agents and activities for the green coffee SC are listed in Table 1. The simulation model implements the Parchment coffee refinement process at the focal company that interprets historical data from 2016 to 2018 fed by the green coffee SC actors under investigation.

Table 1. Agents and activities for the green coffee SC.

Agent	Activities
Cherry-coffee producer Production Agent (CP Agent)	Production
Cherry-coffee producer Delivery Agent (CD Agent)	Satisfy market demand
Parchment-coffee producer Source Agent (PS Agent)	Source Products
Parchment-coffee producer Delivery Agent (PD Agent)	Satisfy market demand
Green-coffee producer Source Agent (GS Agent)	Source Products Verify Products
Green-coffee producer Production Agent (GP Agent)	Plan Production Production
Green -coffee producer Inventory Agent (GI Agent)	Update inventory
Green-coffee producer Reschedule Agent (GR Agent)	Update Plan Production
Green-coffee producer Delivery Agent (GD Agent)	Satisfy market demand
Wholesale Market Agent (WM Agent)	Place order

The decision support agents serve the purposes of deriving useful knowledge to accomplish the decisions: (i) for the Cherry-coffee producer Production Agent (CP Agent), the decision regarding the estimation of Cherry coffee yield obtained at the coffee plantation; (ii) for the Green-coffee producer Source Agent (GS Agent), the Parchment coffee sample verification decision; (iii) for the Green-coffee producer Production Agent (GP Agent), the production plan establishment decision, and (iv) for the Green-coffee producer Reschedule Agent (GR Agent), the plan production update decision. The decision-making rule approach for the CP Agent and the GS Agent is rule-based representation using Fuzzy Logic involving an inference engine with IF-THEN type rules. For the GP Agent, a decision-making rule approach based upon the coupling of IF-THEN fuzzy inference rules and equation-based representation using mixed integer programming with the aim to maximize customer service level is used. Finally, the decision-making rule perspective used for the GR Agent is equation-based representations using mathematical equations once the customer service level falls below the expected level.

In this study, the decision-making rule approach using fuzzy logic is used to generate a knowledge base for the CP Agent, the GS Agent, and the GP Agent. The diverging opinions of the experts are represented as blurred triangular and trapezoidal numbers, which describe the coded knowledge of the expertise of the green coffee SC actors. Accordingly, a Mamdani model codifies the decision criteria related to (i) agricultural practices for growing Cherry coffee in coffee plantations (CP Agent); (ii) the process of sensorial evaluation of coffee beans from which gustatory and olfactory characteristics of the grain are evaluated for quality scoring of a Parchment coffee sample (GS Agent); and (iii) the sorting process scheduling of Parchment coffee entries based on coffee entry quality grading, the percentage of defective coffee beans percentage in the coffee lot (GP Agent). In the Mamdani type model, multiple inputs and outputs represent information using Fuzzy Logic; each input and output variable is represented through a Linguistic variable. The rule base representation is developed according to the IF-THEN type, which constitutes the inference method based on the knowledge base and consequent inference engine. The defuzzification process uses the centroid calculation method.

Fuzzy lLogic is used to mimic the knowledge and expertise of the agricultural field dynamics for coffee growth. We consider variables related to five agricultural practices that are common in any plantation, soil nutrition, control of pests, control of diseases, planting density, and pruning, and two uncertain events that are ever-present in agricultural setting i.e., rainfall and temperature. The consideration of agricultural practices as decision variables by considering two uncertain events for Cherry coffee production makes scenario agricultural field modeling closer to reality, with the simultaneous objectives of maximizing the yield of Cherry coffee obtained at the coffee plantation. Table 2 describes the variables related to agricultural practices, and uncertain events defined in the knowledge database for the CP Agent as Input elements (I), and the operations variables defined as Output elements (O). The knowledge base for coffee growing yield is composed of 1620 inference rules.

Table 2. Variable codification in the knowledge base of the Mamdani type model for coffee growing yield.

Variables	Definition	Measurement Units
Nutrition (*N*)	Agricultural practice related to the transfer of nutrients to the coffee plantation	Number of applications
Rainfall (*R*)	Uncertain event that supplies water to the coffee plantation	mm/month
Control of pests (*CP*)	Agricultural practice that controls pests that affect yield	Number of applications
Control of diseases (*CD*)	Agricultural practice that controls the disease that affects yield	Number of applications
Planting density (*PD*)	Operational variable related to the amount of planted bushes in the coffee plantation	m^2/ha
Pruning (*P*)	Agricultural practice related to the cutting of undergrowth to leave a vegetative cover and prevent erosion	q/ha
Temperature (*T*)	Uncertain event that supplies heat to the coffee plantation	°C
Yield (*Y*)	Linguistic expression that represents the Cherry coffee growth yield obtained at the coffee plantation	q/ha

Fuzzy Logic is also used to mimic the organoleptic evaluation of the Parchment coffee sample. This process comprises the sensorial evaluation of coffee beans, from which the gustatory and olfactory characteristics of the grain are evaluated in order to determine a quality score for two coffee types, Robusta and Parchment. Negative or poor flavors detract from the quality of the coffee. For quality scoring of the Robusta coffee sample, we consider the variables ferment, sour, malodorous, earthy, mold and old.

For quality scoring of the Parchment coffee sample, the following variables are taking into account: aroma; flavor; acidity; body; vinous, fruity, sweetness; green, immatureness; cereal, wood, paper; dry, old; chemical, medicinal; ferment, sour, malodorous; and earthy, mold.

Table 3 describes the variables related to the gustatory and olfactory characteristics defined in the knowledge database for the GS Agent as Input elements (I), and the operations variables defined as Output elements (O). The knowledge base for quality scoring of a Robusta coffee sample and Parchment coffee sample is composed of 4096 inference rules and 96000 inference rules, respectively.

Table 3. Variable codification in the knowledge base of the Mamdani type models for quality scoring of a coffee sample.

Variables	Definition	Measurement Units	Status
	Robusta coffee sample		
Ferment (*F*)	Operational variable related to the fermented taste that detracts from the quality of the coffee	Numerical score	Input
Sour (*S*)	Operational variable related to the sour taste that detracts from the quality of the coffee	Numerical score	Input
Malodorous (*M*)	Operational variable related to the acetic acid smell related to the fermented taste	Numerical score	Input
Earthy (*E*)	Operational variable related to the earthy taste and smell that detract from the quality of the coffee	Numerical score	Input
Mold (*M*)	Operational variable related to the mold taste that detracts from the quality of the coffee	Numerical score	Input
Old (*O*)	Operational variable related to the aged taste that detracts from the quality of the coffee	Numerical score	Input
Robusta Class (*RC*)	Linguistic expression that represents the robusta coffee class obtained from the organoleptic evaluation	Quality score	Output
	Parchment coffee sample		
Aroma (*A*)	Operational variable related to the aromatic impression due to the volatile substances of coffee	Numerical score	Input
Flavor (*F*)	Operational variable related to the balanced impression due to the combination of gustatory and olfactory attributes perceived in coffee	Numerical score	Input
Acidity (*A*)	Operational variable related to the gustatory impression due to organic acids contributing to liveliness, sweetness and fresh-fruit coffee's character	Numerical score	Input
Body (*B*)	Operational variable related to the feeling of fullness and consistency in the mouth, particularly when it is perceived between the tongue and the palate	Numerical score	Input
Vinous, Fruity, Sweetness (*VFS*)	Operational variable related to a pleasing fullness of flavor due to the presence of certain carbohydrates	Numerical score	Input
Green, Immatureness (*GI*)	Operational variable related to the astringent taste that detract from the quality of the coffee	Numerical score	Input
Cereal, Wood, Paper (*CWP*)	Operational variable related to the cereal taste that detract from the quality of the coffee	Numerical score	Input
Dry, Old (*DO*)	Operational variable related to the aged taste that detract from the quality of the coffee	Numerical score	Input
Chemical, Medicinal (*CM*)	Operational variable related to the chemical taste that detract from the quality of the coffee	Numerical score	Input
Ferment, Sour, Malodorous (*FSM*)	Operational variable related to the ferment taste and smell that detract from the quality of the coffee	Numerical score	Input
Earthy, Mold (*EM*)	Operational variable related to the earthy taste and smell that detract from the quality of the coffee	Numerical score	Input
Parchment Class (*PC*)	Linguistic expression that represents the Parchment coffee class obtained from the organoleptic evaluation	Quality score	Output

For the GP Agent, fuzzy logic is used to mimic the sorting process scheduling of coffee entries based on coffee entry quality grading in the dry benefit process. Sorting process scheduling decision is based on the percentage of defective beans resulted from the physical analysis of the Parchment coffee entries necessary to complete the wholesale customer's order. For sorting process scheduling of not-washed Robusta coffee inputs, we consider the variables serious defects, minor defects, pellet, green aspect, and weight. For sorting process scheduling of Robusta coffee inputs, we contemplate the variables humidity, serious defects, minor defects, pellet, green aspect, and weight. Finally, for sorting process scheduling of Parchment coffee inputs, humidity, serious defects, minor defects, pellet, green aspect were taking into consideration. With regards to Output, sift sorting refers to the process by which the coffee is classified by size and shape, pneumatic sorting comprises coffee classification by size and weight, and optical sorting the coffee is classified by a computerized optical system that eliminates undesirable color grains. Table 4 describes the variables related to the wet benefit process defined in the knowledge database for the GP Agent as Input elements (I), and the operations variables defined as Output elements (O). The knowledge base for sorting process scheduling of not-Robusta coffee, Robusta coffee, and the Parchment coffee, is composed of 216 inference rules, 864 inference rules, and 128 inference rules, respectively.

Table 4. Variable codification in the knowledge base of the Mamdani type models for sorting process scheduling of coffee entries.

Variables.	Definition	Measurement Units	Status
Serious defects (*SD*)	Operational variable related to the number of defective coffee beans associated with appearance (black, white, amber, and with irregular spots)	% of defective beans	Input
Minor defects (*MD*)	Operational variable related to the amount of malformed (shell and ear) coffee beans	% of defective beans	Input
Pellet (*P*)	Operational variable related to the number of broken coffee beans	% of defective beans	Input
Green aspect (*GA*)	Operational variable related to the number of immature coffee beans of black-Green color	% of defective beans	Input
Weight (*W*)	Operational variable related to the number of kilograms entering the process schedule	kilograms	Input
Humidity (*H*)	Operational variable related to the water content of the coffee beans	% of humidity	Input
	Not-washed Robusta coffee entry		
Not-washed robusta Schedule 1 (*nrS1*)	Linguistic expression that represents the process schedule: mix, pneumatic sorting, optical sorting, and sift sorting	Number of processes	Output
Not-washed robusta Schedule 2 (*nrS2*); Not-washed robusta Schedule 3 (*nrS3*)	Linguistic expression that represents the process schedule: pneumatic sorting, optical sorting, and sift sorting	Number of processes	Output
	Robusta coffee entry		
Robusta Schedule 1 (*rS1*)	Linguistic expression that represents the process schedule: mix, pneumatic sorting, optical sorting, sift sorting, dry, and dry little	Number of processes	Output
Robusta Schedule 2 (*rS2*); Robusta Schedule 3 (*rS3*); Robusta Schedule 4 (*rS4*)	Linguistic expression that represents the process schedule: pneumatic sorting, optical sorting, sift sorting, dry, and dry little	Number of processes	Output
	Parchment coffee entry		
Parchment Schedule 1 (*pS1*)	Linguistic expression that represents the process schedule: mix, pneumatic sorting, optical sorting, sift sorting, dry, and dry little	Number of processes	Output
Parchment Schedule 2 (*pS2*); Parchment Schedule 3 (*pS3*); Parchment Schedule 4 (*pS4*)	Linguistic expression that represents the process schedule: pneumatic sorting, optical sorting, sift sorting, dry, and dry little	Number of processes	Output

In order to determine the reliability of the aforementioned fuzzy models, a paired *t*-test was applied, which is used to compare the estimated values from each model to real recorded case data. Each record has specific values for each input variable and the output result by the decision-maker. This test produced a confidence interval that includes zero, which shows there is no significant difference between the estimated results and real data, so it can be concluded that the Fuzzy models are valid.

For the GP Agent, the production plan establishment decision is supported by a decision-making rule constructed by equation-based representation using mixed integer programming. Broadly, production planning comprises the determination of the type and quantity of commodities to be produced and resource allocation. The mathematical programming relies in the endeavor made by Sawik [49]. The author proposes a mixed integer programming formulation for customer order assignment over a planning horizon, to maximize service level. The index, parameters, and variables of this model are shown in Tables 5–7, and the description of the problem is explained below.

Table 5. Indices.

Index	Definition
i	Processing stage, $I \in I = (1, \ldots, m)$
j	Wholesaler order, $j \in J = (1, \ldots, n)$
k	Coffee type, $k \in K = (1, \ldots, r)$
t	Planning period, $t \in T = (1, \ldots, h)$

Table 6. Model parameters.

Parameter	Definition
a_j, d_j, s_j	Arrival date, due date, size of order j
b_j	Production lot for order j
c_{it}	Processing time available in period t on each machine in stage i
m_i	Number of identical, parallel machines in stage i
n	Number of customer orders to be scheduled
p_{ij}	Processing time in stage i of each product in order j
$J_i \subseteq J$	$\{j \in J: p_{ij} > 0\}$ subset of wholesaler orders to be processed in stage i
$J1 \subseteq J$	Subset of small wholesaler orders
$J2 \subseteq J$	Subset of large wholesaler orders
J_k	Subset of wholesaler orders for coffee type k

Table 7. Variables.

Variable	Definition
u_j	1, if order j is completed after due date; otherwise $u_j = 0$
x_{jt}	1, if order j is performed in period t; otherwise $x_{jt} = 0$
y_{jt}	Fraction of customer order j to be processed in period t

The green coffee producer can be identified as a flexible flow production system made up of six processing stages in series, and each stage $i \in I = (1, \ldots, m)$ is made up of $m_1 \geq 1$ parallel identical machines (Figure 5). In the system, three types of coffee are produced in a make to- order environment responding directly to wholesale customer-requested orders. Let J be the set of customer orders that are known ahead of a planning horizon. Each order $j \in J$ is described by a triple (a_j, d_j, s_j), where a_j is the order arrival date, d_j is the customer-requested due date, and s_j is the size of order (the number of ordered products of specified type). Each order requires processing in various processing stages; however, some orders may bypass some stages.

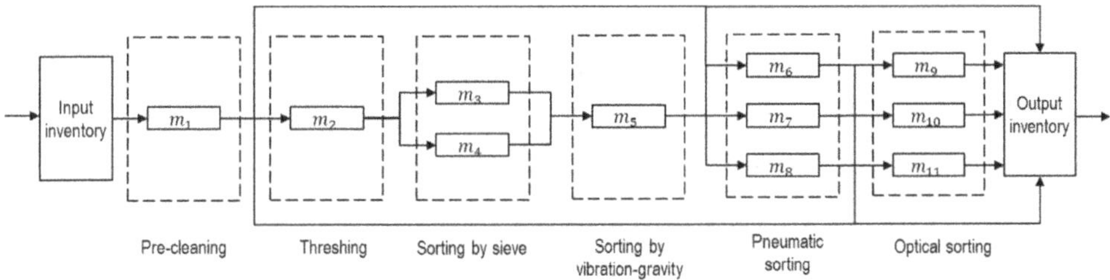

Figure 5. Flexible flow production system of the green coffee producer.

The processing stages are the following: pre-cleaning, threshing, sorting by sieve1, sorting by vibration-gravity, pneumatic sorting, and optical sorting. Let $p_{ij} \geq 0$ be the processing time in stage i of each product in order $j \in J$. The orders are processed and transferred among the stages in lots of various sizes that depend on the ordered product type and let b_j be the size of production lot for order j. The coffee beans are feed into the pre-cleaning machine, m_1, and then, it goes to the threshing machine, m_2; this machine has 2 outputs, good coffee beans, and straw, the straw leaves the system and the coffee continues its way to the next processing stage. The sorting by sieve machines in stage 3, m_3 y m_4, have three outputs, (1) good coffee beans, (2) pellet and (3) straw; the good coffee beans continues its way to enter the next machine, while the broken coffee beans and straw leave the system, separately. The sorting by vibration-gravity machine, m_5, has four outputs, (1) good coffee beans, which passes to the next machine, or failing that, leaves the system as the final product; (2) pellet, (3) dust and (4) dried cherries that leave the system definitively as waste. The pneumatic sorting machines, m_6, m_7 y m_8, have three outputs (1) good coffee beans that leave the system as final product; (2) stain coffee beans that leave the system as waste, and (3) coffee beans that re-enter the machine to be reprocessed. The optical sorting machines, m_9, m_{10} y m_{11}, have two outputs (1) good coffee beans that leave the system as final product; (2) stain coffee beans that leave the system as waste.

The planning horizon consists of h planning periods, and L is the length of each planning period, in this case, working hours per week. Let $T = \{1, \ldots, h\}$ be the set of planning periods and c_{it} the processing time available in period t on each machine in stage i. Customer orders are split into production lots of fixed sizes, each to be processed as a separate job. The following two types of customer orders are considered: (1) small customer order, where each order must be completed in two consecutive time periods, and (2) large customer order, where each order must be completed in four consecutive time periods. In practice, the two types of customer orders are scheduled simultaneously. Denote by $J1 \subseteq J$, and $J2 \subseteq J$, respectively, the subset of small customer orders, and large customer orders.

The mathematical formulation for the initial production schedule for the original customer orders known ahead of the planning horizon is as follows, where all materials are assumed to be available at the beginning, i.e., $a_j = 1$ for each order $j \in J$.

Maximize:

$$1 - \sum\nolimits_{(j \in J)} u_j/n \tag{1}$$

Subject to:

$$x_{jt} + x_{(jt+1)} + x_{(jt+2)} + x_{(jt+3)} \leq 4, \qquad j \in J2, t \in T: a_j \leq t \leq h-3 \tag{2}$$

$$x_{jt} + x_{(jt+1)} + x_{(jt+2)} \leq 3, \qquad j \in J2, t \in T: a_j \leq t \leq h-2 \tag{3}$$

$$x_{jt} + x_{(jt+1)} \leq 2, \qquad j \in J2, t \in T: a_j \leq t \leq h-1 \tag{4}$$

$$x_{jt} + x_{jt'} \leq 1, \qquad j \in J, t \in T, t' \in T: a_j \leq t \leq h-2, t' \geq t+2 \tag{5}$$

$$\sum\nolimits_{(t \in T: t \geq aj)} y_{jt} = 1, \qquad j \in J \tag{6}$$

$$x_{jt} \geq y_{jt}, \qquad j \in J, t \in T: t \geq a_j \tag{7}$$

$$y_{jt} \geq b_j\, x_{jt}/s_j, \qquad j \in J, t \in T: t \geq a_j \tag{8}$$

$$u_j \geq \sum\nolimits_{(t \in T: t > dj)} y_{jt}, \qquad j \in J \tag{9}$$

$$u_j \leq \sum\nolimits_{(t \in T: t > dj)} x_{jt}, \qquad j \in J \tag{10}$$

$$\sum\nolimits_{(j \in J)} p_{ij}\, s_j\, y_{jt} \leq c_{it}\, m_i, \qquad I \in I, t \in T \tag{11}$$

$$\sum\nolimits_{(j \in J: \tau \in T: aj \leq \tau \leq t)} s_j\, y_{j\tau} \geq \sum\nolimits_{(j \in J: dj \leq t)} s_j\, (1 - u_j), \qquad t \in T \tag{12}$$

$$u_j \in \{0, 1\}, \qquad j \in J \tag{13}$$

$$x_{jt} \in \{0, 1\}, \qquad j \in J, t \in T: t \geq a_j \tag{14}$$

$$0 \leq y_{jt} \leq 1, \qquad j \in J, t \in T: t \geq a_j \tag{15}$$

The objective function (1) aims to maximize service level. Each large customer order must be completed in four consecutive time periods and each small customer order must be completed in two consecutive time periods (2)–(5). Each order must be completed (6), each order is allocated among all the periods that are selected for its assignment (7), and the minimum portion of a divisible order allotted to one period is not less than the batch size (8). Regarding tardy order constraints, a tardy order is partly assigned after its due date (9)–(10). The demand for capacity at each processing stage cannot be greater than the maximum available capacity in every period (11). The cumulative production is not less than the cumulative demand minus the tardy demand (12).

In industrial environment agent-based solution adoption, real-time scheduling and rescheduling are becoming increasingly important [17]. The master production schedule has to deal with seasonal fluctuations of demand and to calculate a frame for necessary amounts of overtime, whilst short-term production planning comprises the determination of lot-sizes according to their due dates and the available capacity with minute accuracy [50]. There are two basic elements in this approach [51]: (i) scheduling algorithms are used to generate initial schedules and repair obsolete schedules and (ii) control policies are used to adjust the frequency of repairing a schedule. Considering these control policies from the agent-based modeling and simulation approach, an adaptive agent is capable of modifying them during a simulation based on evolving circumstances [52]. Consequently, the adaptive control approach is based on the control of a set of performance indicators of a system by means of a decision model that analyzes them and selects an appropriate control policy. For SC planning, this approach relates to a partial or even the full change of a previously accepted plan triggered by a new event such as new order arrival, a cancellation of already allocated orders, the availability of a new resource, a failure of existing resources, or changes of the chain objectives [53]. The GR Agent monitors the customer service level and once the indicator falls below the expected level, the Agent updated the production plan. Rescheduling is the process of updating an existing production schedule in response to disruptions or other changes; This includes the machine failures, processing time delays, rush orders, quality problems, and unavailable material [51], in this case, the disturbance relates to delay in the arrival or shortage of materials. The mathematical formulation for the rescheduling algorithm is as follows considering the rescheduling parameters in Table 8. Let t_{mod} be the first planning period immediately after the order modification.

Table 8. Rescheduling parameters.

Parameter	Definition
h'	new planning horizon
E^-	upper limit on maximum earliness
t_{mod}	the planning period immediately following modification of orders
J_{mod}	set of modified orders
J_{old}	subset of orders in J remaining for completion without modification
$J^N{}_{old}$, $J^S{}_{old}$	subset of orders in J_{old}, respectively non-reschedulable, reschedulable
T_{new}	$\{h + 1, \ldots, h'\}$ set of new planning periods
T_{old}	$\{t_{mod}, \ldots, h\}$ subset of remaining planning periods in T
$T^N{}_{old}$	subset of periods in T_{old} with fixed assignment of orders in J_{old}

Prime (') denotes updated parameters after modification of orders.

Step 0. Split the set J_{old} of orders remaining for completion into two disjoint subsets: $J^S{}_{old}$ of reschedulable orders and $J^N{}_{old}$ of fixed, non-reschedulable orders:

$$J^N{}_{old} = \{j \in J_{old}: \sum\nolimits_{(tmod \le t \le tmod + E^*max)} x_{jt} = 1\} \tag{16}$$

$$J^S{}_{old} = J_{old}/J^N{}_{old} \tag{17}$$

Step 1. Set $T^N{}_{old} = \{t_{mod}, \ldots, t_{mod} + E^-\}$

Step 2. Do not change the assignment of non-reschedulable orders $j \in J^N{}_{old}$, i.e.,:

$$y'_{(j,tmod + E^- + 1)} = y_{(j,\, tmod + E^*max + 1)},\ j \in J^N{}_{old} \cap J: x_{(j,tmod + E^-)} = 1 \tag{18}$$

The algorithm is for rescheduling of the remaining customer orders awaiting material supplies [49]. For each order j, product-specific materials are assumed to be unavailable earlier than E^- periods ahead of the order due date d_j. Therefore, each order j cannot be assigned to periods earlier than $d_j - E^-$. In particular, in period tmod product-specific materials are not available for orders due in periods greater than $t_{mod} + E^-$, and hence all such orders can be rescheduled. On the other hand, the unmodified orders with product-specific materials supplied by period t_{mod} are considered non-reschedulable in the algorithm. In the algorithm, the planning horizon is progressively shifted to take into account modifications of the customer orders (changes of order size and/or due date) occurring during the horizon.

3.5. Model Validation

The AnyLogic® Personal Learning Edition multi-method simulation platform that supports not only agent-based general-purpose simulations but also it supports DES modeling was used to build the agent-based simulation model that underpins the decision support system in a HP Workstation with an Intel Zeon CPU operating at 3.40 GHz and equipped with 8 GB RAM. The decision-making rules for the decision support agents, IF-THEN fuzzy inference rules and mixed integer programming, were implemented in Python through the integration AnyLogic®–Python.

To validate the simulation model and to check if it is an adequate representation of the real system, a paired *t*-test was applied, which is used to compare the results from the simulation model to the historical data of the real system. The indicator refers to the amount of good green coffee beans resulting from the sequence of processes, sift sorting, pneumatic sorting, and optical sorting, with an intake of 37,740 kg of Parchment coffee. The results of 10 replicas are shown in Table 9, where X_j is the number of kilograms of good green coffee beans resulting from the sequence of processes (1) sift sorting, (2) pneumatic sorting, and (3) optical sorting, with an intake of 37,740 kg of Parchment coffee, in the real system. In the same way, Y_j is the number of kilograms of good green coffee beans result from the simulation model.

Table 9. Model validation data.

Replicate	X_j	Y_j	$Z_j = X_j - Y_j$	$(Z_j - Z^{-}_{10})^2$
1	27,176.00	27,935.05	−759.05	360,735.09
2	27,901.00	27,811.70	89.30	61,373.56
3	27,348.00	27,734.25	−386.25	51,899.30
4	28,004.00	28,754.03	−750.03	349,980.98
5	27,733.00	27,417.09	315.91	225,007.48
6	27,914.00	28,561.69	−647.69	239,364.85
7	27,682.00	27,270.03	411.97	325,361.82
8	28,412.00	28,461.21	−49.21	11,929.89
9	27,197.00	26,779.30	417.70	331,930.58
10	27,996.00	28,223.02	−227.02	4,703.48
Sum			−1,584.36	1,962,287.05
Average	27,736.30	27,894.74	−158.44	

A 95% confidence interval is constructed using Equations (19)–(21):

$$\begin{gathered} Z^{-}(n) = \Sigma^{n}{}_{(i=1)} Z_i)/n \\ Z^{-}(10) = -1584.36/10 \\ Z^{-}(10) = -158.43 \end{gathered} \tag{19}$$

$$\begin{gathered} \text{Var}^{\wedge}\,[Z^{-}10\,] = \Sigma^{10}{}_{(i=1)}\,[(Z_j - Z^{-}10)^2]/n\,(n - -1) \\ \text{Var}^{\wedge}\,[Z^{-}10\,] = 1{,}962{,}287.05/10\,(10 - -1) \\ \text{Var}^{\wedge}\,[Z^{-}10\,] = 21{,}803.18 \end{gathered} \tag{20}$$

$$\begin{gathered} Z^{-}(n) \pm t_{(n-1,1\text{-}(1-\propto)/2)} \sqrt{\hat{\text{Var}}[Z^{-}n]} \\ -158.43 \pm t_{9,0.975} \sqrt{21{,}803.18} \\ -158.43 \pm 2.26(147.65) \\ (-492.46{,}175.59) \end{gathered} \tag{21}$$

It is observed that the confidence interval includes zero, so it is concluded, with a confidence index of 95% that the difference between the means of the real data and the simulation results is not statistically significant.

The optimal number of replicas of the simulation model was determined using the procedure of estimating the mean $\mu = E\,(x)$ with a specific error. Table 10 shows the results of 10 independent pilot replicas, where each replica represents the number of kilograms of good green coffee beans resulting from the sequence of processes (1) sift sorting, (2) pneumatic sorting, and (3) optical sorting, with an intake of 37,740 kg of Parchment coffee.

Table 10. Simulation results of the pilot replicas.

Replicate	Good Green Coffee Beans (Kilograms)
1	27,900.2004
2	27,907.4386
3	27,905.3017
4	27,889.3183
5	27,895.6686
6	27,853.6233
7	27,934.5803
8	27,896.9872
9	27,863.0423
10	27,889.3104
Average	27,894.7358
Standard deviation	24.9877

The average amount of coffee beans leaving the process, in kilograms, was estimated with an absolute error of $\beta = 15$ kg and a 95% confidence level. The calculations for the number of replications are shown below, where i is the number of replicas, and where the calculation performed must be less than or equal to the absolute error ($\beta = 15$ kg), where $x^{-}_{n} = 27{,}894.73$, $S^2_n = 624.38$, $\beta = 15$ and $\alpha = 0.05$.

$$\begin{array}{c} \min \{i \geq 10\colon t_{(i-1,0.975)} \sqrt{(624.38/i)} \leq 15\} \\ i = 10{:}2.26 \sqrt{(624.38/10)} = 17.87 \geq 15 \\ i = 11{:}2.22 \sqrt{(624.38/11)} = 16.78 \geq 15 \\ i = 12{:}2.20 \sqrt{(624.38/12)} = 15.87 \geq 15 \\ i = 13{:}2.17 \sqrt{(624.38/13)} = 15.09 \geq 15 \\ i = 14{:}2.16 \sqrt{(624.38/14)} = 14.42 \leq 15 \end{array} \quad (22)$$

Therefore, the optimal number of replicas is 14.

4. Results and Discussion

4.1. Simulation Results of the Demand Scenarios

The selection of the demand pattern scenarios responds to the need of the decision-maker for production scheduling during the typical demand for green coffee that declines in the September-December period and peaks in the March-June period. The planning horizon considered was h = 13 weeks, with a length of each planning period of working hours per week. The following three demand patterns were considered.

Increasing, with demand skewed towards the end of the planning horizon, 60 customer orders.

Decreasing, with demand skewed towards the beginning of the planning horizon, 84 customer orders.

Unimodal, where demand peaks in the middle of the planning horizon and falls under the available capacity on the first and last days of the horizon, 63 customer orders.

Figures 6–8 show the initial aggregate production schedule and the estimated production results along with the cumulative aggregate production and cumulative aggregate demand for each case for the increasing demand pattern, the unimodal demand pattern, and the decreasing demand pattern, respectively. The negative values in these figures indicate the tardy demand.

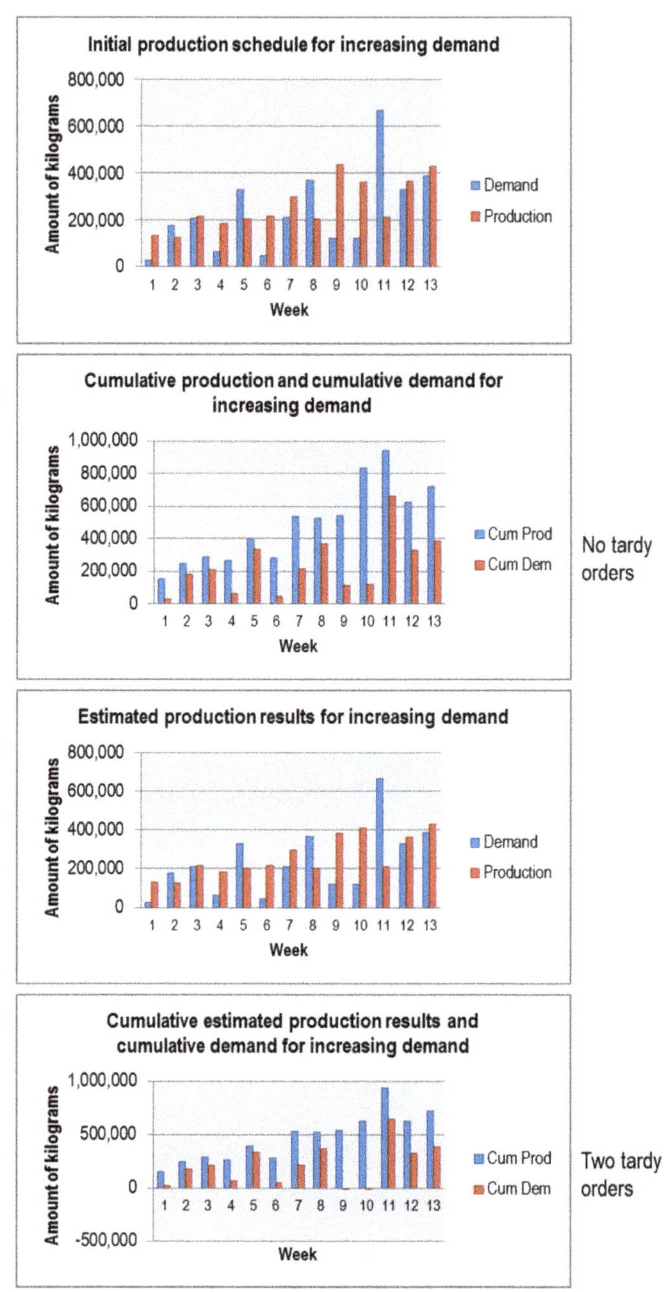

Figure 6. Initial production schedule and estimated production results for increasing demand.

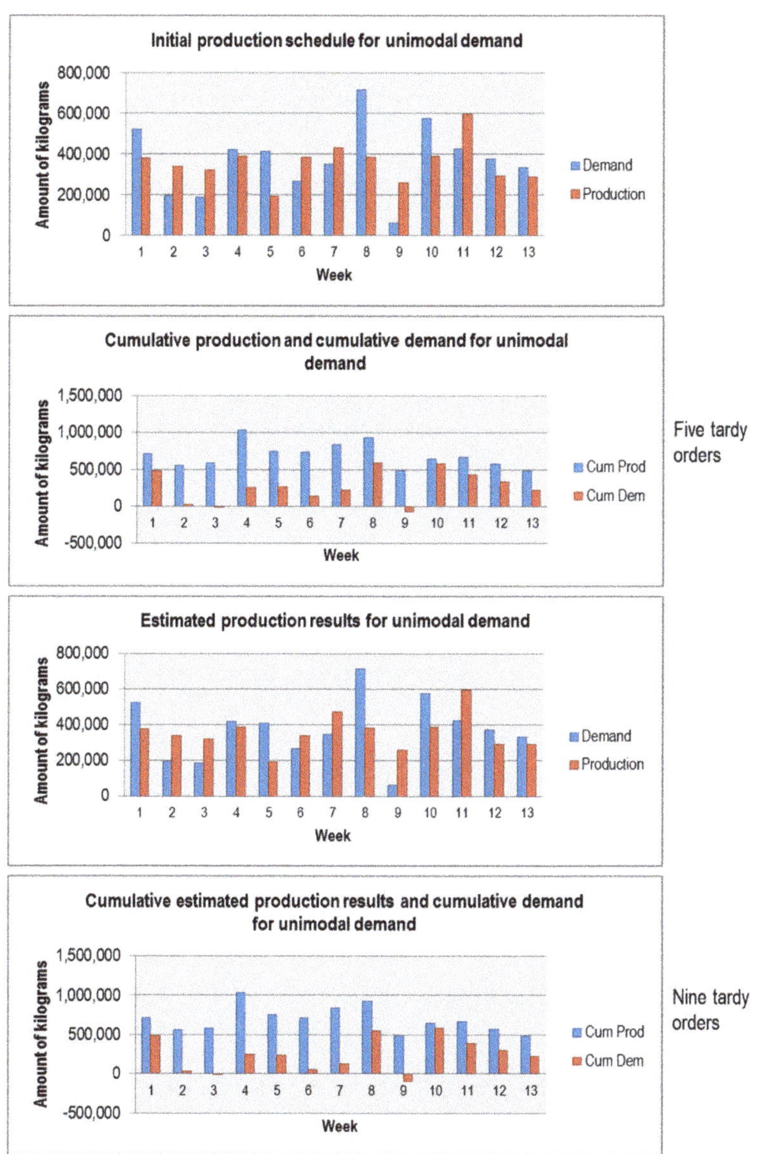

Figure 7. Initial production schedule and estimated production results for unimodal demand.

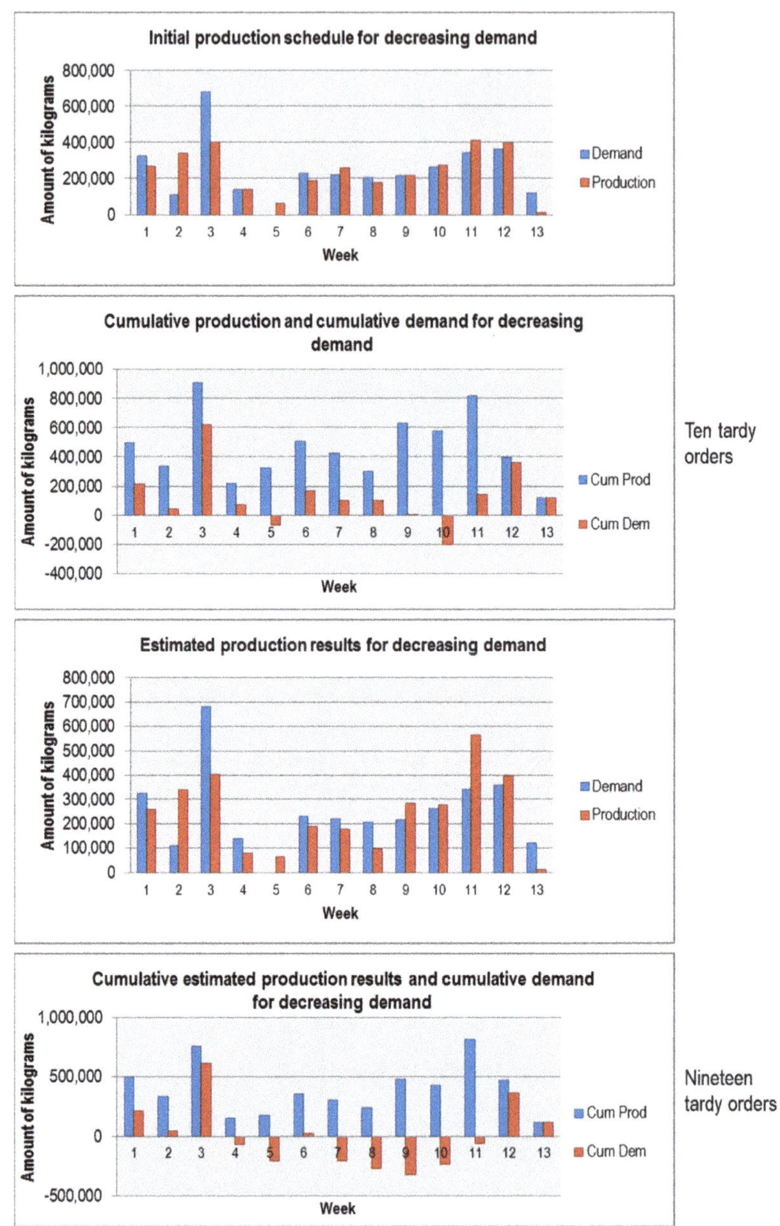

Figure 8. Initial production schedule and estimated production results for decreasing demand.

For the initial aggregate production schedule, all materials are assumed to be available at the beginning of the planning horizon, while for the estimated production results, delay in the arrival of raw materials occurred.

The customer service level results for the increasing demand pattern are as follows (Figure 6): no tardy orders for the initial production schedule, i.e., 100% service level, and two tardy orders for the estimated production results, a 96% service level.

Regarding the unimodal demand pattern, the results are five tardy orders for the initial production schedule, a 92% service level, and nine tardy orders from the estimated production results, an 85% service level. Figure 7 shows that the aggregate production is best leveled over time for the unimodal demand pattern. The study of Sawik [54] for production scheduling in make-to-order manufacturing systems found the aggregate production is best leveled over time for the increasing demand pattern.

4.2. Reactive Scheduling for the Decreasing Demand Scenario

For the decreasing demand pattern, the results indicated the application of the reactive aggregate production scheduling approach in the green coffee SC. For this demand pattern: ten tardy orders for the initial production schedule, an 88% service level, and nineteen tardy orders from the estimated production results, a 77% service level (Figure 8). This value falls below the expected level. Consequently, Figure 9 shows the updated aggregate production schedule with $t_{mod} = 7$ and $E^{-} = 2$, resulting in ten tardy orders and an 88% service level.

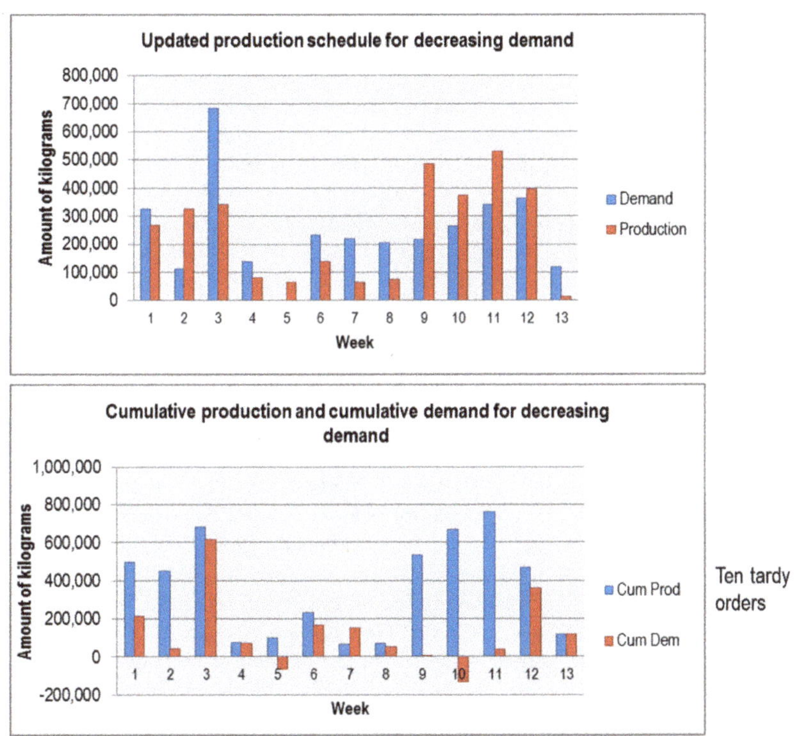

Figure 9. Updated production schedule for decreasing demand and $t_{mod} = 7$ and $E^{-} = 2$

4.3. Discussion

We undertake the analysis of the response surfaces of these fuzzy logic-based decision-making rules. Fuzzy logic is the decision-making rule approach used to generate a knowledge base for the CP Agent, the GS Agent, and the GP Agent as described in Section 3.3. The CP Agent comprises the Cherry coffee-growing yield estimation model, the hectares available for planting and the precipitation and agricultural practices data are used as input. The GS Agent uses the Parchment coffee sample' quality scoring model with the gustatory and olfactory attributes of the sample as input data. The GP Agent comprises the Parchment coffee entries sorting process scheduling model, the percentage of defective coffee beans and the percentage of humidity of the coffee entry are used as input data. Appendix A describes the operating values for the fuzzy sets of the Fuzzy models of Cherry coffee-growing yield; quality scoring of the Robusta coffee sample; quality scoring of the Parchment coffee sample; sorting process scheduling of not-washed Robusta coffee entries; sorting process scheduling of Robusta coffee entries; and sorting process scheduling of Parchment coffee entries.

Regarding the Cherry coffee-growing yield model, the factors that have a significant effect on coffee yield are nutrition, pruning, and control of pests. The factors that have a significant effect from high values are nutrition and pruning. The factor that has a significant effect from low values is control of pests. The factors that do not have a negative effect on coffee yield are rainfall and temperature. The study of Paulo and Furlani Jr. [55] also found that adequate nutrition along with optimal planting density is expected to show high yield responses in coffee plantations.

For the Parchment coffee samples' quality scoring model, ferment, sour, malodorous (FSM) is the factor that has a significant effect on coffee samples' quality score. For the Altura coffee, acidity, chemical taste, flavor, aroma, and body must have high values. For the Extra prima coffee, aroma, acidity, and body must be contained in the medium values. For the Robusta coffee samples' quality scoring model, the malodorous factor has a significant effect on the quality score, from medium values the coffee can be rejected, likewise, when sour is in the high values and ferment in the medium values. The factors that do not have a significant effect on the quality score are old and earthy. The factor that has a significant effect from high values is mold and sour. The presence of both factors, mold and old, has a significant effect on the quality score when they occur at high values. Through the development of a fuzzy expert system for sensorial evaluation of coffee bean attributes to derive quality scoring, Livio and Hodhod [38] found the ranges of the values of the attributes for the lowest and highest quality scores considering Fragrance, flavor, aftertaste, acidity, body, uniformity, balance, clean cup, sweetness, overall, and defects. Also from a fuzzy logic approach, the study of Flores and Pineda [39] conducted a similar analysis with the attributes brew, aroma, taste, aftertaste, and body.

Regarding the Parchment coffee entry sorting process scheduling model, both, Serious defects and minor defects, have a significant effect on sorting process scheduling. The green aspect factor does not have an effect on sorting process scheduling. The pellet factor has a significant effect from low values on sorting process scheduling. The humidity factor has a significant effect from medium to high values and results in complex sorting sequences. For the Robusta coffee entry sorting process scheduling model, both, serious defects and minor defects, have a significant effect from medium to high values on sorting process scheduling. The pellet factor has a significant effect from low to medium values on sorting process scheduling. The humidity factor has a significant effect from high values on sorting process scheduling. According to the a review study of the green coffee processing, Ghosh and Venkatachalapathy [42] asserted that achieving a 12% of humidity of coffee contributes to obtaining acceptable color, size along with the removal of pests for longer safe storage.. The weight factor does not have an effect on sorting process scheduling. Finally, for the not-washed Robusta coffee entry sorting process scheduling model, factors, serious defects, and minor defects have a significant effect from medium to high values on sorting process scheduling and result in complex sorting sequences. Factors, the green aspect, and humidity do not have an effect on sorting process scheduling.

Regarding the demand pattern scenarios analyzed, Figure 10 shows the input inventory of Parchment coffee and the output of green coffee. We noted that the input inventory and output inventory vary similar over time for the unimodal demand pattern, not the case for both the increasing and decreasing demand pattern. The ending input inventory level is at its highest for the increasing demand pattern and its zero for the decreasing demand pattern.

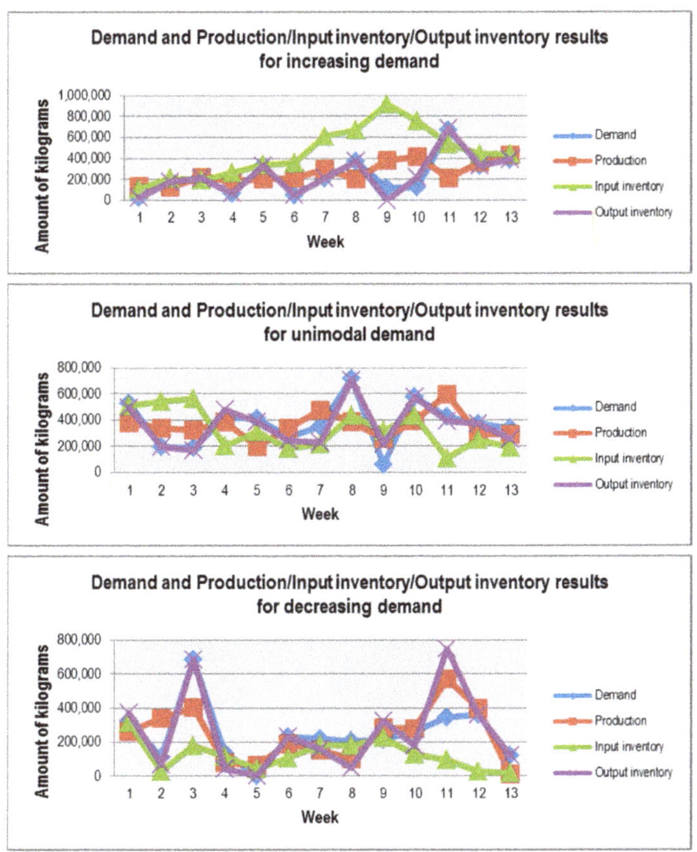

Figure 10. Production/Input inventory/Output inventory results for increasing, unimodal and decreasing demand.

Computational experiments aim to explore use cases for the decision support system to investigate different demand patterns. For the decreasing demand pattern, the results indicated the application of the reactive aggregate production scheduling approach in the green coffee SC due to service level value below the expected level. However, it is important to keep in mind that these results are subject to the assumptions of the model, as well as to the selected parameters and the defined system configurations. Decisions are based on case information, consequently, the limitations relate to this aspect of a case-study- based research. This can lead to situations in which the selection of the decision-making rule approach used for the decision support agents presented in this study results in decisions that could impair the performance of the system.

5. Conclusions and Future Work

In this study, we attempt to tackle the challenge for the ASC regarding the need for logistic systems for planning and scheduling/rescheduling within the agro-industry. To this end, an agent-based model driven decision support system for a Mexican green coffee SC was depicted. Three scenarios of demand patterns were considered to conduct the experiments: increasing demand, unimodal demand, and decreasing demand. A simulation model underpins the decision support system taking into account the use of the SCOR-process oriented approach, a hybrid modeling perspective, agent-based modeling and DES, and an adaptive control approach. A theoretical implication of the use of the hybrid modeling perspective refers to its contribution to achieving model quality as a measure of how the agent-based simulation model appropriately represents the aspects of interest in the green coffee SC, as well as simultaneous execution of events and performance improvement through agent coordination, both benefits of agent-based modeling and simulation with DES combination reported by Zhou et al. [56].

A general perception of the decision-makers within the focal company regarding the functionality of the decision support system the effective response of the system to the disturbance caused by the delay in the arrival or shortage of raw materials, which requires a change in the production schedule. The decision support system can handle these changes while considering scenarios reflecting the context of green coffee production, scenarios where the demand skewed towards the end of the planning horizon, where the demand skewed towards the beginning of the planning, and where demand peaks in the middle of the planning horizon and falls under the available capacity on the first and last days of the horizon. The exhibition of the management process within the green coffee SC context may help practitioners and managers interested in implementing the agent-based modeling and simulation approach in order to increase the possibilities of successful adopting of the reactive aggregate production scheduling.

A social implication for the Cherry coffee producers relates to the determination of the agricultural practices that have a significant effect on coffee yield: nutrition, pruning, and control of pests. Regarding agricultural practices in the study region, Hernández et al. [41] assert that producers generally do not have regulations or recommendations on the use of different varieties of coffee available to them, the culture of soil analysis is practically non-existent, and chemical fertilization is carried out through recommendations from fellow producers, for economic convenience and in some cases, on the recommendation of a commercial agrochemical company. Also, the authors state that there is a culture of control of plantation density with machete, leaving vegetal cover to avoid erosion. However, negative experiences are registered with the use of herbicides both for poisoning in the personnel who apply the product and for the degradation of the soils with frequent use, which has generated a posture of reserved use of the pesticide. Therefore, there are no experiences of pesticide use to control pests or coffee diseases. In regard to the pruning of coffee trees, the practices include sanitary pruning. Finally, the authors report that the producers do not have a multi-year management plan for plant tissue, including renewal.

Future work may consist of including decision support regarding coffee harvesting scheduling and coffee commercialization. The first refers to the decision to harvest ripe Cherry coffee. The cutting of Cherry fruits in mature state results in weight gained in the scale, plus facilitates the wet benefit process, decreases waste when Parchment coffee is processed and the coffee beans gain organoleptic quality.

The second is related to the negotiation in the coffee commercialization among the actors in the chain. The Cherry coffee purchasing parameters include the region of the crop, height and average annual temperature of the coffee plantation, the variety of coffee, the percentage of mature grains, the Parchment coffee yield and the time elapsed since the cut. Parameters for trade-in Parchment coffee include dry Cherry percentage; humidity percentage; yield to green coffee; uniformity of color; the number of defective coffee beans with respect to the weight of the sample analyzed; coffee beans significantly free of improper aromas; and defective beans percentage in the coffee lot.

Author Contributions: A.A.A.-L. was in charge of project administration and the research related to the application of artificial intelligence techniques in the supply chain. M.d.R.P.-S. was in charge of the research project related to the application of the agent modeling approach in the supply chain. M.G.C.-C. was an advisor for the supply chain management of the case study. R.P.-G. proof-read the paper. M.J.d.M.-A. was a collaborator in the use of the simulation platform. J.C.H.-G. was a collaborator in the fieldwork and analysis of the collected data.

Funding: This research received no external funding.

Conflicts of Interest: The authors declare no conflict of interest.

Appendix A

Tables A1–A6 describes the operating values for the fuzzy sets of the Fuzzy models of coffee growing yield; quality scoring of the Robusta coffee sample; quality scoring of the Parchment coffee sample; sorting process scheduling of not-washed Robusta coffee inputs; sorting process scheduling of Robusta coffee inputs; and sorting process scheduling of Parchment coffee inputs.

Table A1. Fuzzy model for coffee growing yield: fuzzy sets and operating intervals.

Coffee Growing						
Input			Knowledge Base	Output		
Variable	Fuzzy Set			Variable	Fuzzy Set	
	Definition	Interval			Definition	Interval
Nutrition (N)	Very low	[1, 1, 1]				
	Low	[1, 2, 2]				
	Appropriate	[2, 3, 3]			Very low	[0, 3, 6, 10]
	High	[3, 4, 4]				
	Very high	[4, 5, 5]				
Rainfall (R)	Low	[600, 600, 820, 1450]				
	Appropriate	[1400, 1500, 1600, 1850]				
	High	[1800, 1941, 2500, 2500]			Low	[7.5, 11, 13, 16]
Control of pests (CP)	Null-minimum	[0, 1, 1]				
	Protection	[1, 2, 2]				
Control of diseases (CD)	Null-minimum	[0, 1, 1]	1620 inference rules	Yield (Y)		
	Protection	[1, 2, 2]				
Planting density (PD)	Low	[0, 1, 1]			Medium	[15, 20, 25, 32]
	Appropriate	[1, 2, 2]				
	High	[2, 3, 3]				
Pruning (P)	Not performed	[0, 0, 0.27]				
	Moderate	[0.2, 0.5, 0.89]				
	Intense	[0.75, 1, 1]			High	[30, 32, 40, 40]
Temperature (T)	Low	[10, 10, 14, 22]				
	Appropriate	[21, 23, 24, 26]				
	High	[26, 28, 50, 50]				

Table A2. Fuzzy model for quality scoring of the Robusta coffee sample: fuzzy sets and operating intervals.

Robusta Coffee Organoleptic Evaluation						
Input			Knowledge Base	Output		
Variable	Fuzzy Set			Variable	Fuzzy Set	
	Definition	Interval			Definition	Interval
Ferment (F)	Not present	[0, 0, 1]				
	Low	[0.8, 1, 2]				
	Medium	[1.8, 2, 3]				
	High	[2.8, 3, 4]			7.2	[7.2, 7.2, 7.31]
Sour (S)	Not present	[0, 0, 1]				
	Low	[0.8, 1, 2]				
	Medium	[1.8, 2, 3]				
	High	[2.8, 3, 4]				
Malodorous (M)	No	[0, 0, 1]				
	Low	[0.8, 1, 2]				
	Medium	[1.8, 2, 3]				
	High	[2.8, 3, 4]	4096 inference rules	Robusta class (RC)	7.3	[7.3, 7.3, 7.41]
Earthy (E)	Not present	[0, 0, 1]				
	Low	[0.8, 1, 2]				
	Medium	[1.8, 2, 3]				
	High	[2.8, 3, 4]				
Mold (M)	Not present	[0, 0, 1]				
	Low	[0.8, 1, 2]				
	Medium	[1.8, 2, 3]				
	High	[2.8, 3, 4]			Rejected	[7.41, 7.41, 7.5]
Old (O)	Not present	[0, 0, 1]				
	Low	[0.8, 1, 2]				
	Medium	[1.8, 2, 3]				
	High	[2.8, 3, 4]				

Table A3. Fuzzy model for quality scoring of the Parchment coffee sample: fuzzy sets and operating intervals.

Parchment Coffee Organoleptic Evaluation						
Input			Knowledge Base	Output		
Variable	Fuzzy Set			Variable	Fuzzy Set	
	Definition	Interval			Definition	Interval
Aroma (A)	Null-little	[0, 2, 2]				
	Very low	[2, 3, 3]				
	Low	[3, 4, 4]				
	Medium	[4, 5, 5]				
	High	[5, 6, 6]			Altura	[2, 2.8, 3,3]
Flavor (F)	Null-little	[0, 2, 2]				
	Very low	[2, 3, 3]				
	Low	[3, 4, 4]				
	Medium	[4, 5, 5]				
	High	[5, 6, 6]				
Acidity (A)	Null-little	[0, 2, 2]				
	Very low	[2, 3, 3]				
	Low	[3, 4, 4]				
	Medium	[4, 5, 5]			Extra prima	[2.8, 3, 3.8, 4]
	High	[5, 6, 6]				
Body (B)	Low	[1, 2, 2]				
	Medium	[2, 3, 3]	96,000 inference rules	Parchment class (PC)		
	High	[3, 4, 4]				
Vinous, Fruity, Sweetness (VFS)	Not present	[0, 1, 1]				
	Low	[1, 1, 2]				
	Medium	[2, 2, 3]				
	High	[3, 3, 4]			Oro	[3.8, 4, 4.8, 5]
Green Immatureness (GI)	Not present	[0, 1, 1]				
	Present	[1, 2, 2]				
Cereal, Wood, Paper (CWP)	Not present	[0, 1, 1]				
	Present	[1, 2, 2]				
Dry, Old (DO)	Not present	[0, 1, 1]				
	Present	[1, 2, 2]				
Chemical, Medicinal (CM)	Not present	[0, 1, 1]				
	Present	[1, 2, 2]			Rejected	[4.8, 5, 5.8, 6]
Ferment, Sour, Malodorous (FSM)	Not present	[0, 1, 1]				
	Present	[1, 2, 2]				
Earthy, Mold (EM)	Not present	[0, 1, 1]				
	Present	[1, 2, 2]				

Table A4. Fuzzy model for process scheduling of not-washed Robusta coffee entries: fuzzy sets and operating intervals.

Sorting Process Scheduling for Not-washed Robusta Coffee Entries						
Input			Knowledge Base	Output		
Variable	Fuzzy Set			Variable	Fuzzy Set	
	Definition	Interval			Definition	Interval
Serious defects (SD)	Normal	[−13.5, −5.58, 9.17, 10.53]		Not-washed robusta Schedule 1 (nwrS1)	Mix	[0, 0.16, 0.33]
	Regular	[10.3, 10.5, 13, 13.5]			Pneumatic	[0.16, 0.33, 0.5]
	Many	[13, 13.5, 22, 22]			Optical	[0.33, 0.5, 0.66]
Minor defects (MD)	Normal	[−7.2, −0.8, 9, 9.2]	216 inference rules		Sift	[0.5, 0.66, 0.83]
	Many	[9, 9.2, 20.23, 20.23]				
Pellet (P)	Normal	[−1.79, 0.106, 3.32, 3.5]		Not-washed robusta Schedule 2 (nwrS2)	Pneumatic	[0, 0.2, 0.4]
	Regular	[3.29, 3.68, 4.5]			Optical	[0.2, 0.4, 0.6]
	Many	[4.3, 4.64, 11.9, 11.9]			Sift	[0.4, 0.6, 0.8]
Green aspect (GA)	Appropriate	[−7.2, −0.8, 13.8, 14.9]				
	Low	[14.7, 15.27, 15.8]				
	Very low	[15.5, 16.4, 16.86]				
	Null-minimum	[16.2, 17.01, 20, 20]		Not-washed robusta Schedule 3 (nwrS3)	Pneumatic	[0, 0.2, 0.4]
Weight (W)	Little	[-1.7×10^4, −7400, 9080, 9180]			Optical	[0.2, 0.4, 0.6]
	Normal	[7250, 8250, 1.44×10^4, 1.45×10^4]			Sift	[0.4, 0.6, 0.8]
	Much	[1.38×10^4, 1.48×10^4, 2.33×10^5, 2.35×10^5]				

Table A5. Fuzzy model for process scheduling of Robusta coffee entries: fuzzy sets and operating intervals.

Sorting Process Scheduling for Robusta Coffee Entries						
Input			Knowledge Base	Output		
Variable	Fuzzy Set			Variable	Fuzzy Set	
	Definition	Interval			Definition	Interval
Humidity (H)	Exceeded	[7.87, 9.47, 10.48, 11.4]	864 inference rules	Robusta Schedule 1 (rS1)	Mix	[0, 0.16, 0.33]
	Appropriate	[10.8, 11, 12.5, 12.75]			Pneumatic	[0.16, 0.33, 0.5]
	Low	[12.5, 12.75, 13]			Optical	[0.33, 0.5, 0.66]
	Null-little	[12.75, 13, 15.7, 17.9]			Sift	[0.5, 0.66, 0.83]
Minor defects (MD)	Normal	[−9, −1, 9.497, 9.81]			Dry	[0.6, 0.83, 1]
	Regular	[9.5, 10, 13, 13]			Dry little	[0.83, 1, 1.16]
	Many	[12.83, 13.2, 20.5, 30]		Robusta Schedule 2 (rS2)	Pneumatic	[0, 0.2, 0.4]
Serious defects (SD)	Normal	[−7.2, −0.8, 10, 10.5]			Optical	[0.2, 0.4, 0.6]
	Many	[10, 10.5, 20.23, 20.23]			Sift	[0.4, 0.6, 0.8]
Pellet (P)	Normal	[−4.814, −1.614, 2.286, 2.536]			Dry	[0.6, 0.8, 1]
	Regular	[2.29, 2.49, 3.49, 3.779]			Dry little	[0.8, 1, 1.2]
	Many	[3.5, 3.75, 10, 10]		Robusta Schedule 3 (rS3)	Pneumatic	[0, 0.2, 0.4]
Green aspect (GA)	Appropriate	[7.32, 8.95, 11, 12.75]			Optical	[0.2, 0.4, 0.6]
	Low	[12.5, 12.75, 13]			Sift	[0.4, 0.6, 0.8]
	Very low	[12.75, 13, 14]			Dry	[0.6, 0.8, 1]
	Null-minimum	[13, 14, 18, 18]			Dry little	[0.8, 1, 1.2]
Weight (W)	Little	[−2988, −188, 8958, 9058]		Robusta Schedule 4 (rS4)	Pneumatic	[0, 0.2, 0.4]
	Normal	[8100, 9100, 8.17×10^4, 8.18×10^4]			Optical	[0.2, 0.4, 0.6]
					Sift	[0.4, 0.6, 0.8]
	Little	[7.98×10^4, 8×10^4, 4.82×10^5, 4.83×10^5]			Dry	[0.6, 0.8, 1]
					Dry little	[0.8, 1, 1.2]

Table A6. Fuzzy model for process scheduling of Parchment coffee entries: fuzzy sets and operating intervals.

Sorting Process Scheduling for Parchment Coffee Entries						
Input			Knowledge Base	Output		
Variable	Fuzzy Set			Variable	Fuzzy Set	
	Definition	Interval			Definition	Interval
Humidity (H)			128 inference rules	Parchment Schedule 1 (pS1)	Mix	[0, 0.16, 0.33]
	Exceeded	[7.87, 9.47, 10.48, 11.4]			Pneumatic	[0.16, 0.33, 0.5]
	Appropriate	[10.8, 11, 12.5, 12.75]			Optical	[0.33, 0.5, 0.66]
					Sift	[0.5, 0.66, 0.83]
					Dry	[0.6, 0.83, 1]
	Low	[12.5, 12.75, 13]			Dry little	[0.83, 1, 1.16]
	Null-little	[12.75, 13, 15.7, 17.9]		Parchment Schedule 2 (pS2)	Pneumatic	[0, 0.2, 0.4]
Serious defects (SD)	Normal	[−9, −1, 2, 2.5]			Optical	[0.2, 0.4, 0.6]
					Sift	[0.4, 0.6, 0.8]
	Many	[2, 2.5, 20.5, 21.4]			Dry	[0.6, 0.8, 1]
Minor defects (MD)	Normal	[−14.4, −8.05, 2, 2.5]			Dry little	[0.8, 1, 1.2]
				Parchment Schedule 3 (pS3)	Pneumatic	[0, 0.2, 0.4]
	Many	[2, 2.5, 22, 22.4]			Optical	[0.2, 0.4, 0.6]
					Sift	[0.4, 0.6, 0.8]
Pellet (P)	Normal	[−3.6, −0.4, 0.5, 0.75]			Dry	[0.6, 0.8, 1]
	Many	[0.5, 0.75, 11, 11]			Dry little	[0.8, 1, 1.2]
Green aspect (GA)	Appropriate	[7.32, 8.95, 11, 12.75]		Parchment Schedule 4 (pS4)	Pneumatic	[0, 0.2, 0.4]
					Optical	[0.2, 0.4, 0.6]
	Low	[12.5, 12.75, 13]			Sift	[0.4, 0.6, 0.8]
	Very low	[12.75, 13, 14]			Dry	[0.6, 0.8, 1]
	Null-minimum	[13, 14, 18, 18]			Dry little	[0.8, 1, 1.2]

References

1. Verdouw, C.N.; Robbemond, R.M.; Verwaart, T.; Wolfert, J.; Beulens, A.J.M. A reference architecture for IoT-based logistic information systems in agri-food supply chains. *Enterp. Inf. Syst.* **2015**, *12*, 755–779. [CrossRef]
2. Macal, C.M.; North, M.J. Tutorial on agent-based modelling and simulation. *J. Simul.* **2010**, *4*, 151–162. [CrossRef]

3. Zhang, W.J.; Xie, S.Q. Agent technology for collaborative process planning: A review. *Int. J. Adv. Manuf. Technol.* **2007**, *32*, 315–325. [CrossRef]
4. Hilletofth, P.; Lättilä, L. Agent based decision support in the supply chain context. *Ind. Manag. Data Syst.* **2012**, *112*, 1217–1235. [CrossRef]
5. Power, D.J.; Sharda, R. Model-driven decision support systems: Concepts and research directions. *Decis. Support Syst.* **2007**, *43*, 1044–1061. [CrossRef]
6. Van der Vorst, J.G.A.J.; Tromp, S.; van der Zee, D.J. Simulation modelling for food supply chain redesign; integrated decision making on product quality, sustainability and logistics. *Int. J. Prod. Res.* **2009**, *47*, 6611–6631. [CrossRef]
7. Yasmine, A.S.L.E.; Ghani, B.A.; Trentesaux, D.; Bouziane, B. Supply Chain Management Using Multi-Agent Systems in the Agri-Food Industry. In *Service Orientation in Holonic and Multi-Agent Manufacturing and Robotics*; Borangiu, T., Trentesaux, D., Thomas, A., Eds.; Springer: Berlin, Germany, 2014; pp. 145–155.
8. Utomo, D.S.; Onggo, B.S.; Eldridge, S. Applications of agent-based modelling and simulation in the agri-food supply chains. *Eur. J. Oper. Res.* **2017**, *269*, 794–805. [CrossRef]
9. Chatfield, D.C.; Hayya, J.C.; Harrison, T.P. A multi-formalism architecture for agent-based, order-centric supply chain simulation. *Simul. Model. Pract. Theory* **2007**, *15*, 153–174. [CrossRef]
10. Macal, C.M. Everything you need to know about agent-based modelling and simulation. *J. Simul.* **2016**, *10*, 144–156. [CrossRef]
11. Van der Vorst, J.G.A.J.; Beulens, A.J.M. Identifying sources of uncertainty to generate supply chain redesign strategies. *Int. J. Phys. Distrib. Logist. Manag.* **2002**, *32*, 409–430. [CrossRef]
12. Suarez-Barraza, M.F.; Miguel-Davila, J.; Vasquez-García, C.F. Supply chain value stream mapping: A new tool of operation management. *Int. J. Qual. Reliab. Manag.* **2016**, *33*, 518–534. [CrossRef]
13. Van der Zee, D.J.; van der Vorst, J.G.A.J. A modeling framework for supply chain simulation: Opportunities for improved decision making. *Decis. Sci.* **2005**, *36*, 65–95. [CrossRef]
14. Bui, T.; Lee, J. An agent-based framework for building decision support systems. *Decis. Support Syst.* **1999**, *25*, 225–237. [CrossRef]
15. Turban, E.; Aronson, J.E.; Liang, T.P. *Decision Support Systems and Intelligent Systems*, 7th ed.; Prentice Hall: Upper Saddle River, NJ, USA, 2005.
16. Jain, L.C.; Lim, C.P.; Nguyen, N.T. Innovations in knowledge processing and decision making in agent-based systems. In *Knowledge Processing and Decision Making in Agent-Based Systems*; Jain, L.C., Nguyen, N.T., Eds.; Springer: Berlin, Germany, 2009; pp. 1–12.
17. Mařík, V.; McFarlane, D. Industrial adoption of agent-based technologies. *IEEE Intell. Syst.* **2005**, *20*, 27–35. [CrossRef]
18. Mattia, A. A multi-dimensional view of agent-based decisions in supply chain management. *Commun. IBIMA* **2012**, *2012*. [CrossRef]
19. Méndez, C.A.; Cerdá, J.; Grossmann, I.E.; Harjunkoski, I.; Fahl, M. State-of-the-art review of optimization methods for short-term scheduling of batch processes. *Comput. Chem. Eng.* **2006**, *30*, 913–946. [CrossRef]
20. Phanden, R.K.; Jain, A.; Verma, R. Integration of process planning and scheduling: A state-of-the-art review. *Int. J. Comput. Integr. Manuf.* **2011**, *24*, 517–534. [CrossRef]
21. Barbati, M.; Bruno, G.; Genovese, A. Applications of agent-based models for optimization problems: A literature review. *Expert Syst. Appl.* **2012**, *39*, 6020–6028. [CrossRef]
22. Lee, J.H.; Kim, C.O. Multi-agent systems applications in manufacturing systems and supply chain management: A review paper. *Int. J. Prod. Res.* **2008**, *46*, 233–265. [CrossRef]
23. Monostori, L.; Váncza, J.; Kumara, S.R.T. Agent-based systems for manufacturing. *CIRP Ann.* **2006**, *55*, 697–720. [CrossRef]
24. Van der Vorst, J.G.A.J.; Beulens, A.J.M.; van Beek, P. Innovations in logistics and ICT in food supply chain networks. In *Innovations in Agri-Food Systems. Product Quality and Consumer Acceptance*; Jongen, W.M.F., Meulenberg, M.T.G., Eds.; Wageningen Academic Publishers: Wageningen, The Netherlands, 2005; pp. 245–292.
25. Ali, J.; Kumar, S. Information and communication technologies (ICTs) and farmers' decision-making across the agricultural supply chain. *Int. J. Inf. Manag.* **2011**, *31*, 149–159. [CrossRef]
26. Borodin, V.; Bourtembourg, J.; Hnaien, F.; Labadie, N. Handling uncertainty in agricultural supply chain management: A state of the art. *Eur. J. Oper. Res.* **2016**, *254*, 348–359. [CrossRef]

27. Higgins, A.J.; Miller, C.J.; Archer, A.A.; Ton, T.; Fletcher, C.S.; McAllister, R.R.J. Challenges of operations research practice in agricultural value chains. *J. Oper. Res. Soc.* **2010**, *61*, 964–973. [CrossRef]
28. Tsolakis, N.K.; Keramydas, C.A.; Toka, A.K.; Aidonis, D.A.; Iakovou, E.T. Agrifood supply chain management: A comprehensive hierarchical decision-making framework and a critical taxonomy. *Biosyst. Eng.* **2014**, *120*, 47–64. [CrossRef]
29. Van der Vorst, J.G.A.J.; van Kooten, O.; Luning, P.A. Towards a diagnostic instrument to identify improvement opportunities for quality controlled logistics in agrifood supply chain networks. *J. Food Syst. Dyn.* **2011**, *2*, 94–105.
30. Handayati, Y.; Simatupang, T.M.; Perdana, T. Value Co-creation in Agri-chains Network: An Agent-Based Simulation. *Procedia Manuf.* **2015**, *4*, 419–428. [CrossRef]
31. Zimon, D.; Domingues, P. Proposal of a concept for improving the sustainable management of supply chains in the textile industry. *Fibres Text. East. Eur.* **2018**, *26*, 8–12. [CrossRef]
32. Killian, B.; Jones, C.; Pratt, L.; Villalobos, A. Is sustainable agriculture a viable strategy to improve farm income in Central America? A case study on coffee. *J. Bus. Res.* **2006**, *59*, 322–330. [CrossRef]
33. Kilian, B.; Rivera, L.; Soto, M.; Navichoc, D. Carbon Footprint across the Coffee Supply Chain: The Case of Costa Rican Coffee Bernard. *J. Agric. Sci. Technol. B* **2013**, *3*, 151–170.
34. Espinosa-Solares, T.; Cruz-Castillo, J.G.; Montesinos-López, O.A.; Hernández-Montes, A. Raw coffee processing yield affected more by cultivar than by harvest date. *J. Agric. Univ. P. R.* **2005**, *89*, 169–180.
35. Bosselmann, A.S.; Dons, K.; Oberthur, T.; Olsen, C.S.; Ræbild, A.; Usma, H. The influence of shade trees on coffee quality in small holder coffee agroforestry systems in Southern Colombia. *Agric. Ecosyst. Environ.* **2009**, *129*, 253–260. [CrossRef]
36. Feria-Morales, A.M. Examining the case of green coffee to illustrate the limitations of grading systems/expert tasters in sensory evaluation for quality control. *Food Qual. Prefer.* **2002**, *13*, 355–367. [CrossRef]
37. Livio, J.; Flores, W.C.; Hodhod, R.; Umphress, D. Smart fuzzy cupper: Employing approximate reasoning to derive coffee bean quality scoring from individual attributes. In Proceedings of the IEEE International Conference on Fuzzy Systems, Rio de Janeiro, Brazil, 8–13 July 2018; pp. 1–7.
38. Livio, J.; Hodhod, R. AI Cupper: A Fuzzy Expert System for Sensorial Evaluation of Coffee Bean Attributes to Derive Quality Scoring. *IEEE Trans. Fuzzy Syst.* **2018**, *26*, 3418–3427. [CrossRef]
39. Flores, W.C.; Pineda, G.M. A type-2 fuzzy logic system approach to train Honduran coffee cuppers. In Proceedings of the 2016 IEEE Latin American Conference on Computational Intelligence, Cartagena, Colombia, 2–4 November 2016; pp. 1–7.
40. FAO Faostat. Available online: http://www.fao.org/faostat/en/#data/QC (accessed on 27 March 2019).
41. Hernández, G.; Escamilla, S.; Velázquez, T.; Martínez, J.L. Análisis de la cadena de suministro del café en el Centro de Veracruz: Situación actual, retos y oportunidades. In *Cafeticultura en la Zona Centro del Estado de Veracruz. Diagnóstico, Productividad y Servicios Ambientales*; López, R., Sosa, V.D.J., Díaz, G., Contreras, H.A., Eds.; Instituto Nacional de Investigaciones Forestales, Agrícolas y Pecuarias: Veracruz, Mexico, 2013; pp. 8–36.
42. Ghosh, P.; Venkatachalapathy, N. Processing and Drying of Coffee—A review. *Int. J. Eng. Res. Technol.* **2014**, *3*, 784–794.
43. Ramalakshmi, K.; Kubra, I.R.; Rao, L.J.M. Physicochemical characteristics of green coffee: Comparison of graded and defective beans. *J. Food Sci.* **2007**, *72*, 333–337. [CrossRef]
44. Lambert, D.M.; García-Dastugue, S.J. An evaluation of process-oriented supply chain management frameworks. *J. Bus. Logist.* **2005**, *26*, 25–51. [CrossRef]
45. Persson, F. SCOR template—A simulation based dynamic supply chain analysis tool. *Int. J. Prod. Econ.* **2011**, *131*, 288–294. [CrossRef]
46. Estampe, D.; Lamouri, S.; Paris, J.L.; Brahim-djelloul, S. A framework for analysing supply chain performance evaluation models. *Int. J. Prod. Econ.* **2013**, *142*, 247–258. [CrossRef]
47. Lockamy, A., III; Mccormack, K. Linking SCOR planning practices to supply chain performance An exploratory study. *Int. J. Oper. Prod. Manag.* **2004**, *24*, 1192–1218. [CrossRef]
48. Bolstorff, P.; Rosenbaum, R. *Supply Chain Excellence: A Handbook for Dramatic Improvement Using the SCOR Model*, 2nd. ed.; AMACOM: New York, NY, USA, 2007.
49. Sawik, T. Integer programming approach to reactive scheduling in make-to-order manufacturing. *Math. Comput. Model.* **2007**, *46*, 1373–1387. [CrossRef]

50. Fleischmann, B.; Meyr, H.; Wagner, M. Advanced Planning. In *Supply Chain Management and Advanced Planning: Concepts, Models, Software and Case Studies*; Stadtler, H., Kilger, C., Eds.; Springer: Berlin, Germay, 2002; pp. 81–106.
51. Vieira, G.E.; Herrmann, J.W.; Lin, E. Rescheduling manufacturing systems: A framework of strategies, policies, and methods. *J. Sched.* **2003**, *6*, 39–62. [CrossRef]
52. Swaminathan, J.M.; Smith, S.F.; Sadeh, N.M. Modeling supply chain dynamics: A multiagent approach. *Decis. Sci.* **1998**, *29*, 607–632. [CrossRef]
53. Rzevski, G.; Andreev, M.; Skobelev, P.; Shveykin, P.; Tugashev, A.; Tsarev, A. Adaptive Planning for Supply Chain Networks. In Proceedings of the Holonic and Multi-Agent Systems for Manufacturing; Springer: Berlin, Germany, 2007; pp. 215–224.
54. Sawik, T. Multi-objective master production scheduling in make-to-order manufacturing. *Int. J. Prod. Res.* **2007**, *45*, 2629–2653. [CrossRef]
55. Paulo, E.M.; Furlani, E., Jr. Yield performance and leaf nutrient levels of coffee cultivars under different plant densities. *Sci. Agric.* **2010**, *67*, 720–726. [CrossRef]
56. Zhou, R.; Lee, H.P.; Nee, A.Y.C. Simulating the generic job shop as a multi-agent system. *Int. J. Intell. Syst. Technol. Appl.* **2008**, *4*, 5–33. [CrossRef]

Article

Spatiotemporal Modeling of the Smart City Residents' Activity with Multi-Agent Systems

Robert Olszewski [1], Piotr Pałka [2,*], Agnieszka Turek [1], Bogna Kietlińska [3], Tadeusz Płatkowski [4] and Marek Borkowski [2]

1 Faculty of Geodesy and Cartography, Warsaw University of Technology, Warsaw 00-661, Poland; robert.olszewski@pw.edu.pl (R.O.); agnieszka.turek@pw.edu.pl (A.T.)

2 Faculty of Electronics and Information Technology, Warsaw University of Technology, Warsaw 00-665, Poland; M.Borkowski@stud.elka.pw.edu.pl

3 Institute of Applied Social Sciences, Warsaw University, Warsaw 00-927, Poland; b.kietlinska@uw.edu.pl

4 Institute of Applied Mathematics, Warsaw University, Warsaw 02-097, Poland; tplatk@mimuw.edu.pl

* Correspondence: p.palka@ia.pw.edu.pl; Tel.: +48-691-241-565

Received: 30 April 2019; Accepted: 16 May 2019; Published: 19 May 2019

Abstract: The article proposes the concept of modeling that uses multi-agent systems of mutual interactions between city residents as well as interactions between residents and spatial objects. Adopting this perspective means treating residents, as well as buildings or other spatial objects, as distinct agents that exchange multifaceted packages of information in a dynamic and non-linear way. The exchanged information may be reinforced or diminished during the process, which may result in changing the social activity of the residents. Utilizing Latour's actor–network theory, the authors developed a model for studying the relationship between demographic and social factors, and the diversified spatial arrangement and the structure of a city. This concept was used to model the level of residents' trust spatiotemporally and, indirectly, to study the level of social (geo)participation in a smart city. The devised system, whose test implementation as an agent-based system was done in the GAMA: agent-based, spatially explicit, modeling and simulation platform, was tested on both model and real data. The results obtained for the model city and the capital of Poland, Warsaw, indicate the significant and interdisciplinary analytical and scientific potential of the authorial methodology in the domain of geospatial science, geospatial data models with multi-agent systems, spatial planning, and applied social sciences.

Keywords: multi-agent systems; smart city development; spatiotemporal modeling; actor–network theory; geoparticipation; social interactions

1. Introduction

Many different disciplines use multi-agent systems as a research tool. One of them is the analysis of social relations in the city, as well as the interaction between residents and spatial objects (the background of the research). The open problem to address is an analysis of various factors that influence changes in the level of residents' social engagement in the process of social participation; above all, changes in the level of mutual trust amongst residents and their trust in social institutions. The multi-agent system (in further parts of the paper, the authors use MAS abbreviation) that models the process of changing the social engagement of residents, proposed by the authors of the article, is the main contribution to the scientific research integrating applied social sciences, geoinformation technologies, and multi-agent systems.

In his 2007 article [1], Michael F. Goodchild introduces the concept of social assembling of spatial information by users identified as specific human agents. Implementing this idea brought about the rapid development of the so-called VGI (volunteered geographic information), manifested by, e.g., social (crowdsourcing) creation of Open Street Map by over 3 million active users around the world.

The purpose of this article is to develop the concept of smart cities' residents as active urban sensors represented by agents in the MAS. Consequently, city residents are considered elements of a specific geospatial multi-agent system. Mutual interactions of the residents, between the residents, as well as the impact the "human agents" have on spatial objects in the long-term influences the activeness of the residents.

It should be emphasized that for this concept, the crucial assumption is that mutual interactions between residents and spatial objects are characteristic of a complex multi-agent system in a smart city. The non-linear exchange of packages of information between individual elements of the system, under special conditions, leads to reinforcing information and "activating" residents in the process of social participation. It is assumed that the residents (represented by agents) also interact with different features of the city, which tend to modify their trust.

In recent decades, the crisis of participatory democracy has been particularly severe in urban centers and areas subject to urbanization. Its outcome is the weakening of the sense that the residents have a real impact on co-creating the vision for a city's development, revitalization of the neglected districts, or spatial order. When interpreting the term "the right to the city," David Harvey [2] emphasizes that it is not only the right to access its resources, but also the right to decide jointly on the direction of the city's development. A smart and sustainable city engages all of its residents in the most critical decision-making processes, making the process of creating spatial order more social, and encouraging the development of participatory and deliberative democracy. An increase in mutual trust among residents and their trust in public institutions are of crucial importance to stimulating the activity. Therefore, the goal of the authors is to model complex social interactions between the urban residents and to establish the level of trust, sense of identity, willingness to participate in the social (geo)participation processes, and the dynamics of changes over time. The devised model has been tested not only on the example of the "model city," but also on the real urban agglomeration of Warsaw, Poland.

The authors intend to develop a concept of a multi-agent decision support system that makes use of game theory, multi-agent systems, and market programming models to support the "weakest link" of the asymmetric urban network's triangle of "municipal authorities" (politics), "business" (urban developers, industrial investments, etc.), and "residents". So that the atomized individuals are transformed into a cooperative urban community, it is necessary to use available means of electronic communication, information and communication technologies (ICT), and geoinformation tools, as well as to revive the Athenian ideas of the (urban) agora that decides on the city through the process of social debate. Such an asymmetry is the basic rule for producing the majority of urban relations described in Latour's actor–network theory [3]. According to this theory, the basic element of all the networks is an actant, meaning a factor influencing all the other factors. In this article, the social sciences' idea of an actant will be synonymous with an agent. This term will apply both to a "human agent" that is a resident characterized by a vector of information specific to his/her age, education, place of residence and work, general health condition, base level of social activity, trust, and so on, as well as a spatial object (district, office, park, and so on) that influences the inhabitants. These factors influence each other, forming a system of actors and networks.

On the basis of this theory, the authors of the article developed a model of interaction between the sensors–actants (both residents and spatial objects that form the urban tissue), which enables the simulation of social interaction processes as well as, indirectly, participatory democracy and the process of social participation of smart cities' residents. The critical element of this process is stimulating the civic activity of the citizens by increasing the level of trust, both in people and institutions.

The mutual trust of residents, as well as the level of the citizens' trust in institutions (e.g., municipal and security authorities, planners, educational institutions, healthcare system, and so forth), is particularly important with regards to implementing the idea of a smart city. While technological development is important to ensure the effectiveness of the process, the development of social interactions among the residents and their joint decision-making on the vision for the city's development is crucial. As stressed in [4], cities of tomorrow need to adopt a holistic model of sustainable urban development. A city is smart when public issues are solved using information and communications technology (ICT) and there is involvement of various types of stakeholders acting in partnership with the municipal authorities (see [5]). The implementation of a smart city is strongly related to the process of social (geo)participation—according to PAS 180: 2014 (Publicly Available Specification under the British Standards Institution), this process means an "effective integration of physical, digital and human systems in the built environment to deliver a sustainable, prosperous and inclusive future for its citizens." Also, the new urban agenda stresses the need to empower all individuals and communities and to promote and broaden inclusive platforms that allow full and meaningful participation in the decision-making and planning process (see [6]). A city can be considered a smart one when it, in parallel, invests in technology and human capital to actively promote sustainable economic development and high quality of life (e.g., it enables natural resources management through civic participation).

The contributions of the paper are (i) development of sociological concepts: Bruno Latour's actor–network theory [3], Edward T. Hall's social distances [7], Erving Goffman's social interactions [8], related to sensors–actants which enables the simulation of social interaction; (ii) implementation of the model using multi-agent methodology [9] in GAMA toolset environment; (iii) spatiotemporal and sociological development of concepts of two smart cities and their implementation in GAMA; and (iv) an illustration of the model's operation in typical situations occurring in cities. The use of these approaches made it possible to model the social activity of residents in a dynamic and non-linear way, as well as to conduct spatiotemporal analysis, and create geospatial data models.

The article consists of six sections. After a short introduction (Section 1) the authors describe related works regarding the analyzed issue and motivate the choice of the methods used (Section 2). Subsequently, the authors discuss the research methodology (Section 3). In this section, the authors discuss the actor–network theory (Section 3.1), which is the basis of our model, as well as the method of city modeling used in this article (Section 3.2), then go to the description of city modeling using agents (Section 3.3) and interactions in which these agents participate (Section 3.4). The next section describes in detail, validates, and calibrates the model city (Section 4). The authors use two scenarios for this purpose: Terra incognita (Section 4.1) and Old Factory revitalization (Section 4.2). This is followed by an analysis of the spatiotemporal model of the city of Warsaw (Section 5) and subsequent sections analyze three scenarios: Parade Square (Section 5.1), "Mordor" on Domaniewska Street (Section 5.2), and Miasteczko Wilanów (Section 5.3). The work ends with a discussion and conclusions (Section 6).

2. Related Works

The interdisciplinary nature of the research undertaken by the authors of this article requires referencing numerous concepts and methods derived from urban planning, spatial planning, sociology, spatial science, as well as mathematics or computer science.

There have been numerous attempts at developing appropriate tools using modern technologies, e.g., geospatial multi-agent system design and integration, agent-based systems, machine learning, data mining, augmented reality, virtual reality, or 3D models, to ensure effective participation of citizens in urban and territorial development decision-making with a game theoretical treatment (see [10,11]). In [12], authors predict that the widespread presence of smartphones will soon mean that citizens will be treated as a network of sensors that the city will use for continuous development. The use of the concept of a personal digital assistant to support a smart-city citizen, which is most often run on smartphones, is described in the paper [13]. Authors propose a software prototype of a personal digital assistant 2.0, which, based on soft computing methods and cognitive computing, improves calendar and mobility management in smart cities. On the other hand, there are many publications on the analysis of social behavior in urban environments by using multi-agent systems or agent-based models. Malleson, in [14], emphasizes the need to combine big data and agent-based modeling tools to analyze a smart city. Karmakharm and Richmond, in [15], include a description of the pedestrian behavior simulation in the event of a threat in the public space. Meanwhile, Sandhu et al., in [16], present the implementation of a model, based on intelligent agents, for controlling streetlights in a smart city. In their study [17], Olszewski, Pałka, and Turek analyze the problem of traffic jams in the office district with regards to car-sharing. Their agent-based model simulates the socio-economic behavior of the employees of the so-called Mordor of Warsaw.

Geo-sensing enables context-aware analyses of physical and social phenomena. Moreover, context-aware analysis can potentially enable a more holistic understanding of spatio-temporal processes [18], where authors discuss the possibilities of integrating spatiotemporal contextual information with human and technical sensor information. Among different types of sensors used to collect such kinds of information, they mention in situ sensors, technical remote sensors, and human agents, discussed by Sagl, Resch, and Blaschke [19]. Resch, in [20], defines human agent data as human-generated measurements. He distinguishes the situation in which humans generates data (subjective observations) and humans that carry "ambient sensors" to measure external parameters. In the literature, there were also attempts made for interpreting data acquired by a "human agent", who uses an interactive location-based service (iLBS) (e.g., to sense cultural-historic facts in the landscape) (see [21]).

Cellular automata (CA) can be used to simulate urban dynamics and land-use changes effectively. Several authors performed simulations of urban development and land-use changes using GIS-based cellular automata (see [22–25]). Li et al., in [26], indicate that using parallel computation techniques can significantly improve the performance of the large-scale urban simulation. Agent-based models are applied to increase the intelligence and flexibility of planning support systems. Saarloos et al., in [27], developed a framework in which an agent organization consists of three types of agents: "interface agents" to improve the user–system interaction; "tool agents" to support the use and management of models; and "domain agents" to provide access to specialized knowledge.

Imottesjo and Kain, in [28], developed a prototype mobile augmented reality (MAR) tool, Urban CoBuilder. The application facilitates participative planning of urban space to increase bottom-up and multi-stakeholder inclusion. Yan Zhang, in [29], prototyped CityMatrix, which is an evidence-based urban decision support system, augmented by artificial intelligence (AI) techniques, including machine learning simulation predictions and optimization of search algorithms. Zhang investigated the strength of these technologies to augment the ability to make better urban decisions. Allen, Regenbrecht, and Abbott, in [30], investigated a smartphone-based augmented reality architecture as a tool for aiding public participation in urban planning by developing a prototype system, which showed 3D virtual representations of proposed architectural designs visualized on top of the existing real-world architecture. The authors investigated whether using a smartphone augmented reality system increases the willingness of the public to participate and the perceived participation in urban planning.

Jing and Hai-xing [31] built a support vector machine (SVM) model to predict the trends of coordinated development. The authors compared the method with an artificial neural network, decision tree, logistic regression, and naïve Bayesian classifier regarding the urban ecosystem coordinated development prediction for the Guanzhong urban agglomeration.

Ultsch, Kretschmer, and Behnisch, in [32], used techniques of machine learning and data mining to discover comprehensible and useful structures in the multivariate municipality data. As Behnisch and Ultsch in [33] indicate, "Urban Data Mining represents a methodological approach that discovers logical, mathematical and partly complex descriptions of urban patterns and regularities inside statistical data".

In the conducted research, multi-agent systems were adopted as a tool for modeling and simulation. It enabled the implementation of the assumptions behind the actor–network theory for modeling social processes in the urban space of a smart city. A multi-agent system (MAS or a "self-organized system") is a computerized system composed of multiple interacting intelligent agents (see [9,34]). Multi-agent systems can solve problems that are difficult or impossible for an individual agent or a monolithic system to solve. The primary assumption of MAS is communication amongst the agents and their autonomy. The notion of agents, as currently used in urban simulation models, is a kind of automaton that mimics the behavior of urban agents in a predetermined way. Portugali, in [35], describes a CogCity (cognitive city) as an urban simulation model that explicitly incorporates in its structure the role of three cognitive processes that typify the behavior of human agents: Information compression, cognitive mapping, and categorization. Moreover, the model CogCity demonstrates the possibility and usefulness of agent-based and cellular automata urban simulation model, which combines top-down and bottom-up processes in one model. Projects from MIT's SENSEable City Lab foster the vision of the real-time city by providing 'a feedback loop between people, their actions, and the city'.

The issue of public opinion formation is the subject of studies conducted by Deffuant, Amblard, and Weisbuch in [36], and Hegselmann and Krause in [37]. These authors consider the issue of social opinion formation through consensus, polarization, and fragmentation. The article investigates various models for the dynamics of continuous opinions by analytical methods as well as by computer simulations. Consequently, the rapid development of advanced technologies (IoT, wearable computing, etc.) forces the process of connecting real-world objects like buildings, roads, household appliances, and human bodies to the Internet via sensors and microprocessor chips that record and transmit data such as sound waves, temperature, movement, and other variables. This supports the development of smart citizens (see [38]).

In [39], Jacobs points to correlation between the urban form and the urban performance, e.g., the quality of life, vibrancy, and safety. Yan Zhang, in [29], takes it a step further and shows the correlation between the urban form and multiple aspects the urban performance. The 17 defined indexes represent 17 aspects of the urban performance of a city district, grouped into four high-level indexes: Density, diversity, proximity, and energy.

Sociological theories have been a vital source of inspiration for the authors of this article. Source studies include the theory of social impact developed by Nowak–Latané, which describes the interaction among members of large groups and the stabilization of opinions in groups [40,41]. Nowak, Szamrej, and Latané in [42] argue that "[modeling] the change of attitudes in a population resulting from the interactive, reciprocal, and recursive operation Bibb Latané's theory of social impact, which specifies principles underlying how individuals are affected by their social environment". Also, the dramaturgical Goffman's theory [8] and Latour's actor–network theory (ANT) [3] have been of crucial importance for the conducted research.

According to Latour's actor–network theory (ANT), an actant, an advanced sensor, is the basic component of all networks; it is a factor influencing other factors. ANT is a theoretical and methodological approach to social theory where everything in the social and natural worlds exists in constantly shifting networks of relationship [3]. It posits that nothing exists outside those relationships. All the factors involved in a social situation are on the same level, and thus, there are no external social forces beyond what and how the network participants are interacting at present. Thus, objects, ideas, processes, and any other relevant factors are seen as just as necessary in creating social situations as humans. Latour distinguishes two types of ties between the actants: Active and passive. The result of an active one is not typical and depends on the mediation between the mediators, meaning that the result is uncertain and variable. In the case of ties between mediations, the situation is stable, and the translation proceeds in a predictable and predetermined manner.

The authors have also been inspired by the studies that take into account the opinion formation model with a "strong leader" (see [43,44]), meaning a leader who significantly influences the molding and modifying of the opinions and attitudes of the residents.

3. Research Methodology

The authors of this article see the relationship between the structure of a city, spatial order, the way residents live, and the level of their social activity. The issue of information asymmetry is of crucial importance for modeling these relationships (see [45]).

3.1. Actor–Network Theory

Depending on the context of the study, a given actant may be divided into a more complex actor–network order (e.g., a city may be analyzed as a system of relations between buildings, districts, authorities, residents, road infrastructure, and so on). The strength of the influence of individual actants—which is a result of various factors, such as the level of mutual trust or the identification with a given place or space—determines the strength of the relations. Furthermore, a number of such forces may impact a single actant at a given moment, each of the forces with suitable power, which is often asymmetrical in relation to others. This lack of symmetry, or an uneven distribution of forces and influences, manifests itself in almost every dimension of the polysemous creation that is a city. Starting from the right to determine, through the levels of capital (social, economic, cultural, and symbolic) of the city's individual users, to planning and urban solutions that may result in, among other things, ghettoization or spatial exclusion of specific groups of residents. However, one may assume that, firstly, an uneven flow of information between particular actants, often conditioned by the previously mentioned level of social trust, underlies each of the urban asymmetries. Secondly, individual asymmetries overlap and form relations with other asymmetry systems, resulting in the production of additional, now much more intricate, networks of mutual influences and interactions. The level of trust is crucial for the activity and social participation of the citizens.

The underlying assumption of the authors in relation to the concept of geospatial multi-agent system design is that modeling of social interactions is a non-linear generalized regression. It is, therefore, assumed that:

- Only selected factors (out of an infinite number of factors) influencing the level of social activity of residents are analyzed in the model. The advantage of this approach is the opportunity to use quantitative models; the disadvantage is the omission of the factors described in the sociological theories of a qualitative nature.
- Modeling the time changes of the sensor–actor system means complex and multiple interactions of individual sensors, which requires considering the iterative approach and simulating long-term processes.
- Following the idea of citizens as sensors, smart city residents are "human agents" that, during multiple interactions, exchange packets of information all the while modifying the parameters that characterize individual elements of the system (actants). According to the actor–network theory, achieving a certain level of parameter "trust" causes the social activation of individual residents.
- Every "human agent" is an autonomous entity with individual goals and information (also conflicting). Every agent strives to achieve its own goals, unattainable without interacting with others. This approach is consistent with the agent programming paradigm, which provided the basis for modeling the system using multi-agent methodology. Also, it is assumed that "human agents" are dynamic objects as interactions change their attributes.

In their research, the authors investigate how the trust of residents change with time: Both the level of mutual trust and the trust in social institutions, which then stimulates the growth of social involvement and social (geo)participation. The level of trust and its changes depend on factors such as, among others, place of residence, type of work, time spent in public transport or public facilities, theatres, as well as the types of building development or the openness of space, and so on. Changing all of the parameters for a population of hundreds of thousands of people requires the use of parallel computing in multi-agent systems and numerical simulations covering millions of calculation epochs, which model decades of a city's functioning. The base level of the residents' identities, the intrinsic idea of deliberative democracy, the so-called strong leaders in the local community, as well as the specific genius loci of the city are all crucial in the process of changing the level of trust and involvement of residents.

3.2. Modeling of the Smart City

The crucial element of studying the development of a city and the way the urban network works using the (broadly defined) game theory is determining whether the knowledge of individual players (advanced sensors) is symmetrical or if there is an informational asymmetry. Multi-agent systems are one of the tools used for modeling the game theory and the theory of market mechanisms.

The models use elements of game theory, emergence, sociology, and multi-agent systems. The model assumes that agents, representing the residents of a city (sensors), move around the city and interact with each other. It takes place during every act of verbal or non-verbal communication. This process involves mutual decoding and simultaneous interpretation of the meaning of symbols used by the other party in communication. Interactions influence and change agent's trust in other residents. The information asymmetry phenomenon is easily modeled in a multi-agent system where every agent (an autonomous software element), while making decisions based on private information and interacting with the remaining agents, has a piece of information whose level may be varied. The simulation is divided into small time quanta (e.g., 15 min), during which the residents interact with each other.

The city is modeled by a system of roads on which the agents move, a set of buildings, including stand-alone and multi-family residential buildings, factories, office buildings, offices, health clinics, schools, and museums; green areas, i.e., boulevards, parks, and water reservoirs. The city is also divided into districts distinguished by a set of general features, which characterize both the district and the people in the district, e.g., the office district is characterized by a significant share of office buildings, and the people in it are blue-collar workers.

3.3. Citizen Modeling—Agents

An agent models a city resident and has a set of features that reflect its social character:

- Age, gender, marital status, and number of children.
- Trust in other residents, a number in the range <0,1> that determines to what extent a resident trusts other people. It is not a pejorative trait; it does not mean naivety, but faith in the capabilities of others.
- Trust in institutions, a number in the range <0,1> that determines to what extent a citizen trusts institutions; that is, governmental agencies, healthcare, or educational institutions.
- Altruism, a number in the range <0,1> that determines to what extent a person wants to work for the society and engage in the social life for the sake of common interests.
- Education, a number in the range <0,1> that determines the degree of education.
- Life satisfaction, a number in the range <0,1> that determines to what extent a citizen is satisfied with her/his life.
- Wealth, a number in the range <0,1> that determines the material status of a citizen.
- Identity, a number in the range <0,1> that determines the emotional connection of the citizen with the city or district in which s/he lives.

Besides the features above, each agent is assigned to a place in which s/he lives (a residential building) and a workplace (an office building or a factory). During the simulation, the agents are moving around the city according to the daily rhythm. Residents navigate the city along a network of roads; eventually, they can go to a demonstration.

Demonstrations take place in the so-called attractors—places that attract social interest and provoke extreme emotions. One such attractor is the Palace of Culture and Science in Warsaw, as its demolition is a continuing matter of dispute. A demonstration causes a clash of extreme emotions of the participants and often results in a change of stance regarding the fate of a given attractor.

The devised model assumes that from Monday to Friday, each agent (human agent) leaves for work in the morning (06:00 to 08:00). An agent stays at work for eight hours and then returns home. Some people go to a governmental agency during work hours (10:00 to 12:00) or to a doctor (09:00 to 10:00). After work, and on the weekends, some of the agents leave the city (18:00 to 23:00). Agents meet and interact when traveling, walking around the city, arriving at work, or places of entertainment.

In addition to ordinary residents, there are also the so-called leaders: Social activists who want to impact on the society so that, following the idea of Václav Havel's "The Power of the Powerless" essay [46], many of the weak gain influence over the molding of the urban fabric through mutual interactions. Leaders are close to controversial events. Positive leaders are those with extreme and affirmative trust in people; they have a beneficial influence over the social trust in others. Negative leaders place extreme and negative trust in people, thereby lowering the trust of people with whom they interact towards others.

3.4. Agents' Interactions

Meetings between residents take place when they are in the same buildings and when the agents are moving around the city. During these meetings, interactions occur, affecting the change of the agents' characteristics. Such interactions are a symptom of human spatial behavior, resulting from the social distances described by Hall in [7]. Generalized Tobler's first law serves as the starting point for modeling the processes of urban interactions in accordance with Latour's actor–network theory. It states that "everything is related to everything else, but near things are more related than distant things." In the modeled process, objects (actors) influencing each other are actants (sensors) understood as the residents of the city, buildings, spatial arrangement, the dominating function of a given city district, and so on. It is important to emphasize that the "nearness" of the actants here (in this context, the residents) can mean not only the distance in the geographical space, but also the similarity of characteristics, shared interests and views, or social media connection. There is also an assumption that interactions may occur through social media or online interactions, without the need for agents to meet physically. Thus, what is taken into account is not the physical distance between the agents (Euclidean distance in the space of a city), but the distance of the social network, which depends on the educational or age difference. The following interactions, two of them defined by Goffman in [8], latter proposed by the authors, are considered:

- Focused interaction: Influences an agent located in a personal distance (45 to 120 cm). It is assumed that the residents participating in focused interaction will pursue their own goals. An interaction occurs through observation, listening, speaking, and acting. The roles of the interaction parties are strictly defined. The model assumes that a leader should be one of the interaction parties.
- Symbolic interaction: Influences an agent located in a personal and social distance (45 to 360 cm). Symbolic interactionism discussed in [47,48] has also been crucial for the research. It is a sociological perspective dealing with the study of interactions taking place as a result of symbols and gestures. Symbolic interactionism is based on the analysis of the mutual interactions processes, understood as the exchange of symbolic meanings. An exchange takes place between conscious partners who are continually interpreting the situation. The interaction between individuals consists of sending, receiving, and interpreting symbols. The devised model assumes that this interaction occurs between all agents.
- Social media interaction: Influences an agent located in a social net distance (age difference <0.05 or educational difference <0.05). It is a process of exchanging opinions and comments among the users of the social network while reading posts, watching movies or images, listening to sound recordings, and conversations.

Additionally, there is an assumption that the mere presence of a citizen in a given district or a building affects her/his characteristics and, in particular, her/his trust in other people. Being in an office, healthcare facility, at a workplace, school, or in an industrial or office district reduces trust (temporarily or permanently). On the other hand, being in a park, museum, on the boulevards, or in a historic or recreational district increases trust (temporarily or permanently). These changes are called the location influence.

The characteristic analyzed by the authors is the change in the level of mutual trust between residents ("human agents") resulting from the multidimensional influence of the actants. The level of their "trust" tends to "equalize" with the interaction of agents representing individual residents, although it is not an immediate or rapid process. The conducted simulations aim to determine how the long-term change of people's trust (mutual and to institutions) over time affects the level of social participation and, indirectly, the development of open civil society and deliberative democracy.

The model includes three groups of actants (agents) interacting with each other in the city:

1. People (residents);
2. Institutions (governmental agencies, businesses, educational facilities, healthcare);
3. Spatial objects (districts characterized by parks, monuments, rivers, and so on).

In the adopted model, the authors of the study take into account the various modifying functions (see Figure 1) of the trust parameter in the interactions between two agents:

- The "linear" function, which is the basis for the formulation of the others. The assumption is that this function, with the interaction of the agent i with the agent j, is as follows:

$$trust^{i}_{People} trust^{i}_{People} + c{\cdot}\Delta_{trust_{People}} \tag{1}$$

$$c k_{edu}{\cdot}\left(1 - edu^{i}\right) + k_{hap}{\cdot}\left(1 - hap^{i}\right) + k_{wea}{\cdot}\left(1 - wea^{i}\right) + k_{age}{\cdot}\left(1 - age^{i}\right) + k \tag{2}$$

$$\Delta_{trust_{People}} trust^{j}_{People} - trust^{i}_{People} \tag{3}$$

where:

k_{edu}—education parameter change modifier;
k_{hap}—satisfaction parameter change modifier;
k_{wea}—wealth parameter change modifier;
k_{age}—age parameter change modifier;
k—change modifier.

- The "reinforcing" function differs from the linear function in the manner of calculating parameter $\Delta_{trust_{People}}$:

$$\Delta_{trust_{People}} tan\left(trust^{j}_{People} - trust^{i}_{People}\right). \tag{4}$$

- The "diminishing" function also differs from the linear function in the manner of calculating parameter $\Delta_{trust_{People}}$:

$$\Delta_{trust_{People}} tgh\left(trust^{j}_{People} - trust^{i}_{People}\right). \tag{5}$$

- A function that uses the fan-idol relation; based on the situation that a person (fan) will imitate another person (idol) when s/he notices a significant similarity of specific features. Then, s/he changes the parameter (trust in people) towards the idol parameter.
- A function that uses the fan-anti-idol relation; based on the situation that a person (fan) will want to distinguish themselves from the other (anti-idol) when s/he notices a significant difference of specific features. Then, s/he changes the parameter (trust in people) in the opposite direction to the anti-idol parameter.

Moreover, it is assumed that the agent's trust that is included in the above dependencies is treated depending on the place where he resides. The trust of the agent staying at the place of residence is increased by 50%, while for the agent staying at the workplace, it is reduced by 20%. In addition, when the agent is in an entertainment location, his trust counts as 10% more. At the same time, the same trust for an agent staying at the governmental agency or health clinic is reduced by 50%. An agent residing in an industrial area (Old Factory or Mordor) also counts as 15% smaller, while when he is in the GreenLand or OldTown districts, it counts as 15% more.

The introduced methodology is used and validated in the following section.

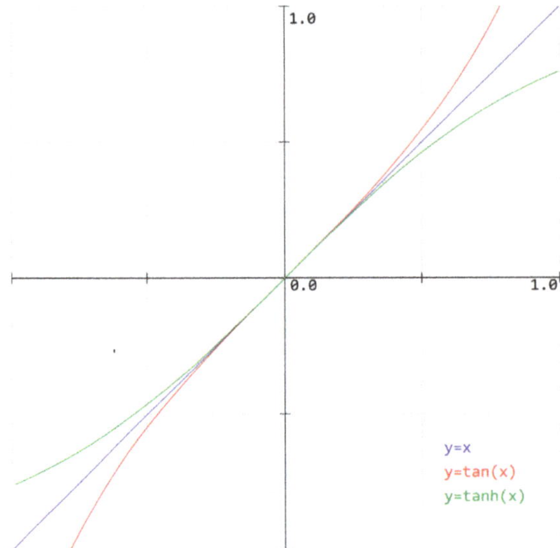

Figure 1. Linear (y = x), reinforcing (y = tan(x)), and diminishing (y = tanh(x)) functions.

4. Model City

To validate the devised model of actant interaction, the authors developed a "model city" comprising seven hexagonal districts with a dominating type of function and structure (Table 1 and Figure 2). One million residents inhabit the city, as illustrated by the dot distribution map (Figure 3). Each of dots, representing 1000 inhabitants, was implemented into the system as an agent with specific characteristics such as age, education, wealth, marital status, number of children, identity (level of identification with the city), trust in people, and trust in institutions. The variations (standard deviation) of the agent characteristics in each district was assumed at 20% for model data. The parameters for agents were drawn according to the normal distribution.

The ArcGIS ESRI (ArcGIS is the name of software developed by Environmental Systems Research Institute; GIS is abbreviation of Geographic Information System) application was used to develop the source spatial database, enabling the preparation of a set of thematic layers (land cover, buildings, communication routes, districts borders, distribution of residents) as shapefiles. These layers were used to build a multi-agent system in the GAMA simulation platform (see [49,50]). GAMA is a modeling and simulation-development environment for building spatially explicit agent-based simulations (see [50]). It is a multiple-application domain platform using a high-level and intuitive agent-based language. With GAMA, users can undertake most of the activities related to modeling, visualizing, and exploring of the simulations using dedicated tools.

Table 1. Model city districts (mean values, in percent); standard deviation is equal to 20%.

No	Name	Population	Trust to People (%)	Trust to Institutions (%)	Altruism (%)	Education (%)	Happiness (%)	Wealth (%)	Identity (%)	Age
1	Greenland	50,000	90	80	70	90	100	90	100	60
2	City Center	250,000	40	60	30	60	60	60	90	50
3	Bedroom Suburb	400,000	40	50	30	70	50	50	20	40
4	Old Town	50,000	80	60	80	80	80	80	100	70
5	Business District	100,000	50	60	20	70	50	60	20	30
6	Old Factory District	150,000	30	20	20	20	10	10	40	60
7	Unspecified Space	0								

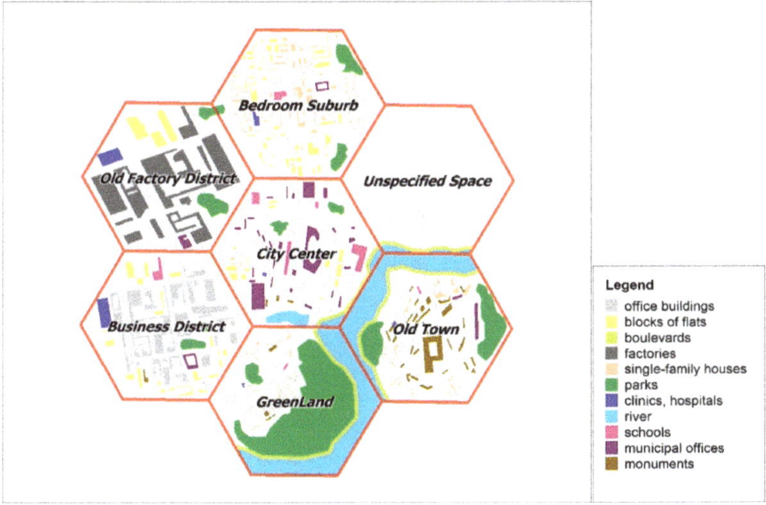

Figure 2. Model city.

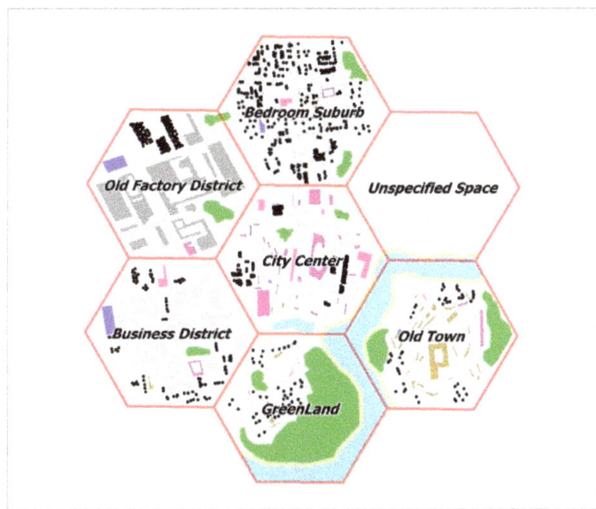

Figure 3. Dot distribution map (each dot represents 1000 inhabitants of the model city).

Thanks to the simulations carried out for the model city, it was possible to calibrate the multi-agent system properly, i.e., to determine the value of individual parameters, which then enabled the use of the devised model for a real urban agglomeration area. For example, 3650 iterations used by the authors correspond to a period of 10 years, during which the level of trust of residents changes significantly (and in a measurable way). The authors repeatedly modified the numerical values of particular factors (e.g., changing the level of trust of individual actants resulting from their mutual interactions) so that the parameterization of the model corresponds to the changes observed in the real cities. Because of the iterative calibration of the system, it was possible to determine the parameters of the model adequate for the research of real metropolises.

The analyses made it possible to check the spatial distribution of changes in the level of trust of the residents of particular districts (Figures 2 and 3) in a long-term (decades-long) process. Thanks to the use of a multi-agent system, it was possible to simulate many years of social processes in computational cycles lasting from several dozen minutes to several hours.

The study also examined the influence of some strong "positive" or "negative" leaders, the impact of the adopted function modifying the traits of the agents (linear, reinforcing, diminishing), as well as specific spatial problems in the model city. In the research, the authors adopted four analytical scenarios:

- City functioning normally;
- The problem of spatial development of the "empty" seventh district;
- The problem of revitalizing a factory going into liquidation;
- The problem of changing the development of the model city's central square (market).

Each of the scenarios is associated with the engagement of a specific group of residents (e.g., the elderly, the less affluent, or the residents of a given part of the city).

Each agent is characterized by the "susceptibility" parameter, which determines the probability of an agent taking part in the social debate, demonstration, or protest. This characteristic is determined on the basis of the agent's other parameters, such as age, education, family, wealth, the level of identification with the city, and so forth. What is key, however, is a given agent's place of residence and the proximity (spatial or social) to the place where a problem, such as the revitalization of a district or the demolition of a controversial town hall, occurs. "Ordinary" agents engage (with a certain probability) in social life after work or on weekends. Only agents representing the "strong leaders" always remain in the conflict places. These agents do not change their level of involvement or trust during the interaction. For positive leaders, it is 1.0, while it is 0.0 for negative ones.

Making use of the devised multi-agent system and the GAMA toolset environment, 3650 computational epochs were carried out. The characteristics of individual agents changed in each iteration because of contacts with individual actants (residents and spatial objects). Using GIS tools, the resulting data were subjected to spatial aggregation analyzing the change in the average level of trust of agents residing in a given district of a model city. First, the authors of the article made calculations, the purpose of which was to check how the particular trust parameter modifying functions works (Figures 4 and 5).

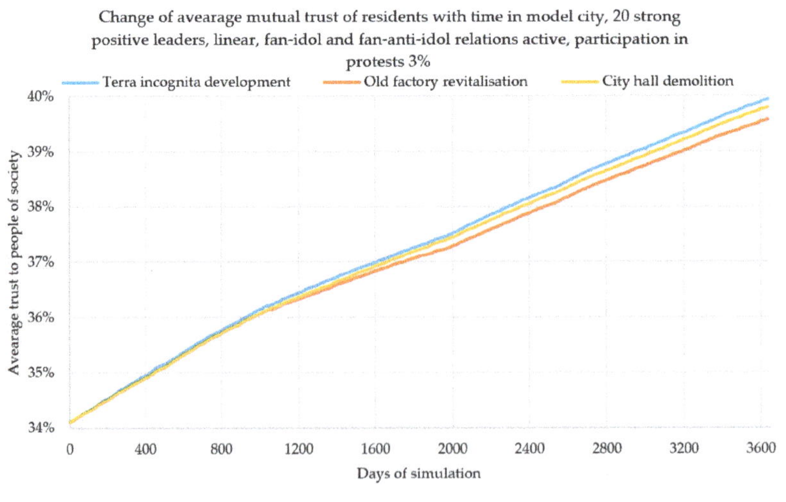

Figure 4. The attractor activity and the level of mutual social trust in a model city through a simulated time of 10 years in the presence of 20 strong positive leaders.

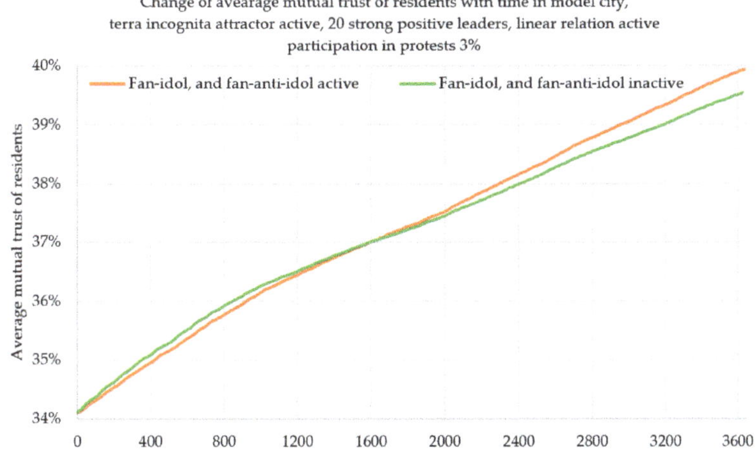

Figure 5. Comparison of the fan-idol and fan-anti-idol relations on the level of mutual social trust in a model city through a simulated time of 10 years in the presence of 20 strong positive leaders.

In Figure 4, one can observe the comparison of the attractor's activity and the level of mutual social trust in a model city through a simulated time of 10 years in the presence of 20 strong positive leaders. All plots rise steadily, but with slightly different slopes. Moreover, in the period of attractors' activeness (days 1200–2400), the curves are moving away from each other. The results indicate that there are differences between individual attractors, which results from various geospatial settings.

In Figure 5, one can observe the comparison of the fan-idol and fan-anti-idol relations on the level of mutual social trust in a model city through a simulated time of 10 years in the presence of 20 strong positive leaders. What is interesting is that up to the 1600-ish day of simulation (four years and four months), the inactivity of the fan-idol and fan-anti-idol relations seems to be stronger. However, after that, the activity of the relations begins to be stronger, and at the end, after 10 years, the mutual trust is 0.4% stronger, compared to the inactivity of the relations.

To sum up, the analysis of various attractors and various functions modifying confidence shows us minor differences, which result from the differences characteristic for a given area, rather than differences in the algorithm used.

The variants of analytical simulations implemented are presented below.

4.1. *Terra Incognita (Unspecified Space)*

Research question: Will the district be built in a "closed" way (gated communities) or (because of the increased level of social (geo)participation and trust) in an "open" way? It is of particular interest to the residents from three marked districts, and young, relatively wealthy, and well-educated people.

Figure 6 shows the comparison between the number of strong leaders and the level of mutual social trust in a model city through a simulated time of 10 years, with fan-idol and fan-anti-idol relations inactive. In the absence of strong leaders, the slope of mutual trust decreases over time and the increase in 10 years is 1.12%. For 20 strong positive leaders, trust increases by 5.44% within the simulation, but the slope is changing; during the manifestation, the slope slightly decreases. It can be explained by the concentration of all the positive leaders in one district of the city. However, in the case of 20 strong negative leaders, the authors observe a slight increase in trust in the beginning, but on the 2000-ish day of simulation (around year 5.5), it begins to drop. Ultimately, after 10 years, trust increases by 1.12%, which is the result of accumulating negative opinions during protests. Similar dependencies, with accuracy to value, occur for other attractors. Finally, the existence of both 20 positive and 20 negative strong leaders results in a different outcome than the non-existence of strong leaders. On the contrary, one can note that the positive leaders have more impact on mutual trust than the negative ones. For this case, the mutual trust after 10 years of simulation increases by 2.94%.

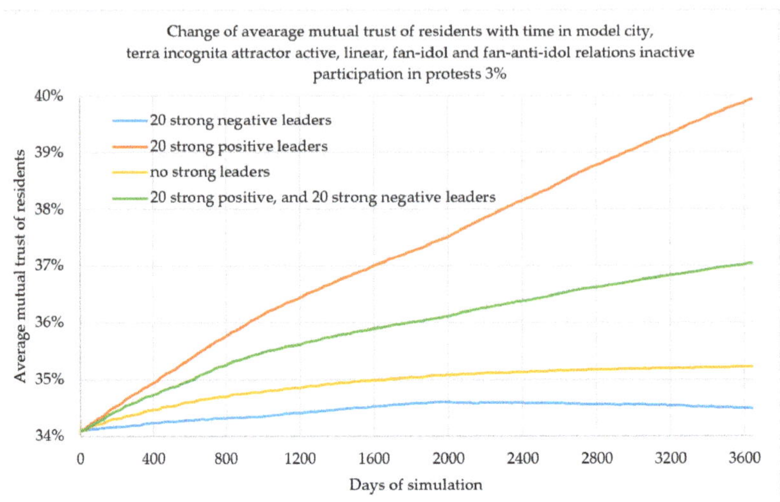

Figure 6. Comparison of the number of strong leaders and the level of mutual social trust in a model city through a simulated time of 10 years, with fan-idol and fan-anti-idol relations inactive.

In Figures 7 and 8, one can see that the changes in the level of trust differ significantly in individual districts of the model city. These changes also have different intensity over time. This process depends not only on the number of strong leaders, but also on the characteristics of residents of particular districts and the level of their involvement in the problem of this attractor. Positive leaders influence a slight increase in trust in Bedroom, while negative leaders considerably reduce the level of trust of the residents of this district. For those who are not very interested in the spatial development of the new district (residents of Old Town and GreenLand), the level of trust decreases both in the presence of positive and negative leaders, although with varying intensity.

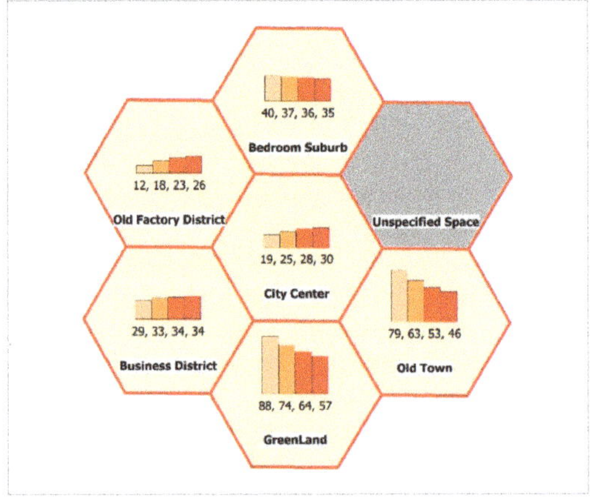

Figure 7. Changes in the level of mutual social trust in a model city through a simulated time of 10 years (20 negative leaders in the "Terra incognita" attractor); iteration 0, 1200, 2400, 3650.

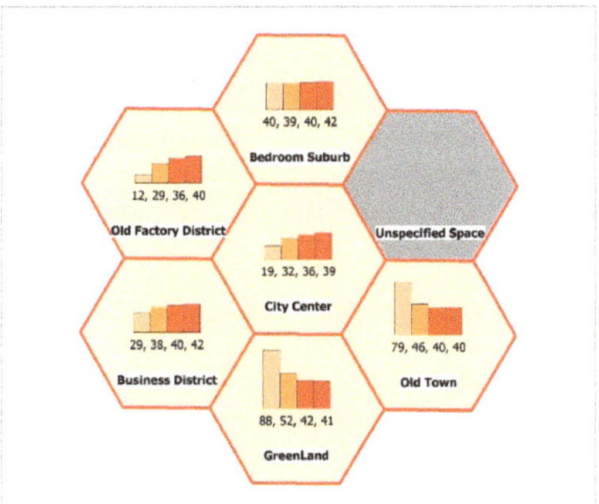

Figure 8. Changes in the level of mutual social trust in a model city through a simulated time of 10 years (20 positive leaders in the "Terra incognita" attractor); iteration 0, 1200, 2400, 3650.

4.2. Old Factory Revitalization

The revitalization of the building after the Steelworks closes down leads to gentrification—and loft spaces for the wealthy. Conducted studies on spatial and descriptive data characterizing the model city have shown that for the relatively poor employees of the closed down Steelworks and the residents of this region (this district and a spatially close part of the bedroom community, center, and business district), this problem is particularly significant (affects negatively). Figure 9 shows a zone of strong and weak impact of the "Old Factory" attractor.

Figure 9. Zone of strong and weak impact of the "Old Factory" attractor.

In Figure 10, one can observe that overall increase in trust by 8% does not mean an even increase of confidence in all districts of the model city. The revitalization of the factory and the creation of residential lofts is especially important for the residents of this district and two neighboring ones (business district and city center). This is also important for some residents of the city who work (or have worked) in a liquidated factory, regardless of where they live. For the inhabitants of the bedroom suburbs, the process of revitalization is of little importance to the wealthy inhabitants of the old town and green land who are not interested in this subject, it even results in lowering the level of social activity. On account of the obtained results, it is possible to state the following:

- A multi-agent system enables simulations of long-term social processes and interactions of the citizens as sensors (both mutual and concerning the urban tissue);
- It is relatively easy to scale the impact of individual factors on the process, which facilitates the development of a model to be used to simulate complex, multi-parameter social processes in real cities;
- The combined use of multi-agent systems, advanced sensors, and GIS tools makes it possible to analyze the interaction of the actants (residents and spatial objects) as well as spatial aggregation and visualization of the results in the form of thematic maps;
- The selection of a spatial database (a digital map of the city) and the spatial distribution of residents and their characteristics make it possible to simulate incredibly complicated processes, e.g., related to spatial development, revitalization, and so on.

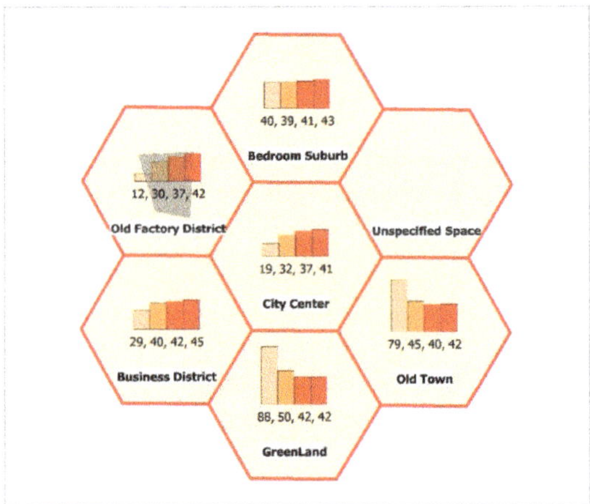

Figure 10. Changes in the level of mutual social trust in a model city through a simulated time of 10 years (20 positive leaders in the "Old Factory" attractor); iteration 0, 1200, 2400, 3650.

5. Spatiotemporal Modeling of the Warsaw Area, Poland

With an appropriately calibrated model and its implementation in the form of a multi-agent system, the authors attempted to conduct research and simulation on real data. The agglomeration of Warsaw in Poland, with its 1,754,000 inhabitants, was the test object (Figure 11). Spatial data used in the study come from the general geographic database, which contains data at the level of accuracy that is equivalent to analogue maps with a scale of 1:250,000. This study was up-to-date in 2016. As with the model city, the distribution of residents was modeled in the form of a dot distribution map (Figure 12), where each of 1754 dots (agents) represents 1000 inhabitants. The data contain a set of characteristics defining the demographic, social, and cultural features of individual residents (Table 2).

The research [51,52] is examining the level of trust of Poles, both in public institutions and each other. It indicates the direct relationship between trust, civic activity, and the level of education. It turns out that "trust increases civic activity only after reaching or exceeding the threshold of secondary education.". In addition, in 2015, at the request of the City Hall of Warsaw, a study was conducted on the quality of life of the residents of Warsaw districts, in which a set of questions was devoted to trust. Also, although the level of trust in friends and family remains at a level close to 90%, confidence in politicians (17%), journalists (34%), and local authorities in Warsaw (35%) still remains very low.

Analytical experiments were carried out for Warsaw by simulating long-term social processes for three selected test areas (Figure 12):

- The central area of the capital (controversy regarding the spatial development of Parade Square);
- The so-called Mordor (with the problem of extreme street congestion during the day and its forlornness at night);
- The so-called Miasteczko Wilanów (socially associated with a "ghetto" for young and wealthy residents born outside of Warsaw, who have a low level of identification with the city).

Table 2. Warsaw (PL) districts (mean values, in percent); standard deviation is equal to 20%. Source: [51].

No	Name	Population	Trust to people (%)	Trust to institutions (%)	Altruism (%)	Education (%)	Happiness (%)	Wealth (%)	Identity (%)	Age
1	Bemowo	120,000	63	83	20	39	12	30	94	60
2	Bialoleka	116,000	63	66	22	29	16	70	90	40
3	Bielany	132,000	64	80	15	33	11	70	93	70
4	Mokotow	218,000	67	79	11	26	5	80	93	100
5	Ochota	84,000	59	75	12	28	5	60	94	80
6	PragaPoludnie	178,000	64	74	19	39	17	60	88	60
7	PragaPolnoc	66,000	61	75	11	22	5	10	86	30
8	Rembertow	24,000	62	72	7	20	4	80	90	60
9	Srodmiescie	118,000	63	74	24	41	11	90	93	90
10	Targowek	124,000	65	81	19	35	10	40	86	30
11	Ursus	58,000	66	84	11	19	3	60	91	70
12	Ursynow	150,000	57	77	23	32	7	80	91	70
13	Wawer	75,000	72	79	18	31	2	60	95	70
14	Wesola	25,000	69	78	11	24	4	70	96	70
15	Wilanow	35,000	73	76	27	48	14	90	94	90
16	Wlochy	41,000	69	83	15	28	3	70	98	70
17	Wola	139,000	69	86	25	37	11	50	93	60
18	Zoliborz	51,000	67	75	17	32	16	80	94	80

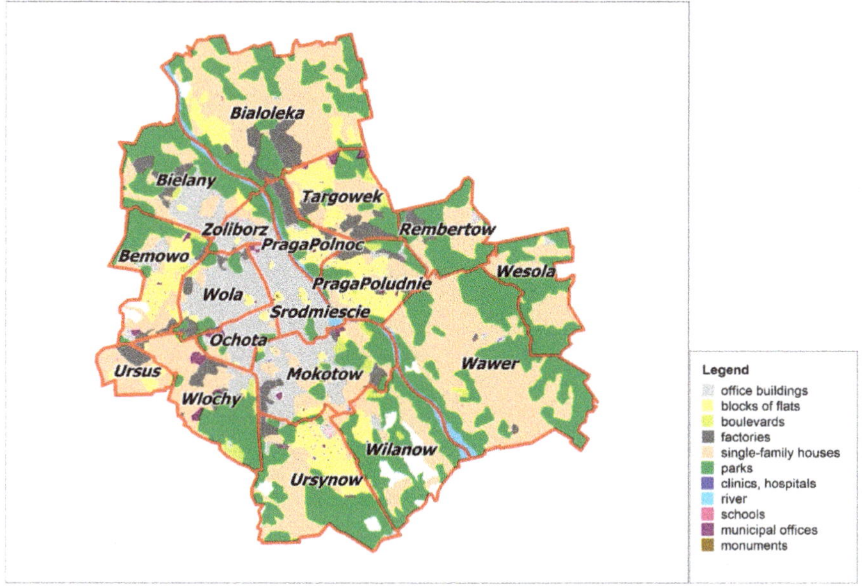

Figure 11. Warsaw (PL) and its districts.

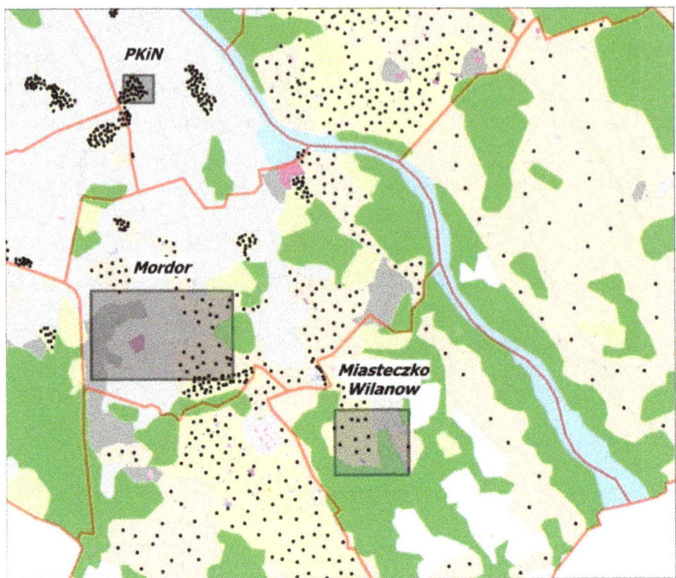

Figure 12. Three attractors in Warsaw and dot distribution map.

5.1. Parade Square (Plac Defilad)

Together with the Palace of Culture and Science, Parade Square has been a source of controversy for years. The background of this conflict is primarily generational and, thus, historical. The problem lies not only in its development, but also in the possible ways of integrating this space into the city.

The seniors born or living in Warsaw declare low trust and distaste for this area. To this group, this place is connected with the period of communist domination, complicated history, and the Soviet Union. They would like to demolish this space along with the Palace of Culture and Science. This group of people has a very high level of identification with the city and its space, especially with the city center. For the youth and people in their prime (up to around 35 years of age), the Palace remains a symbol of Warsaw and the location of many cultural activities, as well as a place for meetings or dating. Young people, in particular, have been trying to revitalize this area for years. They would like to combine the Palace of Culture and Science with space for everyone, where there is a place for greenery, leisure, and a body of water. A manifestation of this is, for example, the so-called Central Park (https://parkcentralny.pl/), which is a project aimed at creating the Green Heart of Warsaw.

In Figure 13, the change of average mutual trust of residents for Parade Square with 20 negative and 20 positive strong leaders can be observed. The mutual trust decreases by 6.8%, which is the opposite of the case of the Model City, where mutual trust grew in a similar situation (20 positive and 20 negative strong leaders). This process is caused by different demographic characteristics of Warsaw (a real urban agglomeration) and the model city.

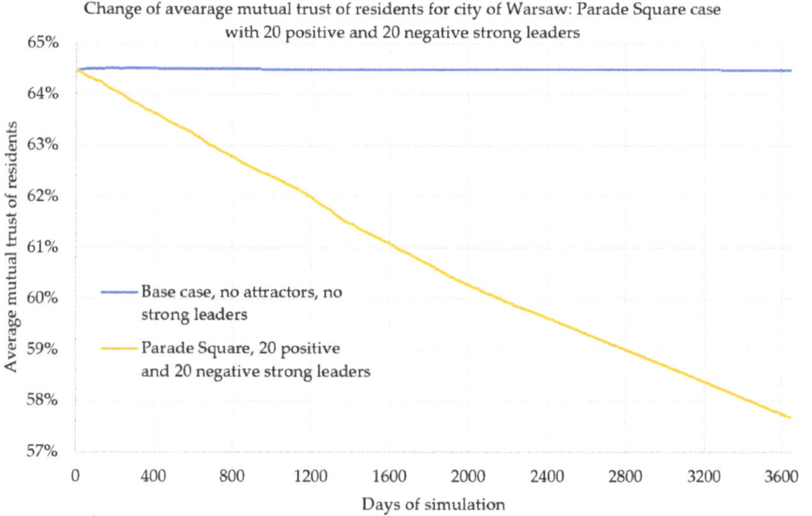

Figure 13. Change in the average mutual trust of residents for Parade Square with 20 negative and 20 positive strong leaders.

5.2. *"Mordor" on Domaniewska Street*

In the 1990s, the process of changing this part of the city from one of industrial function to one of a service function began. At the moment, Domaniewska Street is one of the main streets of the largest office complex in Warsaw. The residents of the city have jokingly named it Mordor, which is the dark land from the novels by J.R.R. Tolkien, because of its traffic-related problems.

The problem with this part of the city is the extreme intensification of office space, lack of greenery, and a limited number of parking spaces. A thick line seems to separate the Mordor of Warsaw from the city (for communicational and architectural reasons, but also mentally). This part of Warsaw has always been associated with a low level of identification with the city. After its change from an industrial area to a service one, the problem of low-level identification and trust remained (which also applies to the attitude of officials and corporations). It is a place where one has to be (work), not where one wants to be. Places that appear in Mordor (restaurants, cafés) are there primarily to serve the corporations; there are no cultural or recreational places (except for fitness centers), and so on.

In this part of Warsaw, all actants have a very low level of trust in public transport and, therefore, in officials as well (employees as well as visitors or residents). For people from the outside, Mordor is a place of very low trust and identification (especially for those who live close to this area, e.g., Mokotów, Ursynów, and so on, neighboring districts). People living in the outlying districts are somewhat indifferent to this place. On the other hand, this place "draws" newcomers who decide to buy flats there. As a result, a sense of new urban identity arises, in a way forged as an opposition to the inhabitants of other parts of Warsaw.

The critical problem of this part of Warsaw is the extreme congestion of streets during rush hour. Almost all of the 100,000 employees of corporations in Mordor commute to work using the company car. An increased mutual trust would allow the rationalization of commuting, e.g., by using carpooling (see [17]).

In Figure 14, one can observe the change of average mutual trust of residents for Mordor with 20 strong negative leaders. The constant drop of mutual trust (−16.06%) can be observed (the drop is more than twofold when compared to Parade Square, see Figure 13). It means that for the city of Warsaw, in the case of strong negative leaders, the level of mutual trust is threatened by a significant reduction. The increase in social trust, and, thus, also the increase in social participation visibility, e.g., through the joint use of company cars leading to increased traffic capacity, therefore requires a strong inspiration of "positive leaders" of social changes in this office area.

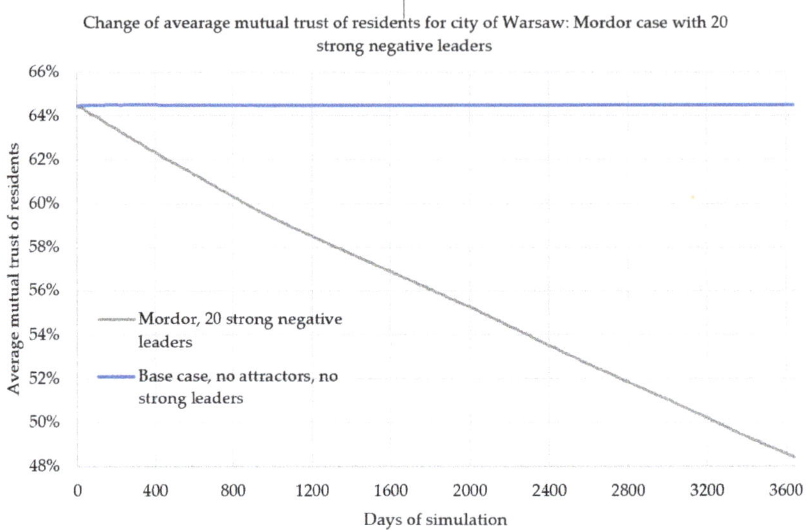

Figure 14. Change in the average mutual trust of residents for Mordor with 20 strong negative leaders.

5.3. Miasteczko Wilanów

This is a recently built district of the city with gated communities inhabited mainly by newcomers with a low sense of identification with Warsaw. The inhabitants of the "old" Wilanów and other Warsaw districts (apart from Białołęka) do not trust the residents of the "new" Wilanów. The relations within this district are also complicated: Those who live in the prestigious buildings built in the very beginning do not trust those who came to live in buildings of a much lower standard and intended for the middle class (a typical class conflict).

What characterizes this place is the lack of kindergartens, schools, sports fields, and so on, as well as the underdeveloped public transport. Also, there is no broadly understood public space or any green areas.

An increased mutual trust and level of identification would not only facilitate the active social participation of residents, but also improve relations between the residents of various parts of Wilanów (the "old" and the "new) and the other districts of the city.

In Figure 15, one can observe the change of average mutual trust of residents for Wilanów with 20 strong positive leaders. In this case, the mutual trust increases steadily up to the level of 73.48% (an increase of 8.94%). What is interesting, for an analogous number of leaders, the drop for Mordor (see Figure 14) is nearly twice the one for Wilanów, which means that Warsaw is a city endangered by the declining mutual trust, and it is more difficult to increase the trust than to decrease it.

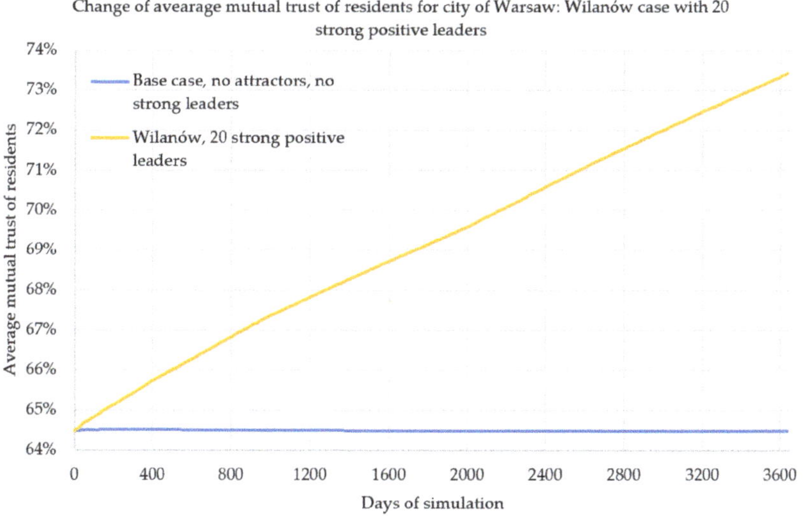

Figure 15. Change in the average mutual trust of residents for Wilanów with 20 strong positive leaders.

6. Conclusions and Future Work

After completing the simulations and accessing the results, it seems that, thanks to the developed concept and the prototype of a multi-agent information system, which uses spatial data, demographic information, and sociological, mathematical, and urban theories, performing complex geospatial big data analyses, geospatial information extraction, and data mining is possible. Because of the idea of "citizens as sensors" represented by "human agents", game theory, information asymmetry, urban morphology, and multi-agent systems, it was possible to model changes in the residents' activity over decades, even in the cases of agglomerations of hundreds of thousands of residents. Therefore, it is a useful tool not only for conducting urban big data analyses in urban studies, spatial science, or applied social science, but also for shaping the smart cities of the future. The analysis of the results for the model city and Warsaw shows that each city has a "potential for mutual trust" that emerges from the distribution of its buildings, road network and, of course, its inhabitants. This potential of social trust can be substantial, causing an increase in mutual trust, as in the case of the model city. It may also be small, as in the case of Warsaw, where it is difficult to increase mutual trust. However, thanks to strong leaders, it is possible to shape the trust and support the process of increasing the activity of residents—active sensors and their social participation in creating a smart city.

Therefore, the following may be stated:

- The idea of "citizens as sensors," expanded with the elements of actor–network theory and multi-agent systems, facilitates spatiotemporal analysis in geospatial data and spatial knowledge acquisition. Increasing the level of mutual trust between residents and their trust in institutions, as well as the sense of local identity results in increased social activity in the (geo)participation process and co-deciding about the city's development.
- Mathematical models (game theory), social theories, ICT tools, geoinformation technologies, and multi-agent systems constitute a tool for modeling spatiotemporal geoinformation structures and enduring social changes. Because of the use of multi-agent systems to model the asymmetry of social relations, it is possible to analyze changes in the level of residents' activity in the model city and real agglomerations.
- Developing a system for a model city makes it possible to experiment with the value of the factors. As a result, one can examine how an increase or a decrease in social trust affects civic engagement

at a given time. Research on real data enables following the initial situation and then applying the results from the model city to, for example, Warsaw. Consequently, one can determine the factors that directly or indirectly influence the increase of participative activities and then strengthen those elements that require reinforcement.

- The system enables measuring the strength of relations between individual human and non-human agents and then, by the modification of selected elements, to reinforce or diminish them.
- The devised concept of a multi-parameter analytical system and the prototype of a geoinformation tool facilitate not only modeling of the social changes, but also their stimulation. Indirectly, it contributes to the development of a virtual urban agora, deliberative participatory democracy, and the reinforcement of the social (geo)participation processes in the smart communities of smart cities.
- The city is a structure of interdependent networks and relationships. Social trust is one of the elements enabling the effective functioning of society. This is a variable that positively influences the consolidation of democracy and formation of citizenship.
- The results of the research suggest that the context does play a significant role in shaping the effect of social trust on social participation.
- The approach proposed by authors may facilitate constructing more and more holistic models of cities.

It is also worth emphasizing that the developed model can be (after minor modifications) also used in other applications, e.g., stimulating the residents' activity to install photovoltaic panels [53], real estate or multilateral negotiations for building plots in distributed multi-agent environment [54].

As future work, it is worth considering to supplement the proposed multi-agent, agent-based model with a game theoretical treatment, in particular to identify possible social dilemmas, such as e.g., public goods, tragedy of the commons and trust dilemma, and their potential impact on the development of public trust in the considered urban agglomeration of groups of interacting individuals with different interests. The authors of the article also plan to expand the model with elements of gamification between residents to model different ways of social activity of city residents; specifically, it is planned to model the social gamification in a smart city, which is likely to stimulate the installation of the photovoltaic panels. A smart city, understood not as intelligent city infrastructure but as a smart, open geoinformation society, is shaped by the "power of the powerless," which can be reinforced.

The developed model is universal; it can be easily parameterized on the basis of any input data, e.g., social, sociological, or economic. The model will be verified and tested in other agglomerations and different cities. These will include cities characterized by a higher baseline level of trust (Scandinavian cities) and culturally different areas (Singapur, Masdar City).

Data Availability: The code of the project in gaml language with included data used to support the findings of this study have been deposited in the git repository: https://gitlab.com/PiotrPowerPalka/smartcitygrowthgama.git.

Author Contributions: Conceptualization, R.O. and P.P.; methodology, P.P, R.O., A.T., and B.K.; software, M.B. and P.P.; validation, A.T and T.P.; formal analysis, R.O. and T.P.; investigation, R.O, P.P., and A.T; resources, A.T.; data curation, P.P and B.K.; writing—original draft preparation, R.O, P.P., A.T., and B.K.; writing—review and editing, A.T.; visualization, R.O, A.T, and P.P.; supervision, R.O.; project administration, A.T.

Funding: This work was supported by the FabSpace 2.0 project funding from the EU 's Horizon 2020 research and innovation program under the grant agreement No. 693210.

Conflicts of Interest: The authors declare that there is no conflict of interest regarding the publication of this paper.

References

1. Goodchild, M.F. Citizens as sensors: The world of volunteered geography. *GeoJournal* **2007**, *69*, 211–221. [CrossRef]
2. Harvey, D. The right to the city. *Int. J. Urban Reg. Res.* **2003**, *27*, 939–941. [CrossRef]

3. Latour, B. *Reassembling the Social: An Introduction to Actor-Network-Theory*; Oxford University Press: Oxford, UK, 2005.
4. European Union Regional Policy. *Cities of Tomorrow—Challenges, Visions, Ways Forward*; European Commission: Brussels, Belgium, 2011.
5. Manville, C.; Cochrane, G.; Cave, J.; Millard, J.; Pederson, J.K.; Thaarup, R.K.; Liebe, A.; Wissner, M.; Massink, R.; Kotterink, B. *Mapping Smart Cities in the EU*; European Parliament: Brussels, Belgium, 2014; Available online: http://www.europarl.europa.eu/RegData/etudes/etudes/join/2014/507480/IPOL-ITRE_ET%282014%29507480_EN.pdf (accessed on 5 May 2019).
6. United Nations Habitat III: New Urban Agenda. Available online: http://habitat3.org/wp-content/uploads/NUA-English.pdf (accessed on 5 May 2019).
7. Hall, E.T. *The Hidden Dimension*; Doubleday & Company, Inc.: Garden City, NY, USA, 1966.
8. Goffman, E. *The Presentation of Self in Everyday Life*; Anchor Books b: New York, NY, USA, 1956.
9. Shoham, Y.; Leyton-Brown, K. *Multiagent Systems: Algorithmic, Gametheoretic, and Logical Foundations*; Cambridge University Press: Cambridge, UK, 2008.
10. Batty, S.E. Game-Theoretic Approaches to Urban Planning and Design. *Environ. Plan. B Urban Anal. City* **1977**, *4*, 211–239. [CrossRef]
11. Tan, R.; Liu, Y.; Zhou, K.; Liao, L.; Tang, W. A game-theory based agent-cellular model for use in urban growth simulation: A case study of the rapidly urbanizing Wuhan area of central China. *Comput. Environ. Urban Syst.* **2015**, *49*, 15–29. [CrossRef]
12. Mostashari, A.; Arnold, F.; Mansouri, M.; Finger, M. Cognitive cities and intelligent urban governance. *Netw. Ind.* **2011**, *13*, 4–7.
13. Kaltenrieder, P.; Altun, T.; D'Onofrio, S.; Portmann, E.; Myrach, T. Personal digital assistant 2.0—A software prototype for cognitive cities. In Proceedings of the 2016 IEEE International Conference on Fuzzy Systems, Vancouver, BC, Canada, 24–29 July 2016.
14. Malleson, N. Big Data, Agent-Based Modeling, and Smart Cities: A Triumvirate to Rival Rome? 2017. Available online: http://www.nickmalleson.co.uk/papers/2017-smart_cities_forecasting.pdf (accessed on 5 May 2019).
15. Karmakharm, T.; Richmond, P. Large Scale Pedestrian Multi-Simulation for a Decision Support Tool. In *EG UK Theory and Practice of Computer Graphics*; The Eurographics Association: Geneve, Switzerland, 2012; pp. 41–44.
16. Sandhu, S.S.; Jain, N.; Gaurav, A.; Narayana Iyenger, N.C.S. Agent based intelligent traffic management system for smart cities. *Int. J. Smart Home* **2015**, *9*, 307–316. [CrossRef]
17. Olszewski, R.; Pałka, P.; Turek, A. Solving "Smart City" Transport Problems by Designing Carpooling Gamification Schemes with Multi-Agent Systems: The Case of the So-Called "Mordor of Warsaw". *Sensors* **2018**, *18*, 141. [CrossRef] [PubMed]
18. Sagl, G.; Blaschke, T.; Beinat, E.; Resch, B. Ubiquitous Geo-Sensing for Context-Aware Analysis: Exploring Relationships between Environmental and Human Dynamics. *Sensors* **2012**, *12*, 9800–9822. [CrossRef]
19. Sagl, G.; Resch, B.; Blaschke, T. Contextual Sensing: Integrating Contextual Information with Human and Technical Geo-Sensor Information for Smart Cities. *Sensors* **2015**, *15*, 17013–17035. [CrossRef] [PubMed]
20. Resch, B. People as Sensors and Collective Sensing-Contextual Observations Complementing Geo-Sensor Network Measurements. In *Progress in Location-Based Services*; Krisp, J.M., Ed.; Springer: Berlin/Heidelberg, Germany, 2013; pp. 391–406.
21. Van Lammeren, R.; Goossen, M.; Roncken, P. Sensing Landscape History with an Interactive Location Based Service. *Sensors* **2009**, *9*, 7217–7233. [CrossRef]
22. Couclelis, H. From cellular automata models to urban models: New principles for model development and implementation. *Environ. Plan. B* **1997**, *24*, 165–174. [CrossRef]
23. Clarke, K.C.; Gaydos, L.J. Loose-coupling a cellular automaton model and GIS: Long-term urban growth prediction for San Fransisco and Washington/Baltimore. *Int. J. Geogr. Inf. Sci.* **1998**, *12*, 699–714. [CrossRef] [PubMed]
24. Wu, T.C.; Hong, B.Y. Simulation of Urban Land Development and Land Use Change Employing GIS with Cellular Automata. In Proceedings of the Second International Conference on Computer Modeling and Simulation, Sanya, China, 22–24 January 2010; pp. 513–516.

25. Hosseinali, F.; Alesheikh, A.A.; Nourian, F. Simulation of Land-Use Development, Using a Risk-Regarding Agent-Based Model. *Adv. Artif. Intell.* **2012**, *2012*. [CrossRef]
26. Li, X.; Zhang, X.; Yeh, A.; Liu, X. Parallel cellular automata for large-scale urban simulation using load-balancing techniques. *Int. J. Geogr. Inf. Sci.* **2010**, *24*, 803–820. [CrossRef]
27. Saarloos, D.J.M.; Arentze, T.A.; Borgers, A.W.J.; Timmermans, H.J.P. A multi-agent paradigm as structuring principle for planning support systems. *Comput. Environ. Urban Syst.* **2008**, *32*, 29–40. [CrossRef]
28. Imottesjo, H.; Kain, J.H. The Urban CoBuilder—A mobile augmented reality tool for crowd-sourced simulation of emergent urban development patterns: Requirements, prototyping and assessment. *Comput. Environ. Urban Syst.* **2018**. [CrossRef]
29. Zhang, Y. CityMatrix—An Urban Decision Support System Augmented by Artificial Intelligence. Master's Thesis, Massachusetts Institute of Technology, Cambridge MA, USA, 23 October 2017. Available online: https://dam-prod.media.mit.edu/x/2017/08/23/ryanz-ms-17.pdf (accessed on 5 May 2019).
30. Allen, M.; Regenbrecht, H.; Abbott, M. Smart-phone augmented reality for public participation in urban planning. In Proceedings of the 23rd Australian Computer-Human Interaction Conference, Canberra, Australia, 28 November–2 December 2011; pp. 11–20.
31. Zhao, J.; Guo, H.-X. Simulation of Coordinated Development of an Urban Ecosystem: A Case Study in Guanzhong Urban Agglomeration. In Proceedings of the International Conference on Management and Service Science, Wuhan, China, 12–14 August 2011. [CrossRef]
32. Ultsch, A.; Kretschmer, O.; Behnisch, M. Systematic Data Mining into Land Consumption in Germany. *CUPUM* **2015**. [CrossRef]
33. Ultsch, A.; Behnisch, M. Are There Cluster of Communities with the Same Dynamic Behaviour? In *Studies in Classification, Data Analysis, and Knowledge Organization*; Springer: Berlin/Heidelberg, Germany, 2010; pp. 445–453.
34. Wooldridge, M. *An Introduction to Multiagent Systems*; John Wiley & Sons: Hoboken, NJ, USA, 2009.
35. Portugali, J. *Complexity, Cognition and the City*; Springer: New York, NY, USA, 2011.
36. Deffuant, G.; Amblard, F.; Weisbuch, G.; Faure, T. How can extremism prevail? A study based on the relative agreement interaction model. *J. Artif. Soc. Soc. Simul.* **2002**, *5*, 4.
37. Hegselmann, R.; Krause, U. Opinion dynamics and bounded confidence models, analysis and simulation. *J. Artif. Soc. Soc. Simul.* **2002**, *5*, 3.
38. Swan, M. Sensor Mania! The Internet of Things, Wearable Computing, Objective Metrics, and the Quantified Self 2.0. *J. Sens. Actuator Netw.* **2012**, *1*, 217–253. [CrossRef]
39. Jacobs, J. *The Death and Life of Great American Cities*; Vintage: New York, NY, USA, 2016.
40. Latané, B. The psychology of social impact. *Am. Psychol.* **1981**, *36*, 343–356. [CrossRef]
41. Latané, B.; Liu, J.H.; Nowak, A.; Bonevento, M.; Zheng, L. Distance Matters: Physical Space and Social Impact. *Personal. Soc. Psychol. Bull.* **1995**, *21*, 795–805. [CrossRef]
42. Nowak, A.; Szamrej, J.; Latané, B. From private attitude to public opinion: A dynamic theory of social impact. *Psychol. Rev.* **1990**, *97*, 362–376. [CrossRef]
43. Hołyst, J.A.; Kacperski, K.; Schweitzer, F. Phase transitions in social impact models of opinion formation. *Phys. A Stat. Mech. Appl.* **2000**, *285*, 199–210. [CrossRef]
44. Hołyst, J.A.; Kacperski, K. Phase transitions as a persistent feature of groups with leaders in models of opinion formation. *Phys. A Stat. Mech. Appl.* **2000**, *287*, 631–643.
45. Kietlińska, B.; Olszewski, R.; Pałka, P.; Turek, A. Urban asymmetries in terms of actor-network theory, social (geo)participation, and multi-agent modeling. In Proceedings of the First European Asymmetry Symposium, Nice, France, 15–16 March 2018.
46. Havel, V. *The Power of the Powerless (Routledge Revivals): Citizens Against the State in Central-eastern Europe*; Routledge: Abingdon, UK, 2009.
47. Blumer, H. *Society as Symbolic Interaction. Human Behavior and Social Process: An Interactionist Approach*; Rose, A.M., Ed.; Houghton-Mifflin: Boston, MA, USA, 1969.
48. Blumer, H. *Symbolic Interactionism. Perspective and Method*; Prentice-Hall: Englewood Cliffs, NJ, USA, 1969.
49. GAMA Website. Available online: https://gama-platform.github.io/ (accessed on 5 May 2019).

50. Grignard, A.; Taillandier, P.; Gaudou, B.; Vo, D.A.; Huynh, N.Q.; Drogoul, A. GAMA 1.6: Advancing the Art of Complex Agent-Based Modeling and Simulation. In Proceedings of the International Conference on Principles and Practice of Multi-Agent Systems, Dunedin, New Zealand, 1–6 December 2013; Springer: Berlin/Heidelberg, Germany, 2013.
51. Quality of Life in Warsaw. Available online: http://jakosczycia.um.warszawa.pl/ (accessed on 10 June 2018).
52. Czapliński, J.; Panek, T. Social Diagnosis 2015: Objective and Subjective Quality of Life in Poland. *Contemp. Econ.* **2015**, *9*, 538.
53. Olszewski, R.; Pałka, P.; Turek, A.; Kamiński, J. Assessing the sustainable development of urban renewable energy systems using multi-agent methodology as a tool for modeling the process of social gamification. *Renew. Sustain. Energy Rev.*. under review.
54. Pałka, P. Multilateral Negotiations in Distributed, Multi-Agent Environment. In Proceedings of the International Conference on Computational Collective Intelligence, Gdynia, Poland, 21–23 September 2011; Springer: Berlin/Heidelberg, Germany, 2011; pp. 80–89.

Article

Scale-Free Features in Collective Robot Foraging

Ilja Rausch *, Yara Khaluf and Pieter Simoens

IDLab, Ghent-University, Technologiepark 126, B-9052 Ghent, Belgium

* Correspondence: ilja.rausch@ugent.be; Tel.: +32-9331-49-75

Received: 19 April 2019; Accepted: 27 June 2019; Published: 30 June 2019

Abstract: In many complex systems observed in nature, properties such as scalability, adaptivity, or rapid information exchange are often accompanied by the presence of features that are scale-free, i.e., that have no characteristic scale. Following this observation, we investigate the existence of scale-free features in artificial collective systems using simulated robot swarms. We implement a large-scale swarm performing the complex task of collective foraging, and demonstrate that several space and time features of the simulated swarm—such as number of communication links or time spent in resting state—spontaneously approach the scale-free property with moderate to strong statistical plausibility. Furthermore, we report strong correlations between the latter observation and swarm performance in terms of the number of retrieved items.

Keywords: agent-based collective intelligence; multi-agent complex systems; scale-free properties; power law distribution; biologically inspired approaches and methods; collective foraging; physics-based simulation; methodologies for agent-based systems; multi-robot simulation

---

1. Introduction

Advances in computation have made it possible to record, simulate, and analyze multi-agent complex systems in nature, such as fish schools, bird flocks, locust swarms, and ant colonies. In many of these collective systems, various attributes were found to be *scale-free* [1], i.e., the attributes do not have a characteristic size or value. Examples of such scale-free features found in biological systems include, among others, (i) asymptotically scale-free correlation lengths of starling flocks [2,3]—the term *asymptotic* refers to the behavior of a variable (in this case spatial correlation) close to a limit (in this case an infinite flock size); (ii) scale-free fluctuations of velocity and orientation correlations in moving bacterial colonies [4]; (iii) time intervals between communication calls that follow a power law—which is the mathematical representation of a scale-free property—in pairs of zebra finches; and (iv) scale-free movement patterns found in models of foraging primates [5] or midge swarms [6].

One of the most prominent findings is that the number of interactions appear to be scale-free in various real-world *networks* of biological and social systems [7–9]. Multi-agent systems benefit from scale-free communication because it enables scalable, fast and efficient information transfer [10–12]. An essential aspect of scale-free networks is that they represent complex topologies in which only a few nodes (called hubs) have a comparably high connectivity degree [7]. This small percentage of highly connected hubs makes scale-free topologies vulnerable to targeted attacks but exceptionally robust to random failures (which are likely to affect the vast majority of nodes that are not hubs) [13]. Furthermore, due to the high connectivity, the network diameter is small, which means that on average, any two nodes can share their information only over a few hops [11], resulting in fast information transfer.

Inspired by the high prevalence of scale-free features in (socio-) biological systems, the aim of the current study is to examine whether scale-free attributes may also spontaneously emerge in *artificial* collective systems. One particularly prominent example of these systems inspired by nature is *swarm robotics*, where robot collaboration is an essential prerequisite for the successful execution of tasks [14–20]. Although accurate understanding and systematic design of swarm robotic systems are considered to be among the greatest challenges of contemporary robotics [21], swarm robotics benefits strongly from the progress made in wireless communication technologies, system integration, machine learning and artificial intelligence (AI). Consequently, artificial cooperative multi-agent systems gain in importance not only in applied science and engineering but also in fundamental research, allowing the shrinkage of the reality gap of detailed modeling and the accurate simulation of distributed biological systems.

Many natural collective behaviors were modeled using artificial collective systems, including bird flocking [22], locust marching [23] or cockroach aggregation [24]. However, among the most prominent challenges is the *collective foraging* task [18]. Extensive efforts were dedicated to the study of the foraging task because of its remarkable prevalence. The foraging behavior is found in various species and subspecies across the world [25,26]. In most cases, its efficient implementation is essential to group survival.

In essence, a multi-agent system performing the foraging task has the goal of collectively retrieving information and other resources from the environment. Different to a single-agent implementation, the benefit of a multi-agent system is its ability to share the accessed resources and information to enhance the overall performance. However, the multi-agent foraging task exhibits a considerable degree of complexity, making its modeling and analysis very demanding [18]. Successful collective foraging often requires a delicate combination of several extensively studied multi-agent sub-behaviors such as deployment [27,28], exploration [29,30], aggregation [31,32] or information sharing [33,34]. Hence, even though collective foraging itself can be considered to be a specific task within a large class of multi-agent problems, it rightfully receives separate attention in numerous contemporary studies [16,35–38].

Moreover, collective foraging is a promising behavior for many real-world applications such as exploration by aerial vehicles [39], underwater monitoring [40], or optimization of electrical networks [41]. Therefore, the foraging performance of artificial multi-agent systems, potentially in combination with other types of AI, is worthwhile investigating in depth. In particular, in robot swarms, various fundamental questions have already been addressed such as the influence of interference [42,43], regulation of information flow [33] or achievement of consensus [44]. Nevertheless, other relevant questions are still open to research, including how does the distribution of individual features change in relation to the input from the environment or social interactions? Is there a connection between particular feature distributions and the performance of the swarm? To address these questions with respect to scale-free properties, we simulate the foraging behavior in a robot swarm and analyze the emergence of scale-free features. For this purpose, the complexity of the foraging task is advantageous as it offers a wide range of features that can be examined for their statistical tendency to be scale-free.

Our goal can be split into the following: (i) investigating the existence of scale-free features in a robot swarm performing the foraging task, (ii) studying the correlation between these features and the swarm performance, (iii) discussing the potential role of feedback mechanisms in the emergence of such scale-free features.

We begin with defining the robot (microscopic) and the swarm (macroscopic) behaviors in Sections 2.1 and 2.2, respectively. The link between these two levels of behaviors is formulated using statistical distributions and elaborated on in Section 2.4. In Section 2.5, we describe the experimental setup. Thereafter, in Section 3 we demonstrate the occurrence of scale-invariant features—such as those related to the communication degree or times spent in foraging or resting—and their correlation with swarm performance. Furthermore, we discuss the present feedback mechanisms that may support the emergence of scale-free features in a sophisticated set of scenario configurations. Lastly, the paper is concluded in Section 4.

2. Methods

2.1. Robot Behavior

We focus on the robot's decision-making process that is defined by the robot's interactions and the robot's individual preferences. The robots are situated in an arena which consists of a nest and a foraging area. Each robot can switch between two states: *resting* and *exploring*. In biology, similar behavior called *forager activation* has been observed in harvester ants, *Pogonomyrmex barbatus*, and reported in several publications [45–47]. In the exploring state, the robot moves around searching for items—which are located only in the foraging area—to retrieve to the nest. The robot moves on a straight line until it encounters another robot or a wall, in which case a *collision avoidance* maneuver is initiated. In the resting state, the robot rests inside the nest and only in this state it is allowed to communicate with the neighbors within its line of sight. Specifically, each robot can broadcast a message about the success or failure of its latest exploration attempt or listen to its neighbors. Received information may either increase or decrease the robot's *probability* to switch from *resting* to *exploring* or vice versa. Probabilities are updated continuously using fixed probability jumps—we refer to those by the term *cues*, as in [14]. In the following, we introduce the different probability cues used in implementing the foraging behavior, in addition to the probabilities determining the switch between the two robot states. We also consider two distinct communication modes defining the duration of information exchange.

2.1.1. State Switching Probabilities

Following [14], there is a minimum duration θ for the robot to stay in a certain state. The purpose of having such a threshold is to ensure that robots can perform the sub-tasks associated with this state for a certain amount of time so that necessary dynamics can take place. For instance, a minimum exploring time θ_e needs to be at least as long as it takes for a robot to reach the most remote items (taking into account the constant linear speed of that robot) [16].

With this in mind, let us formulate the individual response to social and environmental cues in terms of *switching probabilities*. We denote $\{i_e, i_r\} \in \{\mathbb{R}_0^+, \mathbb{R}_0^+\}$ and $\{s_e, s_r\} \in \{\mathbb{R}_0^+, \mathbb{R}_0^+\}$ as the robot's internal (i) and social (s) cues to switch to exploring (e) or resting (r) state, respectively. The probability of a robot to switch from the resting state to the exploring state is denoted by $p_{r\to e}$, whereas the probability to switch from the exploring state to the resting state is denoted by $p_{e\to r}$.

The probabilities are updated iteratively at every simulation time step in a discrete manner as in the following:

$$p_{r\to e}(t+1) = p_{r\to e}(t) + \delta_\eta(t)s_e + \delta_\phi(t)i_e \tag{1}$$

$$p_{e\to r}(t+1) = p_{e\to r}(t) - \delta_\eta(t)s_r - \delta_\phi(t)i_r, \tag{2}$$

where $\delta_\eta(t)$ is the number of 'success' minus 'failure' messages received by the robot from its neighbors at every time step spent in the *resting* state. Additionally, the robot's own experience is characterized using $\delta_\phi(t)$. This is defined for every exploration attempt as follows:

$$\delta_\phi(t) = \begin{cases} +1, & \text{at the time instance of finding an item if } t_{se} < t \leq t_{se} + \theta_e \\ 0, & \text{at all other time instances if } t_{se} < t \leq t_{se} + \theta_e \\ -1, & \text{if } t > t_{se} + \theta_e \text{ and the robot is still exploring} \end{cases} \tag{3}$$

with t_{se} as the time at which the robot started its current exploration. Moreover, $\delta_\phi(t) = 0$ while the robot is resting. In case $p_{e\to r} < 0$ it is truncated to $p_{e\to r} = 0$ and when $p_{e\to r} > 1$ it is truncated to

$p_{e \to r} = 1$; same holds for $p_{r \to e}$. Please note that there is a strict difference between $\delta_\eta(t)$ and $\delta_\phi(t)$: $\delta_\eta(t)$ may be non-zero only when the robot is resting inside the nest because it is computed based on the information broadcast by the neighbors which can only be received in the nest (i.e., when the robot is in the resting state). Whereas, $\delta_\phi(t)$ may be non-zero only when the robot is exploring—it is computed based on the robot's own foraging experience. Table 1 lists the parameters relevant for the computation of the switching probabilities.

Table 1. An overview of parameters defining the switching probabilities.

Description	Symbol
Probability to switch from resting to exploring	$p_{r \to e}$
Probability to switch from exploring to resting	$p_{e \to r}$
Number of 'success' minus 'failure' messages received	$\delta_\eta(t)$
Impact of minimum exploring time θ_e	$\delta_\phi(t)$
Social cues for exploring and resting, respectively	s_e, s_r
Internal cues for exploring and resting, respectively	i_e, i_r

2.1.2. Communication Modes

We focus on the local communication between the robots and their influence on the global swarm behavior. A common approach is to restrict robot communication *only* to the area within the nest. This approach is inspired by several natural systems, in which the communication takes place mainly inside the nest or hive, such as in the case of ants or honey bees [14,47–50]. Moreover, this approach accommodates two relevant properties of foraging systems: (i) it is common that the foraging area is significantly larger than the nest area, and hence, individual encountering rates outside the nest are negligibly low; (ii) high density of individuals within the nest leads to more accurate information about the environment due to the high encounter rate of individuals that explored different, distant parts of the foraging area.

Regarding particular communication strategies, it is common to let robots broadcast the last exploration result only *once*, namely when the robots switch to the *resting* state. Henceforth, we will refer to this approach as the *discontinuous communication mode* (DCM), because after broadcasting the message once, the active communication of the robot is interrupted and is limited to listening. In contrast, we use the term *continuous communication mode* (CCM) to refer to the mode in which robots continue broadcasting the result of their last foraging attempt at every time step until they switch back to the *exploring* state. As we will see later, the difference between these two modes does not have substantial impact on swarm performance. However, it has a significant impact on the statistical distribution of various system features for which we study the scale-free property.

2.2. *Swarm Behavior*

At the macroscopic level, global behavior emerges as a result of complex interactions between the robots as well as between robots and their environment. The quality of such global behavior is evaluated with respect to quantifiable objectives. In the present study, we define the swarm performance in terms of three quantitative measures:

1. the total number of collected items at the end of the experiment N_{coll}
2. the average number of collected items per time spent in collision avoidance $\omega_{ca} = \frac{N_{coll}}{T_{ca}}$, where T_{ca} is the total duration of all collision avoidance events in the swarm throughout the experiment. One collision avoidance event includes slowing down and turning around until the angle to the closest possible obstacle is $> 25°$. In essence, ω_{ca} reflects the trade-off between coordination and interference.

3. the average number of collected items per time spent exploring $\omega_e = \frac{N_{coll}}{T_e}$, where T_e is the aggregate time that all individuals spent in the exploring state.

2.3. Measured Features

There is a large variety of features that could potentially have scale-free character in collective foraging. We investigate such features categorizing them into, *space* and *time* features. An overview of the measured features is given in Table 2. Space features are mostly related to the inter-robot communication according to their distribution in the arena. This is a distribution that changes over time while the robots are in motion. Among the space features, the robot's communication degree d is the most important and evident. It is defined as the number of communication links to neighbors within the robot's communication range. However, in dynamic topologies—where robots move around and neighbor lists are constantly updated—the communication degree d changes frequently. Hence, we track additionally the change of the communication degree Δd of a robot whenever fellow robots enter or leave its communication range. Beside the communication degree, we analyze space features that reflect the foraging progress such as the difference between the number of *received success* and *failure* messages, denoted by the *critical degree* $d_{c,rec}$. Similarly, we include features that reflect the success-degree of a particular individual by measuring the difference of *success* to *failure* messages *sent* by that individual, denoted by $d_{c,sent}$.

Table 2. An overview of the investigated space and time features.

Description	Symbol
Space features	
Degree	d
Change of degree	Δd
Critical degree (sent, received)	$d_{c,sent}$, $d_{c,rec}$
Time features	
Foraging time	τ_f
Homing time	τ_h
Resting time	τ_r
Collision avoidance time	τ_{ca}

With respect to *time* features, we note that in swarm robotics the individuals are commonly subject to physical interference. Robots interfere with each other or with obstacles as a result of finite-size effects influencing the dynamics of the collective behavior [42,43,51]. Therefore, we investigate the time spent on collision avoidance, denoted by τ_{ca}. Additionally, we study time features that are related to the robot's exploring time τ_e. This time can be split into foraging time τ_f, i.e., the time spent on searching for items, and homing time τ_h, i.e., the time spent on returning to the nest. While a long foraging time effectively increases the probability of finding items, long homing times indicate overcrowding close to the nest. Finally, another relevant time feature is the resting time τ_r that includes the duration of robot interaction within the nest.

2.4. Data Analysis

In complex systems such as swarm robotics, the statistical analysis of relevant system properties paves the way to mathematical modeling, useful simplifications, or inference of long-term behaviors. Consequently, it helps in defining the link between the individual robot behavior and the emergent global swarm behavior, referred to as the micro-macro link [52]. In our study, we focus on how the collective foraging behavior can be related to the scale-freeness of a set of individual and global features. The main statistical characteristic of scale-free features is that they are distributed according to a power law [1,53]. Therefore, to identify scale-free features in our simulated swarms, it is of central importance to measure the statistical distribution of these features and to perform a sound power law fitting procedure.

2.4.1. Power Law Fitting Procedures

To verify whether a feature is scale-free, we use a set of techniques that are described in [53–55] for fitting its distribution by the *power law distribution*. The power law distribution takes the form of a straight line on a log-log scale of $p(x)$. However, most real-world data displays significant fluctuations due to randomness. When fitting power law to the data, random fluctuations are considered by the statistical value p which represents the *goodness-of-fit*. When $p < 0.1$ the power law fit can be considered to be unreliable [54]. Furthermore, power law behavior emerges mostly only in the tail of the distribution, i.e., for higher values of x above a statistically determined lower bound x_{min} [53]. Please note that this effectively reduces data set to fit by the power law, which is important to keep in mind by considering a ratio of the total number of data points to the points that satisfy the condition $x > x_{min}$. Finally, there are several other statistical distributions that may resemble the characteristic straight-line tendency of a power law on a log-log plot. Hence, for a sound statistical analysis it is important to compare the power law fit to other statistical models [54–56]. More precisely, the power law fitting procedure can be summarized by the following three steps:

1. Using maximum likelihood estimation, fit the data by the power law distribution

$$p\left(x, x_{min}, \alpha\right) = \frac{\alpha - 1}{x_{min}} \left(\frac{x}{x_{min}}\right)^{-\alpha}, \tag{4}$$

   where α is the scaling parameter and x_{min} is the lower bound. In particular, α and x_{min} are estimated using procedures described in [54].
2. Apply Kolmogorov-Smirnov statistic to carry out the *goodness-of-fit* tests and verify the above results. Here, essentially, a large set of synthetic data is generated from a power law distribution (with α and x_{min} found in 1.) and their distances to their respective fits are compared to the distance between the empiric data and the best-fit found in step 1. The outcome of this procedure is the p-ratio which estimates the contribution of random fluctuations. If $p < 0.1$ the power law fit found in step 1 is very likely to be due to inherent randomness in the empiric data. Moreover, the p-value is unreliable when the data set is too small. Therefore, the percentage of data points $N_{data,pl}$ that lie above x_{min} should be 10% or higher.
3. Finally, even if $p > 0.1$, the power law fit might be not the only model that fits the data well. Consequently, complete the above steps for a set of other potential distributions including exponential or lognormal distributions. Then, compare the resulting best fits to the one obtained for the power law by computing the ratio R. The latter is defined as the log likelihood of the power law over the log likelihood of another distribution. If $R > 0$, power law is the statistically superior fit. Although this last step still does not give us the certainty that the data is power law distributed, it makes the hypothesis more plausible. For this step we used an open-access Python toolbox [56].

2.4.2. Quality Ratio ρ_q

Given a high quantity of empiric data sets, it is useful to find an automated way for the evaluation of the power law fits. For the analysis of our experiments, we introduce a quality ratio ρ_q which we use as a practical estimate of the plausibility of a (truncated) power law fit based on the well-known rigorous statistical tests described above. The quality ratio ρ_q includes the three criteria discussed in Section 2.4.1:

p-value, $N_{data,pl}$ and the number of likelihood-ratio tests resulting in $R > 0$. We account for these criteria by defining ρ_q as the product of $\rho_{q,p}$, $\rho_{q,data}$ and $\rho_{q,lrt}$:

1. First, we begin with the p-value. As mentioned above, the linear shape of the data distribution on a log-log plot can be mainly attributed to random fluctuations if $p < 0.1$. Taking this into account, we design ρ_q to be a binary piecewise function evaluating the *goodness-of-fit* in terms of the p-value:

$$\rho_{q,p} = \begin{cases} 1.0, & \text{if } p \leq 0.1 \\ 0.0, & \text{otherwise.} \end{cases} \tag{5}$$

   This way, we take into account the possibility that random fluctuations may be present but as soon as $p > 0.1$ we do not assign the precise value of p to the ranking of the fit. The reason is that random fluctuations might be present even if the data is in fact power law distributed. In that case the p-value could be very low even if the data is in fact power law distributed. In general, it might be more substantial to consider the size of the fitted data set and to compare the power law fit to other important distributions [54,56].
2. Second, $\rho_{q,data}$, denotes the ratio of the data which is fit by the (truncated) power law $N_{data,pl}$ to the total number of data points $N_{data,tot}$.

$$\rho_{q,data} = \frac{N_{data,pl}}{N_{data,tot}}, \tag{6}$$

3. Third, $\rho_{q,lrt}$ represents the fraction of likelihood-ratio-tests in which the (truncated) power law fit proved to be statistically more plausible than other distributions. To include the quality of the (truncated) power law fit as compared to other distributions, we count how many times $n_{lrt,pl}$ we obtained $R > 0$ from the likelihood-ratio tests. We compare the power law fit to *six* distributions: *truncated power law, exponential, stretched exponential, lognormal, positive lognormal* and *normal*; all of them are implemented in [56] (except the normal distribution). Hence, we use the piecewise function,

$$\rho_{q,lrt} = \begin{cases} 0, & \text{if at least one likelihood-ratio test yields } R < 0 \\ \frac{n_{lrt,pl}+1}{7}, & \text{otherwise.} \end{cases} \tag{7}$$

   where we added 1 to $n_{lrt,pl}$ to account for the possibility that the likelihood-ratio test yields $R \approx 0$, in which case the support for the power law fit is neither strengthened nor weakened.
   Please note that in Equation (7) we set $\rho_{q,lrt} = 0$ if at least one distribution is a more reliable model than the power law. However, it is important to remember that our simulated systems are meant to include real-world attributes (e.g., finite-size effects, physical interference, line-of-sight interruptions during communication) and therefore deviate from ideal systems. Consequently, the assumption of power law (i.e., scale-free) distribution might be distorted and needs to be corrected. The deviation is often particularly distinct in the heavy tail. Therefore, one common correction technique is to consider the power law distribution with an exponential cutoff (also known as *truncated power law*) [57]:

$$p\left(x\right) = \frac{\lambda\left(\lambda x\right)^{-\alpha}}{\Gamma\left(1-\alpha, x_{min}\lambda\right)} e^{-\lambda x}, \tag{8}$$

   where λ is the scaling parameter of the exponential decay and $\Gamma(y, z)$ is the upper incomplete gamma function. While Equation (4) directly implies that the feature is scale-free, Equation (8) describes an asymptotic scale-freeness in the limit $\lambda x \to 0$. This equation approaches the power law distribution asymptotically for $\lambda x \to 0$ and the exponential distribution for $x\lambda \gg 1$, respectively. Thus, accepting

that our systems are significantly constrained within physical boundaries we can slightly soften the criteria given by Equation (7) in the following way:
If the truncated power law passes more likelihood-ratio tests than the power law fit, i.e., if $n_{lrt,tpl} > n_{lrt,pl}$, we consider the success-ratio of the former. In short:

$$\rho_{q,lrt} = \begin{cases} 0, & \text{if at least one likelihood-ratio test yields } R < 0 \\ \frac{n_{lrt,tpl}+1}{7}, & \text{if } n_{lrt,tpl} > n_{lrt,pl} \\ \frac{n_{lrt,pl}+1}{7}, & \text{otherwise.} \end{cases} \tag{9}$$

Finally, including all the above criteria, we define the quality ratio:

$$\rho_q = \rho_{q,p} \cdot \rho_{q,data} \cdot \rho_{q,lrt}. \tag{10}$$

Consequently, we obtain $\rho_q = 0$ if $p \leq 0.1$ or $R < 0$. Conversely, $\rho_q = 1$ in the case of $p > 0.1$, $N_{data,pl} = N_{data,tot}$ and $n_{lrt} = 6$, which is an unlikely but nevertheless possible scenario. Using this ranking, we can link the quality of a fit to a quantifiable value and describe the support for the (truncated) power law as illustrated in Table 3.

Table 3. Classification of the power law fit quality with respect to the quality ratio ρ_q used in our study.

Support for the (Truncated) Power Law	Numerical Value of ρ_q
No support	$\rho_q = 0$
Weak	$0 < \rho_q \leq \frac{0.1}{7} \approx 0.0143$
Moderate	$\frac{0.1}{7} < \rho_q \leq \frac{0.5}{7} \approx 0.0714$
Strong	$\frac{0.5}{7} < \rho_q$

The denominator value represents the total number of considered distributions (i.e., the power law and the six alternative distributions we compare it to). The lower limit of the 'moderate' classification corresponds to the case with $\rho_{q,p} = 1$, $\rho_{q,data} = 0.1$ and $\rho_{q,lrt} = \frac{1}{7}$—i.e., at least 10% of the data is included in the fit and none of the alternative distributions is a statistically better fit than the power law. The upper limit considers the case with $\rho_{q,p} = 1$ and $\rho_{q,data} \cdot \rho_{q,lrt} = \frac{0.5}{7}$—i.e., either the fit includes a high number of data or the power law is statistically a better fit than other distributions. Please note that ρ_q multiplicatively combines standard power law fitting techniques [53–56] into a quantitative estimate of the quality of the (truncated) power law fit.

It is important to emphasize that even if the hypothesis of the data following the power law distribution is found to be plausible using the above statistical analysis, care needs to be taken when interpreting this observation. Firstly, there is still no guarantee that the data is in fact power law distributed and although our rigorous analysis includes several common distributions, other non-obvious distributions may prove to be a better fit. Secondly, the power law fit may be valid only for a small fraction of data. However, as the power law behavior is commonly found for a subset of data, namely at the tail of the distribution, the group that displays power law (i.e., scale-free) behavior includes individuals that stand out from the rest of the swarm by having features with values that are significantly above average. The way in which such individuals impact the global swarm performance remains an open question worth investigating.

2.4.3. Correlation Measures

To examine the presence of correlations between the support for the power law distribution (i.e., the value of ρ_q) and the swarm performance it is important to use an appropriate correlation measure. One of the most prominent correlation measures is the *Pearson* correlation coefficient [58,59]. It evaluates the quality of a linear association between two distributions. In essence, it calculates the covariance of the mean values of two distributions, over the root of their standard deviations. It is closely related to linear regression and does not require the data to be normally distributed. Despite its mathematical simplicity it is an appropriate correlation measure for many distributions and, therefore, is widely used [60–63].

However, one could argue that the Pearson correlation coefficient is not ideal for skewed distributions with strong outliers. Popular alternatives are the *Spearman's rank* and the *Kendall's tau* correlation coefficients [62–65]. Both are based on generating ranked distributions by assigning a rank to each variable with respect to its value. The correlation coefficient is then given as a measure of the association between the two ranked distributions. Consequently, both correlation metrics are robust to outliers and suitable for non-linear distributions.

Although both correlation measures commonly return very similar results, Kendall's tau handles *ties* (i.e., cases in which there is no difference between the ranks) in a mathematically more straightforward way. More precisely, Kendall's tau returns the density difference between *concordant* and *discordant* pairs. Consider two vectors of length n, $(x_1, x_2, ..., x_n)$ and $(y_1, y_2, ..., y_n)$. Concordant pairs are pairs of data points that satisfy $sgn(x_i - x_j)sgn(y_i - y_j) > 0$ (where $sgn\ (z)$ is a sign-function equal to +1 if $z > 0$, -1 if $z < 0$ and 0 if $z = 0$); similarly, discordant pairs satisfy $sgn(x_i - x_j)sgn(y_i - y_j) < 0$. Furthermore, ties are pairs for which $x_i = x_j$ or $y_i = y_j$. Hence, with n_c (n_d) as the number of concordant (discordant) pairs, respectively, and n_x (n_y) as the number of ties in x (y), respectively, the Kendall's tau (also known as the Kendall's tau-b) is given by [66]:

$$\tau_{Kendall} = \frac{n_c - n_d}{\sqrt{n_c + n_d + n_x}\sqrt{n_c + n_c + n_y}}, \tag{11}$$

with

$$n_c = \sum_{i,j} \delta_{i,j}^{(c)}, \ n_d = \sum_{i,j} \delta_{i,j}^{(d)}, \ n_x = \sum_{i,j} \delta_{i,j}^{(x)}, \ n_y = \sum_{i,j} \delta_{i,j}^{(y)}, \tag{12}$$

where

$$\begin{aligned} \delta_{i,j}^{(c)} &= \begin{cases} 1, & \text{if } sgn(x_i - x_j)sgn(y_i - y_j) > 0 \\ 0, & \text{else} \end{cases} \\ \delta_{i,j}^{(x)} &= \begin{cases} 1, & \text{if } x_i = x_j \\ 0, & \text{else} \end{cases} \end{aligned} \tag{13}$$

and similarly, for $\delta_{i,j}^{(d)}$ and $\delta_{i,j}^{(y)}$.

2.5. Simulation Setup

We designed and implemented a set of physics-based simulations using the state-of-the-art simulator for large-scale swarms, ARGoS [15]. An overview of all parameter values used in our simulations is given in Table 4. The simulations are conducted in a square-shaped arena, which is confined within four walls, each being of the length of 50 m. The arena is divided into two regions: (i) the nest A_n: it is the gray 10×50 m^2 area in Figure 1a, and (ii) the foraging area A_f: it is the white 40×50 m^2 area in Figure 1a.

The items are scattered uniformly over the foraging area, and keep reappearing after robots retrieve them to the nest—as in [14]—with constant probability. This prevents the system from drifting into an absorbing state in which there are no items left to recover.

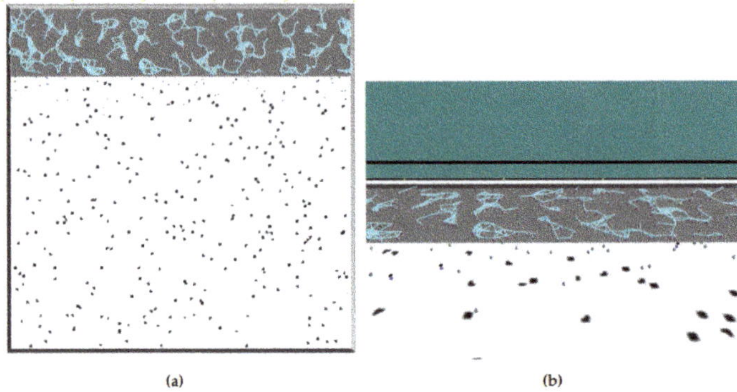

Figure 1. Illustrations of the arena. (**a**) A snapshot from a simulation in ARGoS. Gray area: nest; white area: foraging field; black dots: items; blue objects: Footbots; light-blue lines: communication (range-and-bearing) links. (**b**) 3D view on the same arena. In both figures, the communication links are formed *only* for *resting* robots *inside* the nest, as in our experiments moving robots neither broadcast nor listen to any messages.

Table 4. Robot and arena parameters used for the simulation setup.

Parameter	Value
Robot parameters	
Type	Footbot
Proximity sensor range r_{prox}	0.1 m
Range-and-bearing sensor range r_{rab}	1.25 m
Maximum moving speed	$1\ \frac{m}{s}$
No-turn threshold α_{nt}	10°
Soft-turn threshold α_{st}	30°
Hard-turn threshold α_{ht}	90°
$s_e, i_e, i_r, s_r \in \{0.0, 0.01, 0.5, 0.9\}$	
Arena parameters	
Total area A	50×50 m^2
Nest area A_n	10×50 m^2
Foraging area A_f	40×50 m^2
Number of robots N_{robots}	950
Number of items N_{items}	300
Radius of an item r_{item}	0.2 m
Total experiment duration T	10^4 ts

A phototaxis behavior is used to assist the robots in leaving and re-visiting the nest. For that purpose, light beacons are positioned equidistantly at the nest wall (yellow dots at the bottom of Figure 1a). Their light is perceived by the robots' light sensors. Each robot is programmed to move *away* from the beacons when it needs to *leave* the nest and *towards* the beacons when it needs to *return*. We use a homogeneous swarm of Footbots (see http://www.swarmanoid.org/swarmanoid_hardware.php) in our simulations, and the communication radius of the robots is set such that the fraction of the circular communication area around the robot is 0.982 % of the nest area, which is close to the fraction used in [14]. For better readability, we will limit our discussion of the robot states to only *resting* and *exploring*. While the former is distinct, the latter is composed of further states of which only *foraging* and *homing* are relevant because they are the most time consuming (for a detailed list of the robot states please see Supplementary Material Section 2).

At the beginning of the simulation, each robot switches from resting to exploring with a probability of 0.01. Consequently, within the first 500 time steps (ts) most robots leave the nest. After another ≈ 500 ts most of the swarm returns, with or without an item. Although this behavior is subsequently repeated several times, the number of simultaneously switching robots gradually decreases, and the switching rate from resting to exploring (or vice versa) approaches a constant limit. In most cases, the system approached an equilibrium after $5 \cdot 10^3$ ts (for the given arena, item density, and swarm size). Our measurements of the system features begin from that time instance on-wards and the experiment proceeds for another $T = 10^4$ ts.

Furthermore, to conduct a solid statistical study we use large-scale swarms with $N_{robots} = 950$ units which is up to an order of magnitude higher than what is commonly used [14,16,36,42]. We selected the value of N_{robots} by running preliminary experiments, in which we observed for this particular swarm size—under the given arena and item density—a maximum in swarm performance.

Finally, in our experiments, the most important means of influencing the swarm dynamics is by adjusting the numerical values of the internal and social cues—i_e, i_r and s_e, s_r, respectively—at the start of each experiment. We consider a spectrum of 256 distinct scenario configurations, which differ by the 4-tuples $a = (s_e, i_e, i_r, s_r)$ drawn from:

$$\Omega := \{a : s_e, i_e, i_r, s_r \in \{0.0, 0.01, 0.5, 0.9\}\}. \tag{14}$$

The rationale behind the choice of these parameter values is to include four fundamentally different kinds of cue impact on swarm dynamics: (i) none (ii) low (iii) intermediate and (iv) high. Please note that any additional value in the set a greatly increases the associated computational and analytic effort—as the number of scenarios scales with $dim(a)^4$. However, based on preliminary results, additional values would offer potentially little informative gain (at the current stage) because the swarm dynamics would be similar to a mix of the dynamics generated by the above values.

3. Results and Discussion

We performed simulations with all combinations of cues and communication modes. Each simulation was repeated with 30 random seeds and the data analysis procedure was carried out as discussed in the previous section.

3.1. Presence of Power Law Distributed Features

The analysis of our simulation data shows that in most scenarios there was only weak or no statistical support for the (truncated) power law distribution (see Figure 2A). In particular, in roughly half of all scenarios no power distributed features were found. This observation suggests that in the present system, scale-free features are rare. Nevertheless, we found 245 + 71 = 316 (truncated) power law distributions with moderate or strong statistical plausibility for different features in various scenario configurations and for both communication modes, DCM and CCM. Thus, our findings are in line with a recent study showing that scale-free networks may occur rarely but across different areas [55].

As the scatter plots in Figure 2 show, most of the distributions with weak or moderate support for power law are concentrated below or close to the average values of swarm performance while the distributions with strong support for power law are associated with above-average performance in terms of N_{coll} and ω_{ca}. Swarm performance was measured using (i) the number of items retrieved by the robots,

N_{coll}, (ii) the average number of collected items per time spent on collision avoidance ω_{ca}, and (iii) the average number of collected items per time spent exploring ω_e. Figure 3 shows the values recorded for these three metrics under both continuous and DCMs and for the entire range of 256 scenario (cues) configurations, respectively. Repeating performance patterns can be observed over different sets of configurations. The regions over which these patterns emerge are (from left to right): (i) all scenarios with $s_e = 0$ and $i_e = 0$, i.e., constant $p_{r \to e}$ (blue region with a left tilted mesh in Figure 3), (ii) all scenarios with $s_e = 0$ and $i_e > 0$, i.e., no social and only internal influence on $p_{r \to e}$. This region is henceforth referred to as NS_e (shown in orange, no mesh), (iii) all scenarios with $s_e = 0.01$, i.e., low social impact on $p_{r \to e}$ (green region with vertical mesh, henceforth denoted as LS_e), and (iv) all scenarios with $s_e = 0.5$ or $s_e = 0.9$, i.e., high social influence on $p_{r \to e}$ (red region with right tilted mesh, henceforth denoted as HS_e). Please note that in all four regions $p_{e \to r}$ is altered in the same way, i.e., for s_r and i_r all values from $\{0.0, 0.01, 0.5, 0.9\}$ are included. The best swarm performance in terms of N_{coll} and ω_{ca} emerges when the influence of internal cues on the swarm dynamics is negligible compared to social cues, i.e., when $s_e \sum_t |\delta_\eta(t)| \gg i_e \sum_t |\delta_\phi(t)|$ and $s_r \sum_t |\delta_\eta(t)| \gg i_r \sum_t |\delta_\phi(t)|$.

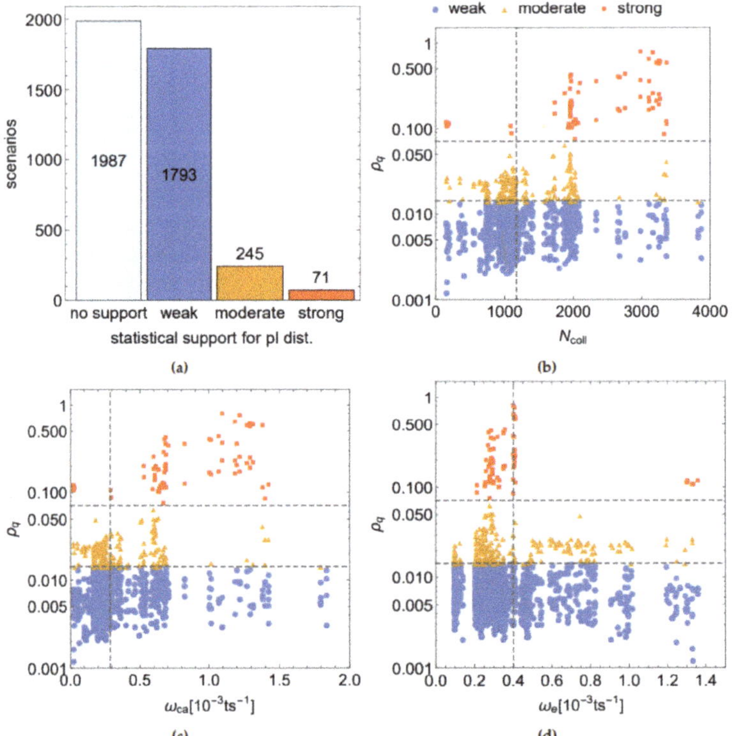

Figure 2. (**a**) Feature data sets obtained from simulations, sorted by the type of statistical support for a corresponding power law fit. The classifications follow Table 3. (**b–d**) Log-linear scatter plots relating the power law fit quality ratio ρ_q to the swarm performance in terms of N_{coll}, ω_{ca} and ω_e, respectively. The vertical dashed lines indicate the mean performance values while the horizontal dashed lines separate the quality categorizations taken from Table 3.

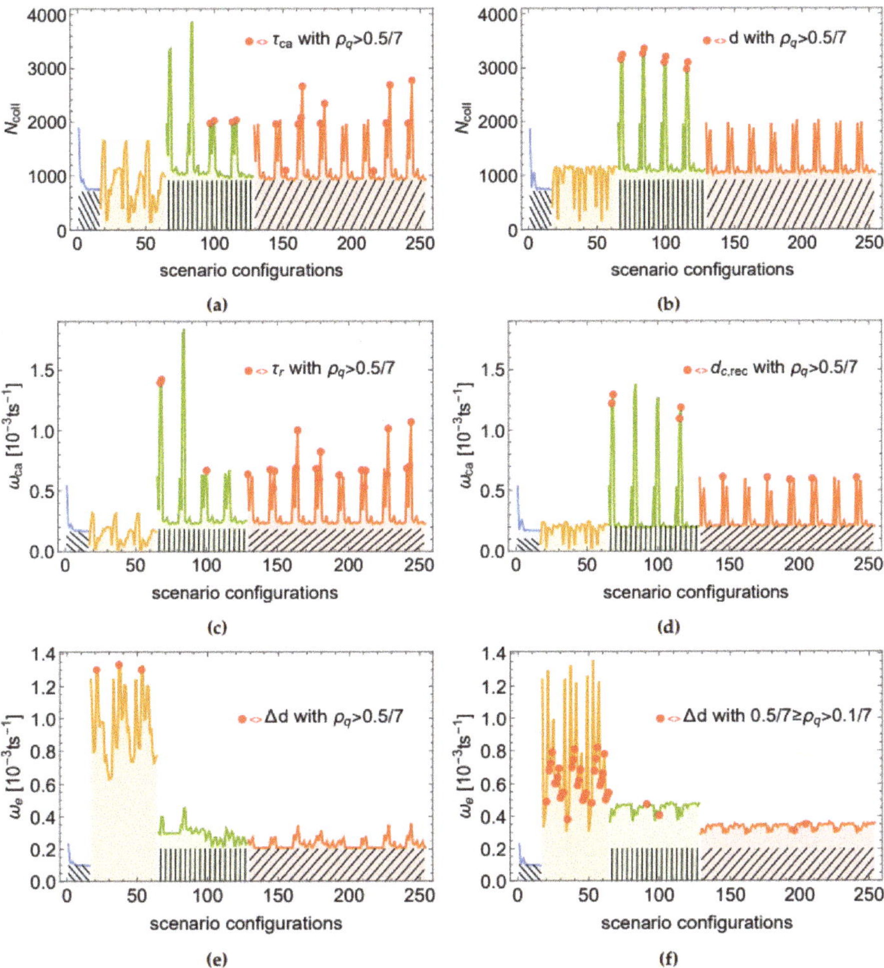

Figure 3. Swarm performance in terms of (**a**,**b**) N_{coll}; (**c**,**d**) ω_{ca} and (**e**,**f**) ω_e, respectively. For each performance measure, 256 scenario configurations were implemented (i.e., with all cue values from Equation (14)), using one of two communication modes: DCM (left) and CCM (right). The x-axis represents the IDs of the scenario configurations. The colors and the mesh patterns highlight regions that display different dynamics. Apart from (**f**), in all plots the red dots mark the scenarios in which the feature mentioned in the inset demonstrated a high value of ρ_q, i.e., there was a strong support for the distribution to be power law. In (**f**), the red dots mark the scenarios with moderate support. See Supplementary Material Section 3 for combined plots of N_{coll} and d distribution in CCM over the complete set of 256 scenario configurations.

The best performance levels in terms of N_{coll} and ω_{ca} were reached over the LS_e region. For instance, the maxima of N_{coll} and ω_{ca} correspond to the scenario configurations in which $s_e = 0.01$, $i_r = 0$ and $s_r \geq 0.5$. For the same configurations, (truncated) power law distributions of space features were found in the CCM (examples shown in Figure 4). Contrary to CCM, in the DCM the robot interactions are interrupted. These interruptions may explain why, in DCM, space features such as communication degree tend to not follow a power law distribution (weak overall support for the presence of a power law behavior). Nevertheless, we found fits with moderate to strong support for (truncated) power law to time features, such as τ_r and τ_{ca}, demonstrated in Figure 5. The best power law fits of the DCM correspond to the peaks in swarm performance in terms of N_{coll} and ω_{ca} over the HS_e regions.

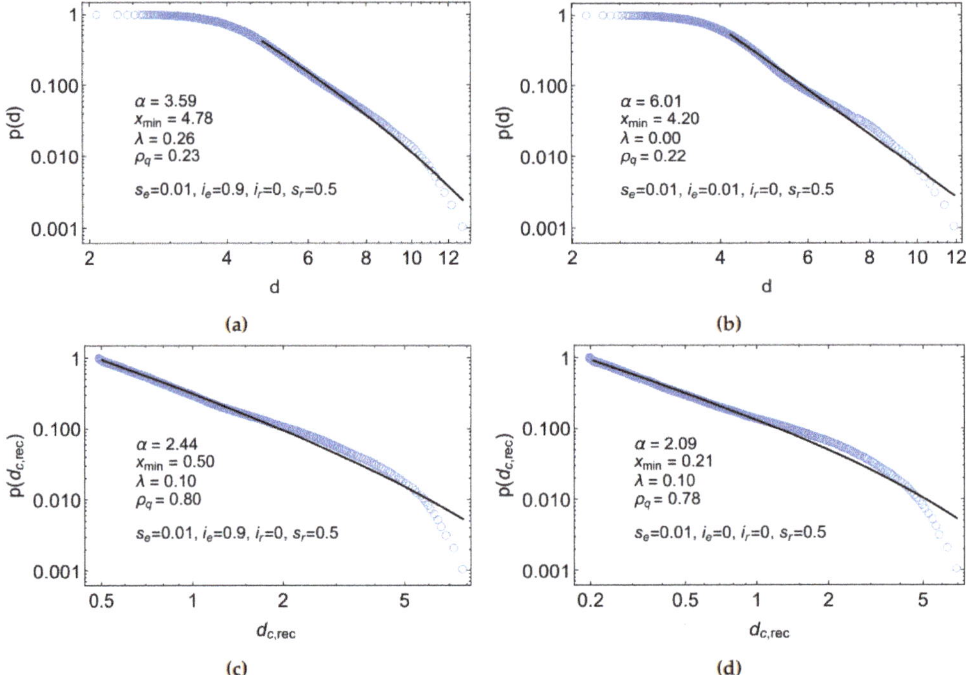

Figure 4. Log-log scale plots of the degree d (top) and the critical degree $d_{c,rec}$ (bottom) distributions in CCM. The black lines represent the corresponding truncated power law fits. The insets show the fit parameters as well as the scenario configurations. The plots (a) and (b) differ by the scenario configurations shown in the insets; similarly for (c) and (d). These scenarios are among the top five swarm performances with respect to N_{coll}. Please note that λx_{min} is relatively small, i.e., power law is a good fit for x close to x_{min}. See Supplementary Material Section 3 for plots of d in CCM over all 256 scenario configurations.

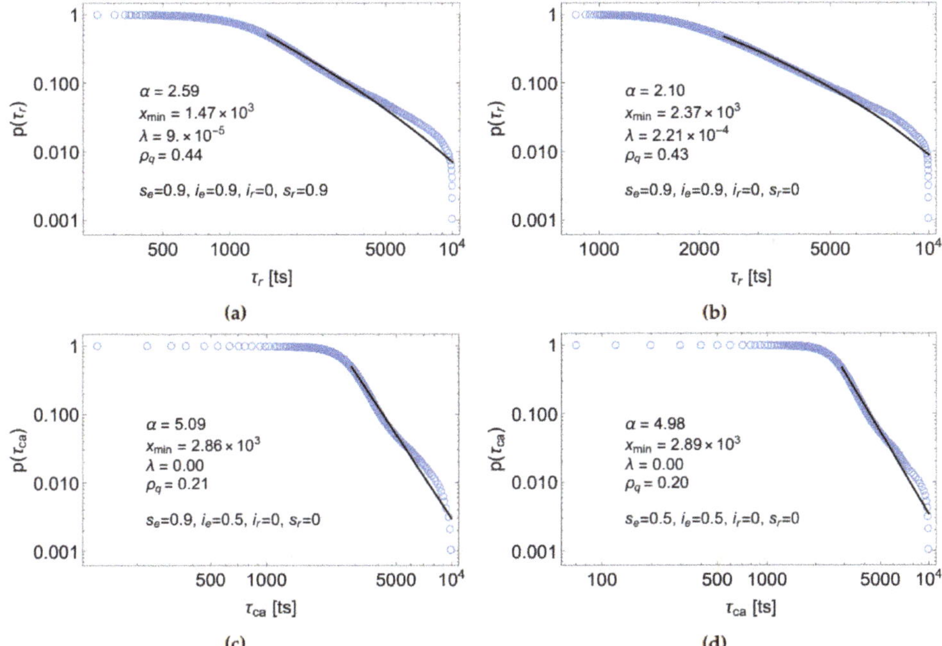

Figure 5. Log-log scale plots of the resting time τ_r (top) and the collision avoidance time τ_{ca} (bottom) distributions in DCM. The black lines represent the corresponding truncated power law fits. The insets show the fit parameters as well as the scenario configurations. The plots (**a**,**b**) differ by the scenario configurations shown in the insets; similarly for (**c**,**d**). These scenarios belong to a subset of the best swarm performance with respect to ω_{ca}. Please note that $\lambda = 0$ for the fits of τ_{ca}, indicating better support for the power law fit than for truncated power law.

The third performance measure, i.e., ω_e, reached its best values over the NS_e region. Its maxima correspond to cases where $s_e = 0$ and $s_r = 0$. Interestingly, for these scenario configurations we found fits with moderate to strong support for (truncated) power law to the data of Δd, i.e., the change of the average communication degree of the robot (examples shown in Figure 6). This is an interesting finding because it indicates that a communication feature may be power law distributed also in those scenarios in which the swarm tries to minimize the number of foraging robots and maximize the number of resting ones. Moreover, in most Δd distributions with strong or moderate support for the power law, the fit includes only 10–20% of data points. The reason for the relatively low ratio of power law fitted data is that the tail of the distribution is likely to represent by the fraction of robots that rest or move close to the border between the nest and the foraging area.

In general, the findings suggest that internal cues (in the absence of social cues) keep robots at the edge of minimal activity while social cues (in the absence of internal cues) drive the robots towards maximal activity.

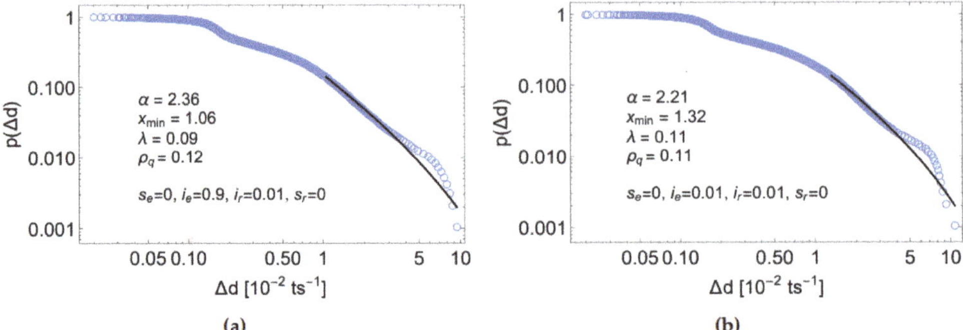

Figure 6. Log-log scale plots of Δd per 100 ts in CCM. The black lines represent the corresponding truncated power law fits. The insets show the fit parameters as well as the scenario configurations. The plots (**a**,**b**) differ by the scenario configurations shown in the insets. These scenarios are among the best swarm performances with respect to ω_e.

3.2. Correlation with Swarm Performance

In the previous section we have illustrated that swarm performance is likely to reach its peaks over cue configurations that include asymptotically scale-free space or time features (see Figure 3). In this section, we analyze this observation statistically, using correlation measures such as the Pearson and Kendall's tau rank correlation coefficients introduced in Section 2.4.3. However, note that both correlation measures have strengths and shortcomings. On the one hand, while the Pearson correlation coefficient is widely used and has an elegant mathematical form, it is sensible to outliers and may not be appropriate for non-linear distributions. On the other hand, Kendall's tau is suitable for non-linear distributions as well as robust to outliers. Nevertheless, reducing the values to ranks may disregard the significance of the variable's value being far from the average. In particular, replacing the real value of the quality ratio ρ_q by its rank leads to loss of information about the extent to which ρ_q represents the quality of the power law distribution. Moreover, following the definition of Kendall's tau in Equations (11)–(15), each difference between data point pairs is assigned the same weight which may not always be appropriate. For instance, consider the ranked swarm performance in terms of N_{coll} and the corresponding distribution of ρ_q in CCM for d in Figure 7a and for $d_{c,sent}$ in Figure 7b, respectively. In both cases, Kendall's tau defined by Equations (11)–(15) returns values indicating no correlation (i.e., $\tau_{Kendall} = 0.02$ and $\tau_{Kendall} = -0.04$, respectively). However, as evident in Figure 7, both cases show different dynamics, with ρ_q for d following N_{coll} more closely than for $d_{c,sent}$. The main reason is that the dominant fluctuations of ρ_q close to zero are assigned the same weight (i.e., rank step 1) as the more permanent increase of ρ_q for high values of N_{coll}. Similar considerations hold for the other features and the DCM. To account for this type of behavior, we use a generalization of Equation (15) that weights the ranking steps by a parameter κ, which is relative to the average change, such that:

$$\delta_{i,j}^{(c)} = \begin{cases} \kappa, & \text{if } sgn(x_i - x_j)sgn(y_i - y_j) > 0 \\ 0, & \text{else} \end{cases}$$
$$\delta_{i,j}^{(x)} = \begin{cases} \kappa, & \text{if } x_i = x_j \\ 0, & \text{else} \end{cases} \tag{15}$$

and similarly, for $\delta_{i,j}^{(d)}$ and $\delta_{i,j}^{(y)}$. The weight parameter κ is given by

$$\kappa = \frac{1}{2}\left(\frac{|x_i - x_j|}{\mu_x} + \frac{|y_i - y_j|}{\mu_y}\right), \tag{16}$$

where μ_x and μ_y are averages over all $|x_i - x_j|$ and $|y_i - y_j|$, respectively. For each i, j pair, Equation (16) considers the data distances of both distributions, normalized by their respective mean distances. Consequently, κ does not favor any distribution and weights each ranking step relative to other distances. Please note that in general, there is no correlation metric that is perfectly adequate for all types of studies and data distributions; it is thus common to consider appropriate modifications [67–70]. In the present case, for $\kappa = 1$ we obtain the standard Kendall's tau rank correlation coefficient described in Section 2.4.3. However, by implementing Equation (16), the correlation coefficient is less sensitive to fluctuations than the standard Kendall's tau, while still being more robust to outliers and non-linearity than the Pearson correlation measure. Therefore, in the following we will use this modified Kendall's tau rank correlation coefficient to investigate the presence of correlations between ρ_q and swarm performance.

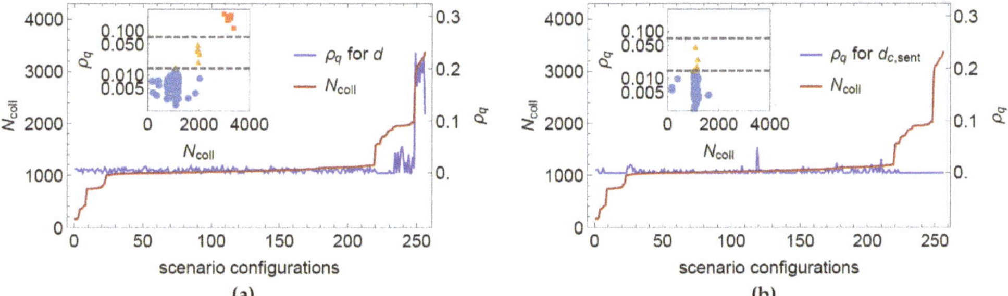

Figure 7. Ranked distribution of N_{coll} (dark red, left y-axis). For the same cue configurations, the CCM distributions of ρ_q (right y-axis) for (**a**) d and (**b**) $d_{c,sent}$ are shown in blue, respectively. The insets depict the corresponding scatter plots with data points representing weak (circles), moderate (triangles) and strong (squares) support for power law distribution; gray lines indicate the onsets of the different support classifications (similar to Figure 2).

The correlations are shown for all features in Table 5 between the three measures of the swarm performance and the feature 'scale-freeness' quantified by ρ_q. We found strong correlations of the scale-free property of various features with the swarm performance. In particular, high correlations exist for τ_{ca}, τ_r, Δd in DCM; and, additionally, for $d, d_{c,rec}$ in CCM. Remarkably, for those features for which we found moderate or high correlation values (highlighted in blue in Table 5), most high-quality power law fits appear in the same scenarios as the highest swarm performance peaks. The red dots in Figure 3 illustrate this finding by highlighting the scenarios in which the quality ratio is $\rho_q > \frac{0.5}{7}$. Moreover, the swarm tends to demonstrate low performance with respect to N_{coll} and ω_{ca} for those scenarios in which ω_e is highest, the latter being well correlated with Δd.

Table 5. Correlation coefficients quantifying correlations of ρ_q with N_{coll}, ω_{ca} or ω_e.

Property	Correlation with N_{coll}	Correlation with ω_{ca}	Correlation with ω_e
	Discontinuous communication mode (DCM)		
d	0.11	0.17	−0.03
Δd	−0.19	−0.33	0.64
$d_{c,sent}$	0.03	−0.05	−0.03
$d_{c,rec}$	0.14	0.29	−0.51
τ_{ca}	0.76	0.80	0.06
τ_r	0.74	0.77	0.08
τ_f	0.27	0.26	−0.20
τ_h	0.22	0.28	−0.44
all	0.40	0.42	0.02
	Continuous communication mode (CCM)		
d	0.50	0.56	0.10
Δd	−0.02	−0.16	0.51
$d_{c,sent}$	−0.15	−0.39	0.54
$d_{c,rec}$	0.44	0.44	0.15
τ_{ca}	0.75	0.80	−0.13
τ_r	0.59	0.67	−0.02
τ_f	0.26	0.29	−0.24
τ_h	0.11	0.17	−0.27
all	0.40	0.41	0.05

A high correlation coefficient means: the better the quality of (truncated) power law distribution, the higher the likelihood that the swarm performed well. Cells highlighted in blue show moderate or strong correlations.

The correlation coefficients confirm the observation, supported by the data shown in Figures 2 and 3, that most power law distributions with strong support (i.e., high ρ_q) appear in scenarios with peak swarm performance. To further examine this observation, we consider the correlations of the swarm performance with different ρ_q support classifications (based on Table 3). As Table 6 shows, there are moderate and strong positive correlations between features with strong support for power law distribution and swarm performance in terms of N_{coll} and ω_{ca} for both communication modes. This suggests that the observation of scale-free features is more likely in scenarios in which the agents are more successful in retrieving a high number of food items.

Table 6. Correlation coefficients between ρ_q and N_{coll}, ω_{ca} or ω_e for different power law support classifications.

Support for pl	Correlation with N_{coll}	Correlation with ω_{ca}	Correlation with ω_e
	Discontinuous communication mode (DCM)		
weak	0.18	0.18	−0.16
moderate	0.10	0.08	0.21
strong	0.31	0.40	−0.14
moderate + strong	0.50	0.53	0.22
	Continuous communication mode (CCM)		
weak	0.10	0.11	0.03
moderate	0.37	0.33	0.03
strong	0.66	0.63	−0.25
moderate + strong	0.74	0.76	−0.02

Correlation of the swarm performance with different categories of power law distribution support. Cells highlighted in blue show moderate or strong correlation values.

3.3. The Role of Feedback Mechanisms in the Emergence of Scale-Free Features

An attribute of complex systems that is widely known to support the emergence of scale-free characteristics is the presence of (positive and negative) feedback loops [1,53,71]. We specify the feedback effect to be positive or negative based on the individual response to the information input from its neighborhood. Hence, we refer to the feedback mechanism as *positive* feedback if it pushes the individuals to the same state as the state of the majority, whereas *negative* feedback pushes them away from it.

Most scale-free features were found in scenarios in which (i) the robot behavior was dominated by social interactions, (ii) the swarm attempted to balance positive and negative feedback and (iii) the swarm displayed a tendency towards active exploration. In particular, in CCM, the first 17 scenarios sorted by ρ_q in descending order were found over the LS_e region and with $i_r = 0$. Similarly, in DCM, the first 28 scenarios were found over the HS_e region and with $i_r = 0$. To understand this, it is necessary to consider in more detail the impact of each cue on swarm dynamics and the feedback mechanisms.

For conciseness, we focus our system analysis on CCM and its most relevant set of parameter configurations. Similar conclusions hold for DCM. In particular, we can simplify our analysis based on the repeating patterns of swarm performance (see Figure 3) and the following observations: (i) The swarm performance is qualitatively very similar between the cue values 0.5 and 0.9 (for all four cues). Thus, we focus, in the following, on $\{a : s_e, i_e, i_r, s_r \in \{0.0, 0.01, 0.5\}\}$. (ii) The cue i_e has a negligible impact on the foraging dynamics when $s_e > 0$. By neglecting scenarios in which $i_e \neq 0$, except those with $s_e = 0$, we can further shorten the set of relevant scenarios. Finally, (iii) there are significant differences in the dynamics between scenario configurations with $i_r = 0$ and those with $i_r > 0$ but negligible differences between $i_r = 0.01$ and $i_r = 0.5, 0.9$. Thus, we focus, in the following, only on scenarios with either $i_r = 0$ or $i_r = 0.01$. Figure 8 shows the final set of 24 scenarios relevant to the discussion below.

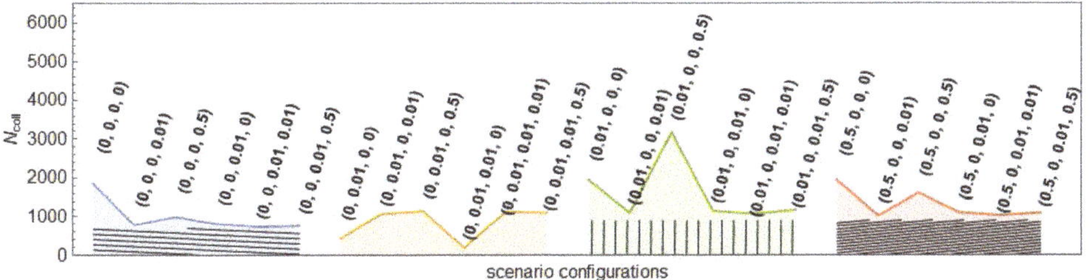

Figure 8. Number of collected items for a selected set of 24 scenario configurations a in CCM. The data labels show the corresponding cue values of $a = (s_e, i_e, i_r, s_r)$.

Please note that (ii) and (iii) are consequences of the internal cues i_e and i_r acting only on *exploring* robots. In the exploring state, the crucial parameter is $p_{e \to r}$ because it defines the probability to stop exploring and change to resting. A non-zero value of the internal cue i_r has a substantial impact on dynamics as it alters $p_{e \to r}$ after each exploration attempt. As the likelihood of finding and retrieving a food item is low, i_r mostly reduces $p_{e \to r}$. The more $p_{e \to r}$ is lowered by i_r, the less likely the robot is to find a food item during the next exploration attempt. Thus, i_r has a strong inhibitory influence on the swarm's exploration activity. Consequently, there is a significant difference in swarm performance between the scenario configurations with $i_r = 0$ and those with $i_r > 0$. As the swarm actively attempts to explore the environment and collect food items, the influence of $i_r > 0$ can be considered an important driver of negative feedback. In contrast, $p_{r \to e}$ acts only on *resting* robots. Consequently, any change of $p_{r \to e}$ through i_e is easily distorted by s_e, i.e., the social interactions with the neighborhood of the resting robot. Hence, only when $s_e = 0$ there is an inhibitory impact of i_e on swarm dynamics (similar to i_r, due to the scarcity of food items), otherwise i_e is negligible. In short, when the probability of finding a food item is low, with $i_r = 0$ and $s_e > 0$ enabling the swarm to significantly damp the feedback mechanisms that drive the swarm towards inactivity.

Next, we consider the particular contributions of social cues s_e and s_r. Social interactions represent a direct form of feedback loops, enabling the swarm to drift towards an absorbing state (e.g., uninterrupted resting or exploring) or maintain a balance between positive and negative feedback. In general, note that high values of s_e often lead to $p_{r \to e} = 0$ due to the high probability of encountering a robot with a failed exploration (due to the low density of food items). With such relatively high values of s_e, $p_{r \to e}$ can be reduced to zero within a few time steps. By contrast, with $s_e = 0.01$, $p_{r \to e}$ does not fluctuate as strongly. Similar considerations hold for s_r and $p_{e \to r}$. In terms of active exploration, i.e., long exploration times and high number of retrieved items, it is beneficial for the swarm to have robots with $p_{r \to e} = 1$ and $p_{e \to r} = 0$. Indeed, in the present system we observe that the swarm approaches such behavior for $s_r = 0.5$ and $s_e = 0.01$ (i.e., over the LS_e region). More importantly, such balance of s_r and s_e allows positive and negative feedback loops to coexist with the positive feedback being slightly more dominant. Due to this feedback coexistence, a robot that happened to be surrounded by unsuccessful neighbors will tend to have low $p_{r \to e}$ and high $p_{e \to r}$, i.e., its resting time τ_r will increase (together with its d or $d_{c,rec}$) and vice versa. Over time, such dynamics will result in robots that are increasingly inactive (with increasingly higher τ_r, d or $d_{c,rec}$) and robots that are increasingly active (with increasingly lower τ_r, d or $d_{c,rec}$). When the majority tends towards active exploration, the inactive group of robots experiences negative feedback and while the active group is subject to positive feedback. The prevalence of the positive feedback decreases the number of consistently resting robots significantly below the number of consistently exploring ones. Ultimately, this leads to skewed or heavy tailed distributions, such as the power law and, consequently, to the emergence of scale-free features. Similar considerations apply to the DCM over the HS_e region. The difference is that in DCM each robot can broadcast its exploration result only once. Thus, s_e needs to have high values for dynamics similar to CCM to emerge.

To illustrate the above considerations, let us examine the scenario configuration of CCM with $s_e = 0.01$, $i_e = 0.9$, $i_r = 0.0$, and $s_r = 0.5$ (see Figure 4a), in which a high performance value of N_{coll} was observed (i.e., this scenario is similar to the peak in Figure 8). The value of s_r has a high impact on $p_{e \to r}$ (the probability to transition to resting). For example, if a robot receives at least two 'success' messages more than 'failure' messages—i.e., if $\delta_\eta(t_r) \geq 2$ in Equation (2)—its $p_{e \to r}$ drops to zero. When $p_{e \to r} = 0$ and $i_r = 0.0$, the robot will stop exploring only if it finds an item. During the subsequent resting, this robot is likely to cause one of its neighbors to reach $p_{e \to r} = 0$, which repeats an analogous cycle of events. The corresponding dynamics can be translated in terms of the positive feedback pushing the robots out of the nest and increasing the number of robots in the foraging area (i.e., in the exploring state). In the long term, due to the positive feedback, the swarm drifts towards the absorbing state in which all robots have $p_{e \to r} = 0$ and $p_{r \to e} = 1.0$. In the short term, while most robots is exploring, some robots remain in the nest, e.g., due to crowding at the entrance of the nest. Those robots have a higher number of neighbors because the nest is significantly smaller than the foraging area. Therefore, during this crowding behavior, the swarm experiences the coexistence of positive and negative feedback loops. A specific balance between these feedback loops may lead to the emergence of scale-free features such as the space feature d (for which, indeed, the above mentioned scenario configuration has one of the best truncated power law fits with $\rho_q \approx 0.23$, shown in Figure 4a). Similar considerations hold for other CCM examples presented in Figure 4 or the DCM examples in Figure 5.

The above example demonstrates positive feedback regarding the *exploring state*. However, under some configurations, positive feedback can also be observed around the *resting state*. For example, during the crowding behavior in the nest, a robot which is surrounded by a high number of resting neighbors is likely to get 'stuck' and be unable to leave the nest. This robot will eventually switch to the resting state and broadcast a 'failure' message. Neighbors that receive this message will decrease their probability to explore $p_{r \to e}$, and, through physical interference, lower the neighboring robots' chances to leave the nest. Consequently, positive feedback at the resting state may lead, in the long term, to an increase of the average communication degree of the resting robots and the emergence of power law distributed features, alongside the occurrence of outstanding robots whose features such as Δd (examples shown in Figure 6), τ_r or τ_{ca} exhibit exceptionally above-average values.

4. Conclusions

In this paper, we have investigated the interplay between scale-free features and swarm dynamics of a foraging swarm. Our results demonstrate that in the studied system (i) several space and time features tend to be asymptotically scale-free for multiple parameter configurations; (ii) the emergence of scale-free features can be attributed to the presence of positive/negative feedback mechanisms. Furthermore, (iii) in several cases the swarm performance is moderately or strongly correlated with the tendency of space and time features to follow the power law distribution—which is the mathematical backbone of the scale-free property.

This study serves as a first step towards a better understanding of the interplay between the presence of scale-free features and the swarm behavior in terms of collective performance. Although our results do not indicate a causal relationship, we found conclusive evidence for a close connection between scale-free features and swarm performance. Moreover, our analysis of power law distributed features shows a strong link between the microscopic behavior of robots determined by specific cues and the macroscopic behavior of the entire swarm exhibiting peak performance. However, care needs to be taken when considering cases where the power law fit includes only a small fraction of data, as focusing on a small subgroup that plausibly displays scale-free features may disregard a significant piece of information about the global swarm behavior.

Please note that the presented exploratory study was conducted with an emphasis on *whether* scale-free features may emerge autonomously in artificial multi-agent systems, without focusing on *why* they do so. Hence, more sophisticated work is needed to precisely understand the exact causes for the emergent scale-free characteristics in our systems. For instance, strong feedback mechanisms may push the system close to a critical point at which a phase-transition occurs. In case of a *continuous* phase-transition, the latter is known to be associated with the emergence of scale-free features [1,53]. In fact, using approximations (such as the assumption of a well-mixed system) it could be shown analytically that the social or internal cues can be used as control parameters, moving the system between its phases (e.g., phases in which the number of resting robots is minimized or maximized). However, if the system approaching a phase-transition is the cause for the emergence of scale-free features in our experiments, we expect to find a correlation length (i.e., the distance over which one robot influences another, directly or indirectly) that is longer than the size of the system (e.g., the length of the nest). In contrast, our preliminary analysis indicated the opposite behavior: as we approached those scenarios in which scale-free characteristics were observed, the correlation length decreased below the system size. In general, a detailed finite-size-scaling analysis is necessary to explain our findings more thoroughly as well as reveal which (physical) boundaries are most relevant and what impact they have on the system dynamics.

The canonical foraging task continues drawing scientific attention due its importance and prevalence in nature as well as artificial systems. In addition, the complexity of collective foraging as a combination of several sub-behaviors allows the modeling and analysis of a large number of scenarios and examine various features. For these reasons, we focused exclusively on the foraging behavior. However, it would be interesting to extend the scope and investigate other multi-agent tasks, such as aggregation or flocking, from the same perspective as in our study thereby broadening the understanding of scale-free phenomena in artificial collective systems.

Supplementary Materials: The following are available at http://www.mdpi.com/2076-3417/9/13/2667/s1.

Author Contributions: Conceptualization, Y.K.; Formal analysis, I.R.; Funding acquisition, P.S.; Investigation, I.R.; Methodology, I.R. and Y.K.; Project administration, P.S.; Resources, P.S.; Supervision, Y.K. and P.S.; Validation, I.R.; Visualization, I.R.; Writing—original draft, I.R.; Writing—review and editing, I.R., Y.K. and P.S.

Funding: This research received no external funding.

Conflicts of Interest: The authors declare no conflict of interest.

References

1. Khaluf, Y.; Ferrante, E.; Simoens, P.; Huepe, C. Scale invariance in natural and artificial collective systems: A review. *J. R. Soc. Interface* **2017**, *14*, 20170662. [CrossRef] [PubMed]
2. Cavagna, A.; Cimarelli, A.; Giardina, I.; Parisi, G.; Santagati, R.; Stefanini, F.; Viale, M. Scale-free correlations in starling flocks. *Proc. Natl. Acad. Sci. USA* **2010**, *107*, 11865–11870. [CrossRef] [PubMed]
3. Huepe, C.; Ferrante, E.; Wenseleers, T.; Turgut, A.E. Scale-Free Correlations in Flocking Systems with Position-Based Interactions. *J. Stat. Phys.* **2015**, *158*, 549–562. [CrossRef]
4. Chen, X.; Dong, X.; Be'er, A.; Swinney, H.L.; Zhang, H.P. Scale-Invariant Correlations in Dynamic Bacterial Clusters. *Phys. Rev. Lett.* **2012**, *108*, 148101. [CrossRef] [PubMed]
5. Boyer, D.; Ramos-Fernández, G.; Miramontes, O.; Mateos, J.L.; Cocho, G.; Larralde, H.; Ramos, H.; Rojas, F. Scale-free foraging by primates emerges from their interaction with a complex environment. *Proc. R. Soc. B Biol. Sci.* **2006**, *273*, 1743–1750. [CrossRef] [PubMed]
6. Reynolds, A.M.; Ouellette, N.T. Swarm dynamics may give rise to Lévy flights. *Sci. Rep.* **2016**, *6*, 30515. [CrossRef] [PubMed]
7. Albert, R.; Barabási, A.L. Statistical mechanics of complex networks. *Rev. Mod. Phys.* **2002**, *74*, 47–97. [CrossRef]
8. Fewell, J.H. Social Insect Networks. *Science* **2003**, *301*, 1867–1870. [CrossRef]
9. Kasthurirathna, D.; Piraveenan, M. Emergence of scale-free characteristics in socio-ecological systems with bounded rationality. *Sci. Rep.* **2015**, *5*, 10448. [CrossRef]
10. Goh, K.I.; Kahng, B.; Kim, D. Universal Behavior of Load Distribution in Scale-Free Networks. *Phys. Rev. Lett.* **2001**, *87*, 278701. [CrossRef]
11. Cohen, R.; Havlin, S. Scale-Free Networks Are Ultrasmall. *Phys. Rev. Lett.* **2003**, *90*, 058701. [CrossRef] [PubMed]
12. Thivierge, J.P. Scale-free and economical features of functional connectivity in neuronal networks. *Phys. Rev. E* **2014**, *90*, 022721. [CrossRef] [PubMed]
13. Albert, R.; Jeong, H.; Barabási, A.L. Error and attack tolerance of complex networks. *Nature* **2000**, *406*, 378–382. [CrossRef] [PubMed]
14. Liu, W.; Winfield, A.F.T.; Sa, J.; Chen, J.; Dou, L. Towards Energy Optimization: Emergent Task Allocation in a Swarm of Foraging Robots. *Adapt. Behav.* **2007**, *15*, 289–305. [CrossRef]
15. Pinciroli, C.; Trianni, V.; O'Grady, R.; Pini, G.; Brutschy, A.; Brambilla, M.; Mathews, N.; Ferrante, E.; Di Caro, G.; Ducatelle, F.; et al. ARGoS: A modular, parallel, multi-engine simulator for multi-robot systems. *Swarm Intell.* **2012**, *6*, 271–295. [CrossRef]
16. Pini, G.; Brutschy, A.; Pinciroli, C.; Dorigo, M.; Birattari, M. Autonomous task partitioning in robot foraging: An approach based on cost estimation. *Adapt. Behav.* **2013**, *21*, 118–136. [CrossRef]
17. Brambilla, M.; Ferrante, E.; Birattari, M.; Dorigo, M. Swarm robotics: A review from the swarm engineering perspective. *Swarm Intell.* **2013**, *7*, 1–41. [CrossRef]
18. Bayındır, L. A Review of Swarm Robotics Tasks. *Neurocomput.* **2016**, *172*, 292–321. [CrossRef]
19. Khaluf, Y.; Dorigo, M. Modeling robot swarms using integrals of birth-death processes. *ACM Trans. Auton. Adapt. Syst. (TAAS)* **2016**, *11*, 8. [CrossRef]

20. Khaluf, Y.; Pinciroli, C.; Valentini, G.; Hamann, H. The impact of agent density on scalability in collective systems: Noise-induced versus majority-based bistability. *Swarm Intell.* **2017**, *11*, 155–179. [CrossRef]
21. Yang, G.Z.; Bellingham, J.; Dupont, P.E.; Fischer, P.; Floridi, L.; Full, R.; Jacobstein, N.; Kumar, V.; McNutt, M.; Merrifield, R.; et al. The grand challenges of Science Robotics. *Sci. Robot.* **2018**, *3*. [CrossRef]
22. Ferrante, E.; Turgut, A.E.; Huepe, C.; Stranieri, A.; Pinciroli, C.; Dorigo, M. Self-organized flocking with a mobile robot swarm: A novel motion control method. *Adapt. Behav.* **2012**, *20*, 460–477. [CrossRef]
23. Khaluf, Y.; Rausch, I.; Simoens, P. The Impact of Interaction Models on the Coherence of Collective Decision-Making: A Case Study with Simulated Locusts. In *Swarm Intelligence*; Dorigo, M., Birattari, M., Blum, C., Christensen, A.L., Reina, A., Trianni, V., Eds.; Springer International Publishing: Cham, Switzerland, 2018; pp. 252–263.
24. Correll, N.; Martinoli, A. Modeling and designing self-organized aggregation in a swarm of miniature robots. *Int. J. Robot. Res.* **2011**, *30*, 615–626. [CrossRef]
25. Deneubourg, J.L.; Goss, S. Collective patterns and decision-making. *Ethol. Ecol. Evol.* **1989**, *1*, 295–311. [CrossRef]
26. Sumpter, D.J. The principles of collective animal behaviour. *Philos. Trans. R. Soc. B Biol. Sci.* **2005**, *361*, 5–22. [CrossRef] [PubMed]
27. Batalin, M.A.; Sukhatme, G.S. Spreading Out: A Local Approach to Multi-robot Coverage. In *Distributed Autonomous Robotic Systems 5*; Asama, H., Arai, T., Fukuda, T., Hasegawa, T., Eds.; Springer: Tokyo, Japan, 2002; pp. 373–382.
28. Ren, H.; Tse, Z.T.H. Investigation of Optimal Deployment Problem in Three-Dimensional Space Coverage for Swarm Robotic System. In *Social Robotics*; Ge, S.S., Khatib, O., Cabibihan, J.J., Simmons, R., Williams, M.A., Eds.; Springer: Berlin/Heidelberg, Germany, 2012; pp. 468–474.
29. Mclurkin, J.; Smith, J. Distributed algorithms for dispersion in indoor environments using a swarm of autonomous mobile robots. In *Distributed Autonomous Robotic Systems 6*; Asama, H., Alami, R., Chatila, R., Eds.; Springer: Tokyo, Japan, 2007; pp. 399–408.
30. Sharma, S.; Shukla, A.; Tiwari, R. Multi robot area exploration using nature inspired algorithm. *Biol. Inspired Cogn. Archit.* **2016**, *18*, 80 – 94. [CrossRef]
31. Schmickl, T.; Crailsheim, K. A Navigation Algorithm for Swarm Robotics Inspired by Slime Mold Aggregation. In *Swarm Robotics*; Şahin, E., Spears, W.M., Winfield, A.F.T., Eds.; Springer: Berlin/Heidelberg, Germany, 2007; pp. 1–13.
32. Gauci, M.; Chen, J.; Dodd, T.J.; Groß, R. Evolving Aggregation Behaviors in Multi-Robot Systems with Binary Sensors. In *Distributed Autonomous Robotic Systems*; Ani Hsieh, M., Chirikjian, G., Eds.; Springer: Berlin/Heidelberg, Germnay, 2014; pp. 355–367.
33. Pitonakova, L.; Crowder, R.; Bullock, S. Information flow principles for plasticity in foraging robot swarms. *Swarm Intell.* **2016**, *10*, 33–63. [CrossRef]
34. Bullock, S.; Crowder, R.; Pitonakova, L. Task Allocation in Foraging Robot Swarms: The Role of Information Sharing. In Proceedings of the The 2018 Conference on Artificial Life: A Hybrid of the European Conference on Artificial Life (ECAL) and the International Conference on the Synthesis and Simulation of Living Systems (ALIFE), Cancun, Mexico, 4–8 July 2016; pp. 306–313.
35. Ostergaard, E.H.; Sukhatme, G.S.; Matari, M.J. Emergent bucket brigading: A simple mechanisms for improving performance in multi-robot constrained-space foraging tasks. In Proceedings of the Fifth International Conference on Autonomous Agents, Montreal, QC, Canada, 28 May–1 June 2001; pp. 29–30.
36. Hamann, H.; Wörn, H. An Analytical and Spatial Model of Foraging in a Swarm of Robots. In *Swarm Robotics: Second International Workshop, SAB 2006, Rome, Italy, September 30–October 1, 2006, Revised Selected Papers*; Şahin, E., Spears, W.M., Winfield, A.F.T., Eds.; Springer: Berlin/Heidelberg, Germany, 2007; pp. 43–55.
37. Hoff, N.R.; Sagoff, A.; Wood, R.J.; Nagpal, R. Two foraging algorithms for robot swarms using only local communication. In Proceedings of the 2010 IEEE International Conference on Robotics and Biomimetics, Tianjin, China, 14–18 December 2010; pp. 123–130.

38. Hecker, J.P.; Letendre, K.; Stolleis, K.; Washington, D.; Moses, M.E. Formica ex Machina: Ant Swarm Foraging from Physical to Virtual and Back Again. In *Swarm Intelligence*; Dorigo, M., Birattari, M., Blum, C., Christensen, A.L., Engelbrecht, A.P., Groß, R., Stützle, T., Eds.; Springer: Berlin/Heidelberg, Germany, 2012; pp. 252–259.
39. Castañeda Cisneros, J.; Pomares Hernandez, S.E.; Perez Cruz, J.R.; Rodríguez-Henríquez, L.M.; Gonzalez Bernal, J.A. Data-Foraging-Oriented Reconnaissance Based on Bio-Inspired Indirect Communication for Aerial Vehicles. *Appl. Sci.* **2017**, *7*. [CrossRef]
40. Wang, H.; Li, Y.; Chang, T.; Chang, S.; Fan, Y. Event-Driven Sensor Deployment in an Underwater Environment Using a Distributed Hybrid Fish Swarm Optimization Algorithm. *Appl. Sci.* **2018**, *8*, 1638. [CrossRef]
41. Hernández-Ocaña, B.; Hernández-Torruco, J.; Chávez-Bosquez, O.; Calva-Yáñez, M.B.; Portilla-Flores, E.A. Bacterial Foraging-Based Algorithm for Optimizing the Power Generation of an Isolated Microgrid. *Appl. Sci.* **2019**, *9*, 1261. [CrossRef]
42. Lerman, K.; Galstyan, A. Mathematical Model of Foraging in a Group of Robots: Effect of Interference. *Auton. Robot.* **2002**, *13*, 127–141. [CrossRef]
43. Khaluf, Y.; Birattari, M.; Rammig, F. Analysis of long-term swarm performance based on short-term experiments. *Soft Comput.* **2016**, *20*, 37–48. [CrossRef]
44. Hoff, N.; Wood, R.; Nagpal, R. Distributed colony-level algorithm switching for robot swarm foraging. In *Distributed Autonomous Robotic Systems*; Springer: Berlin/Heidelberg, Germany, 2013; pp. 417–430.
45. Schafer, R.J.; Holmes, S.; Gordon, D.M. Forager activation and food availability in harvester ants. *Anim. Behav.* **2006**, *71*, 815 – 822. [CrossRef]
46. Prabhakar, B.; Dektar, K.N.; Gordon, D.M. The Regulation of Ant Colony Foraging Activity without Spatial Information. *PLoS Comput. Biol.* **2012**, *8*, 1–7. [CrossRef] [PubMed]
47. Pinter-Wollman, N.; Bala, A.; Merrell, A.; Queirolo, J.; Stumpe, M.; Holmes, S.; Gordon, D. Harvester ants use interactions to regulate forager activation and availability. *Anim. Behav.* **2013**. *86*, 197–207. [CrossRef]
48. Seeley, T.D.; Visscher, P.K.; Schlegel, T.; Hogan, P.M.; Franks, N.R.; Marshall, J.A.R. Stop Signals Provide Cross Inhibition in Collective Decision-Making by Honeybee Swarms. *Science* **2012**, *335*, 108–111. [CrossRef]
49. Reina, A.; Valentini, G.; Fernández-Oto, C.; Dorigo, M.; Trianni, V. A Design Pattern for Decentralised Decision Making. *PLoS ONE* **2015**, *10*, 0140950. [CrossRef]
50. Valentini, G.; Ferrante, E.; Hamann, H.; Dorigo, M. Collective decision with 100Kilobots: Speed versus accuracy in binary discrimination problems. *Auton. Agents Multi-Agent Syst.* **2016**, *30*, 553–580. [CrossRef]
51. Khaluf, Y.; Birattari, M.; Rammig, F. Probabilistic analysis of long-term swarm performance under spatial interferences. In Proceedings of the International Conference on Theory and Practice of Natural Computing, Cáceres, Spain, 3–5 December 2013; pp. 121–132.
52. Hamann, H.; Valentini, G.; Khaluf, Y.; Dorigo, M. Derivation of a micro-macro link for collective decision-making systems. In *Proceedings of the International Conference on Parallel Problem Solving from Nature*; Springer: Cham, Switzerland, 2014; pp. 181–190.
53. Newman, M.E. Power laws, Pareto distributions and Zipf's law. *Contemp. Phys.* **2005**, *46*, 323–351. [CrossRef]
54. Clauset, A.; Shalizi, C.R.; Newman, M.E.J. Power-Law Distributions in Empirical Data. *SIAM Rev.* **2009**, *51*, 661–703. [CrossRef]
55. Broido, A.D.; Clauset, A. Scale-free networks are rare. *Nat. Comm.* **2019**, *10*, 1017. [CrossRef] [PubMed]
56. Alstott, J.; Bullmore, E.; Plenz, D. Powerlaw: A Python Package for Analysis of Heavy-Tailed Distributions. *PLoS ONE* **2014**, *9*, e85777. [CrossRef] [PubMed]
57. Viswanathan, G.M.; Afanasyev, V.; Buldyrev, S.; Murphy, E.; Prince, P.; Stanley, H.E. Lévy flight search patterns of wandering albatrosses. *Nature* **1996**, *381*, 413–415. [CrossRef]
58. Galton, F. *The Journal of the Anthropological Institute of Great Britain and Ireland*; Number v. 15; Anthropological Institute of Great Britain and Ireland: London, UK, 1886.
59. Pearson, K. *Proceedings of the Royal Society of London*; Number v. 58; Taylor & Francis: Abingdon, UK, 1895.
60. Lee Rodgers, J.; Nicewander, W.A. Thirteen ways to look at the correlation coefficient. *Am. Stat.* **1988**, *42*, 59–66. [CrossRef]

61. Kraaikamp, F.D.C.; Meester, H.L.L. *A Modern Introduction to Probability and Statistics*; Springer: Berlin/Heidelberg, Germany, 2005.
62. McDonald, J.H. *Handbook of Biological Statistics*; Sparky House Publishing: Baltimore, MD, USA, 2009; Volume 2.
63. Wang, Y.; Li, Y.; Cao, H.; Xiong, M.; Shugart, Y.Y.; Jin, L. Efficient test for nonlinear dependence of two continuous variables. *BMC Bioinform.* **2015**, *16*, 260. [CrossRef] [PubMed]
64. Spearman, C. The Proof and Measurement of Association between Two Things. *Am. J. Psychol.* **1904**, *15*, 72–101. [CrossRef]
65. Kendall, M.G. A new measure of rank correlation. *Biometrika* **1938**, *30*, 81–93. [CrossRef]
66. Christensen, D. Fast algorithms for the calculation of Kendall's τ. *Comput. Stat.* **2005**, *20*, 51–62. [CrossRef]
67. Blest, D.C. Theory & Methods: Rank Correlation—An Alternative Measure. *Aust. N. Z. J. Stat.* **2000**, *42*, 101–111.
68. Darken, P.F.; Holtzman, G.I.; Smith, E.P.; Zipper, C.E. Detecting changes in trends in water quality using modified Kendall's tau. *Environmetrics* **2000**, *11*, 423–434. [CrossRef]
69. Emond, E.J.; Mason, D.W. A new rank correlation coefficient with application to the consensus ranking problem. *J. Multi-Criteria Decis. Anal.* **2002**, *11*, 17–28. [CrossRef]
70. Sengupta, D.; Maulik, U.; Bandyopadhyay, S. Entropy steered Kendall's tau measure for a fair Rank Aggregation. In Proceedings of the 2011 2nd National Conference on Emerging Trends and Applications in Computer Science, Shillong, India, 4–5 March 2011; pp. 1–5.
71. Khaluf, Y.; Hamann, H. On the definition of self-organizing systems: Relevance of positive/negative feedback and fluctuations. In Proceedings of the Swarm Intelligence: 10th International Conference, ANTS 2016, Brussels, Belgium, 7–9 September 2016; Springer: New York, NY, USA, 2016; Volume 9882, p. 298.

Article

Situated Psychological Agents: A Methodology for Educational Games

Michela Ponticorvo [1,*], Elena Dell'Aquila [2], Davide Marocco [1] and Orazio Miglino [1,3]

1 Natural and Artificial Cognition Lab, University of Naples "Federico II", 80133 Naples, Italy; davide.marocco@unina.it (D.M.); orazio.miglino@unina.it (O.M.)
2 Department of Political Sciences, University of Naples "Federico II", 80133 Naples, Italy; elena.dellaquila@unina.it
3 Institute of Cognitive Sciences and Technologies, National Research Council, 00185 Rome, Italy
* Correspondence: michela.ponticorvo@unina.it; Tel.: +39-081-2535465

Received: 4 October 2019; Accepted: 12 November 2019; Published: 14 November 2019

Featured Application: The present paper introduces a methodology based on situated psychological agents that can be fruitfully applied to design and implement educational games, as it permits to represent the flows inside the game on the educational, psychological, and pedagogical level while detailing agents' features at a psychological level.

Abstract: In recent years, the ever-increasing need for valid and effective training to acquire competences in multiform contexts has led to a wide diffusion of educational games (EG). In spite of their diffusion, there is still a need to reflect on the design process that should embed the games' pedagogical potential and the instructional process in the entertainment scope. Moreover, as building EG, especially in digital environments, is an enterprise that involves specialists with different expertise, it can be useful to have a shared methodology that is easily understandable and usable by many users. In this paper, we propose to use situated psychological agents (SPA) as a methodology to design and build effective EG and show how to represent games in terms of SPA and their interactions by diagrams and describe different examples of how this approach has been applied.

Keywords: educational games; game design; situated psychological agents; education; competences

1. Introduction

Education is a key step and challenge in every society as successfully preparing future citizens in terms of knowledge, skills, and competences strongly affects the competitiveness at an individual and collective level. Not by chance, European Commission is tracing indications to strengthen human capital [1] so that everyone may have a key set of competencies that allow personal fulfillment and include transversal skills, such as digital competence and entrepreneurship competence.

Competences are more easily acquired through pedagogical models that favor the active involvement of the learner in the acquisition process [2–4] thus, opening the way to innovative educational strategies, including educational games (EG) [5–8]. Moreover, the introduction of information and communication technologies has led to a revolution in education concerning different aspects, for example, the tools that can be used for education, the places where education can happen, the possibility to interact with an incomparable higher amount of learning resources and educational figures. ICT has brought to the evolution of new approaches such as technology-enhanced learning and game-based learning [9,10].

It means, for example, that a learner, child or adult, can now access not only books, one of the most used learning sources for a very long time, but also additional multimedia contents, simulations, and social media to obtain information. The world where education takes place is no more limited to the physical classroom but is expanded to cover potentially an unlimited world, in terms of space and time, that is totally or partially virtual. If we consider the tools that have been introduced in the educational context, it seems that, in spite of their wide diffusion, there is still a need to reflect on the methodology to design, implement, and use them in a learning scenario.

In more detail, it can be a useful reflection on the methodology to design and implement educational games, as an effective methodology should permit to have a shared formal representation of the main game elements, of their connections and their interactions.

In what follows, we will propose a methodology that meets this requirement, which is based on the situated psychological agents (SPA) approach, connecting it to the EG design process.

2. Educational and Serious Games: The Design Process

Serious games are games that educate, train, and inform, to use the title of the highly cited paper by Michael and Chen [5]. Since the first book by Abt [11], which referred to card and board games, serious games have become digital and have strongly affirmed their educational potentials [12,13]. This fact forces to critically reflect on this kind of games, as they intervene in the learning process in a way that has not been faced yet in traditional learning theory and typical game theory as well. Many remarkable theoretical frameworks have been proposed, also recently, to satisfy this need. Among these, it is useful to cite the learning mechanics-game mechanics model [14], which draws a set of pre-defined game mechanics and pedagogical elements abstracted from the literature and connects them to identify the main pedagogical and entertainment features of a game. This work highlights the fundamental game mechanics and how they are translated into learning mechanics. Also noteworthy is the activity theory-based model of serious games [15] that provides a useful representation of EG, with game elements, their connections and their contribution to pedagogical goals achievement. These works have the value of trying to answer the question of how the concrete components of the game should be structured to support learning and which elements are crucial to address the design process.

The design process, indeed, is extremely important in determining EG success.

The starting point of each designing enterprise is to clarify what is the goal to be reached. In the case of games, the questions are "What is the goal of the game? What the player can do?". In the case of entertainment games, some typical objectives are:

Erase: The player must "eliminate" the opponent, such as chess and checkers game;

Solve: The player must find a solution to a puzzle or answer a question: Examples are Cluedo or Trivial Pursuit;

Chase: The player runs towards or away from someone or something. One famous example is the video game Super Mario Bros;

Build: The player has to build something: A house, a city, an empire, as in The Sims, Civilization, and Age of Empires.

On the other hand, if our goal is to build EG, where the educational aspect is crucial, the designers have to specify also "What is the learning goal of the game? How it can be achieved by actively involving the player?" In this case, help can come from learning theory. For example, Bloom's updated taxonomy [16,17], which is commonly used to describe learning goals and includes remembering, understanding, applying, analyzing, evaluating, and creating, can help focalize what the EG aims at leading the player to. After the learning and game objectives are defined, the design must define three fundamental and interconnected levels: The shell level, the core level, and the educational level. It represents a multi-level approach for design, where there are two concentric levels, the shell, and core level and a ubiquitous one, the educational level, also named evaluation and tutoring level [2]. The level interconnections are represented in Figure 1.

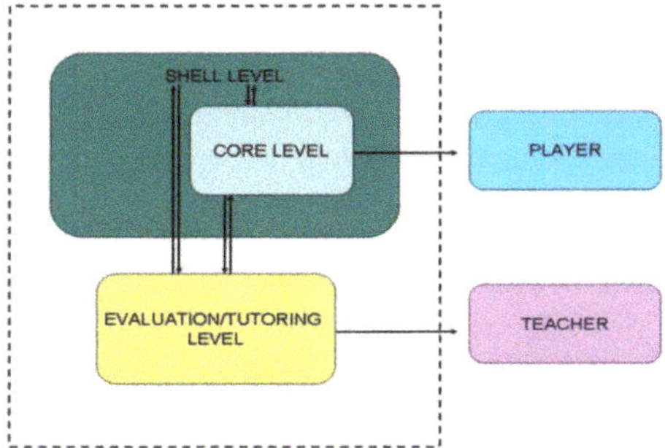

Figure 1. A multi-level approach for game design with shell, core, and educational level. Shell and core levels can be found in every kind of game, whereas the educational level characterizes educational games.

2.1. The Shell (Game Narrative) and Core (Game Mechanics) Levels

The shell and the core levels are present in every game, and in almost every cultural product as well. The shell level represents the visible content that is immediately accessible to the player. It frames the game dynamics within the core level. The educational level, even if it is present in many entertainment games, is explicitly characterized in educational and serious games, as it allows, on the teacher's side, to understand if and how the player/learner has acquired the concepts conveyed by the EG and, in some cases, directly intervenes in the learning process.

At the shell level, superficial and visible, we find the game narrative. EGs, like many other cultural products, are expressed through a narrative metaphor. It is, therefore, important to define who are the characters, what actions they can perform, what interactions are possible between characters, and the environment within which those actions take place. If we adopt a theatrical jargon, the plot, the scenario, the roles, the setting are aspects to be defined.

It is widely recognized that narration is a key aspect in human cognition [18], and it is, therefore, possible to find it in a wide variety of cultural products, such as fairy tales, movies, news, to cite some. Games, as cultural products, share this feature and then narration is present in games too [19]. As an example, the characters can be two armies in the chess game, the scenario can be a futuristic world in a videogame, the plot can be an interaction between relatives in a role-playing game. In the well-known game of Monopoly, for example, a pair of dice are rolled to move a player's piece around the board. Buying and trading properties mean to represent real estate trading that strongly helps to engage the player in the negotiation. The shell level, where narrative resides, keeps a hidden level with specific mechanisms and rules: This hidden level is called the core. Adopting a term that is commonly used in the context of videogame creation and development, this deep level is the game engine [20]. The game engine allows implementing core functionalities related to game dynamics, for example, related to physics, animation, artificial intelligence, etc.

These levels interact: One level can have strong effects on the other. The narrative provides a framework where the hidden content lives, as it was in a shell, as suggested by the name.

In the context of EG, the shell level is essential to provide a semantic context to the educational activities, whereas the core level is related to skills, abilities, competences to be transferred, and to the relevant learning objectives. It is interesting to note, as hinted at before, that the concepts of core and shell levels are present in every kind of game, not just in educational games.

2.2. The Educational Level

In EG, a relevant role is played by the educational level, which includes evaluation and tutoring activities, with the explicit educational goal to allow students to accomplish specific learning outcomes. It is, therefore, important to pay attention to the design process of such a key function. Together with this, all design decisions at all levels should be harmonized in order to provide a meaningful learning tool, as shown in Figure 1.

At the educational level, the evaluation function analyzes players' game performances relative to the specified training objectives and provides the players and the trainer with important information and data about the learning process. At this level, we find learning analytics, which is the measurement, collection, analysis, and reporting of data about learners, intending to improve the learning process as well as the environment in which it occurs. Despite some challenges that can derive from the effort of introducing learning analytics in EG, nonetheless, studies report that this effort can be useful to achieve greater effectiveness and measurements of progress in learning [21–23]. Paraphrasing Siemens's words [24], learning analytics is the use of intelligent, learner produced data, and analysis models to discover information and social connections for predicting and advising people's learning. From the teacher's perspective, this level is fundamental because it supplies specific tools and functions to support the training process.

3. Agents in EG and the SPA Approach

The ubiquitous presence of interacting artificial and real actors at each level, together with the importance of the narrative, recalls the theatrical metaphor already presented for the shell level. From the educational point of view, this metaphor is extremely powerful to represent interactions between the various actors of the educational process in EG. However, the theatrical metaphor effectively applies to all kinds of educational games only when agent's conception and design is based and inspired to psychological models, as they ultimately make choices, take decisions, and act within the environment they live in [2].

Indeed, the various actors populating the different stages of an EG, at the shell level (users, learners), core level (interactions between actors), and educational level (trainers, educators, tutors) can be represented as agents with different features and functions. If we think of EG, it is evident that the people involved in the learning process are a key element both on the educational and game side. Almost every educational situation is characterized by interactions between the learner, at the center, and the people involved in the whole educational processes, both in formal (teachers, educational designers, tutors etc.) [25] and informal contexts (parents, peers, etc.) [26–28]. Nevertheless, the kinds of interactions that specify the educational settings can be varied and show specific nuance that every methodology aiming at modeling the educational process should take into account.

By looking at Figures 2 and 3 we can see two different implementations of an educational process. The first one is usually observable in children who learn with a teacher through multisensory experience. It is characterized by well structured educational materials [29–31], e.g., a Montessori-inspired classroom [32] and a well-structured environment. In this case, the teacher can be modeled as an agent that we can call generically trainer and directly affects learners' activities. In this view, the environment within which the learner acts as the playground of the learning process. The trainer provides external guidance and support during the play, thus allowing a full understanding at the cognitive level.

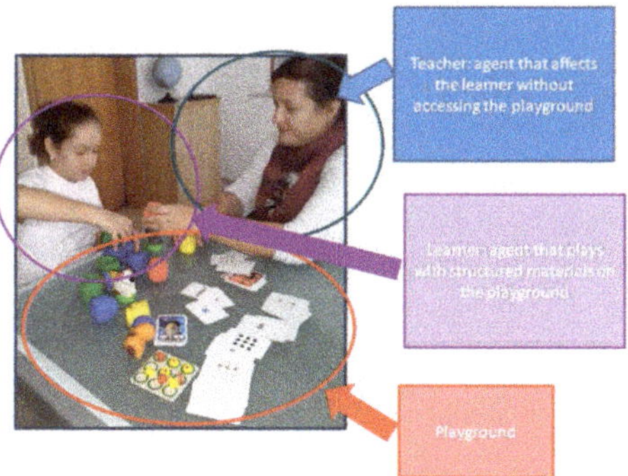

Figure 2. Learning interactions in a Montessori-type environment.

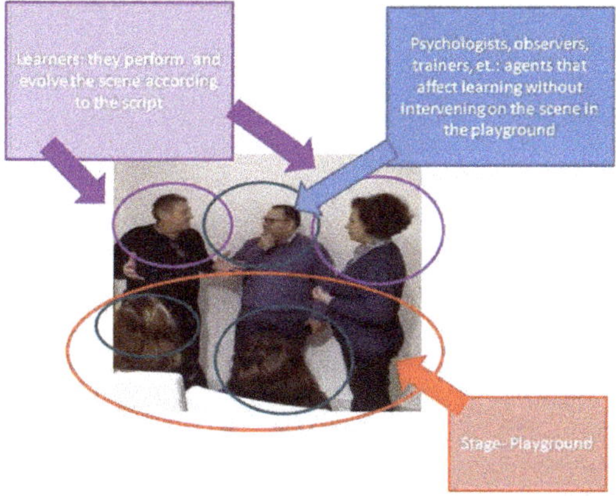

Figure 3. Moreno role-playing games.

The second situation comes from Moreno's role-playing games [33] where the learner acts exactly as an actor, by evolving the scene on the stage according to a given script.

Role-playing games simulate a social situation in which users are asked to cover and interpret specific roles to develop a certain competence, such as effective communication or negotiation [34–38]. Here, learners, as actors, according to a specific script, perform and develop their actions on a stage. In this way, the stage represents the playground where the learning takes place [29]. Behind the stage, the psychologist, the trainer, or the observers, which all can be seen as agents interacting with the learner, can provide guidance and support, affecting the learning environment, though not directly intervening on the playground.

These interactions always happen inside the game and can be partially or completely virtual if the agents are ruled by artificial intelligence [39–41].

However, the relevant elements of these learning situations are useful to define a more general methodology:

The playground: A space (physical and/or conceptual) that delimitates the actions of one or more learners. The playground is defined by the narrative structure. It can contain objects (physical and/or conceptual) that can be manipulated by the learner.

Learners: The learners can act in the playground, changing its state directly. They can be considered agents that are situated, immersed, in a scene of the play, and can select autonomously the actions that modify the playground in the function of their psychology, including cognition, emotion, etc.

Trainers: Teachers or people who have educational, training, or assessment functions affect directly or indirectly the learners but cannot modify the playground state.

In the design process of EG, it is, therefore, necessary to keep in mind the following elements:

(1) definition of the narrative structure with the necessary agents;
(2) definition of the actions and interactions modalities of the agents;
(3) definition of the agents' control system, whether human-controlled or guided by a set of rules or AI systems.

Learners and other characters present at the shell level, and therefore, belonging to the narrative of the game, can be called on-stage agents (OSA), as they directly interact and affect the core level according to their specific endowment. The BSA can interact with the game indirectly, by affecting OSA actions and are mainly present at the educational level. This distinction was firstly introduced by Dell'Aquila and colleagues [2] and Ponticorvo and colleagues [29]. Moreover, it can be also adopted at two different levels: 1) The first, related to the educational material (in EG interacting elements can be conceptualized in the form of agents), 2) and the second, to the learning scenario, where learners are conceived as agents interacting with other agents (real or artificial), thus defining the educational environment.

Taken together, the description of the design process and the focus on interacting agents and playground, are the main elements of the SPA approach for educational games, which allows addressing the EG design both at a high level of abstraction and at a high level of detail.

Therefore, the SPA are agents with different characteristics: OSA can directly act on the playground or BSA if they externally interact with the OSAs within a well-defined educational process. They are situated, as they are present and somehow "immersed" in the educational process, being in the playground or in the overall narrative structure. They are psychological, as they are endowed with cognitive and emotional features: In the case of human agents, it is automatic that agents have a psychological characterization. In the case of AI-controlled agents, it is possible to take inspiration from psychological theories and models to define their psychological characteristic and behaviors. It is useful to underline that the agents share the same context. Thus, there is a shared meaning between the actors involved in the learning situation.

The SPA approach, at the shell level, identifies the game characters, their characteristics, and interactions. At the core level, each agent is accurately defined as to its sensory and action endowments, i.e., what the agent perceives and what action it can perform. These actions must follow the game rules that are defined both by setting constraints and by the agent actions defined in the core level itself.

SPA can be useful for EG design because all the interacting entities within the game can be represented as agents, some immersed in the playground, and some not. Both the players and the roles that guide the learning process from backstage (psychologists, teachers, or trainers) can be conceived as agents. Players become OSA, and contour figures become agents with specific functions, from supervision or score recording to observation, tutoring, advising or mentoring. Considering the different levels described in Section 2, we can say that, at the shell level, an EG is a mise-en-scene of a plot by one or more agents interacting in a well and formally defined setting. On the core level, actions performed by OSA directly modify the game state, whereas BSA supports OSA at the educational level. It is possible to identify a clear separation between the shell level and the core one: The visible dimension can be conceived through traditional narrative techniques, and the core level, expressed in terms of SPA, implies the formal definition of the various game components that we have introduced.

The Educational Level in the SPA Approach

In this section, we will focus on the educational level in the SPA approach. BSA are the main characters at the educational level: They may have the function to support the learners involved in EG, mainly as OSA. BSA does not intervene directly in the playground but provides what is required to enhance the learning process. These agents cover specific roles, functional to the achievement of educational goals. Educational or learning goals are inspired by a specific learning theory, such as the already cited work by Bloom or Kolb [42] who emphasized concrete experience, active experimentation, reflective observation, and abstract conceptualization. A meaningful learning process is characterized by the presence of feedback, as giving (and receiving) feedback is essential to understand how close learners are to the defined learning goals. Feedback, together with debriefing are regarded as the most important element for maximizing the learning process [43], as they guide learners through a reflective process about their learning [44], offer a space for giving personal meaning to the learning experience [45], and help to relate this learning experience to real-life contexts. In the SPA approach, feedback is provided by BSA, and in the case of digital games, it can come both from real or artificial tutors. In digital games, their role is essential to provide learners with short feedback cycles through which they can get continuous and immediate information regarding the effect of their actions on the game interactions. Conversely, in traditional educational approaches where teachers generally have to mark students' work using conventional means (i.e., manually), there is a significant delay until students can receive the appropriate information regarding some aspects of their task. Digital EG can help to reduce such delays almost to zero. Moreover, feedback is offered throughout the full game session. A very important moment for delivering feedback is at the end of the game, during the debriefing phase, when the learners receive feedback about the overall performance. It is also a process which gives the opportunity to analyze what dynamics occurred during the game, what went wrong, and was achieved, and share experiences with other people, making it possible to compare different perspectives from other players or from other people involved in the learning process, such as tutors.

The SPA approach to developing EG opens a way to adopt software based on artificial intelligence systems to model the interactions between OSA and BSA and allows to conceive these complex interactions between agents as finalized to a meaningful learning process through feedback and debriefing activities. The tutor can be a human being, but also a virtual entity, thanks to artificial intelligence. In both cases, it is crucial that the tutor observes and traces behaviors, actions, reactions of learners during the game, similarly to what happens when a student performs a task or takes a test in face to face situations to create a learner's profile. To create such a profile, educational games can rely on a wide amount of data available, even more than in real life-oriented tasks. Digital games offer a system able to record every single action performed by the players, the time required by each action, as well as not effective choices made. Thanks to all this information, both real and virtual tutors can operate various analyses, to understand the cognitive state of the learner, thus implementing learning analytics. It is reasonable to hypothesize that real and virtual tutors can be even more effective when they jointly operate, as the virtual tutor can record a significant amount of data and provide immediate feedback, which is impossible to achieve from a human tutor, and the human tutor can supervise and actively guide the learning process in such a way that is, at the moment, very difficult, if not impossible, to achieve by a virtual agent.

The tutoring agents, both human or artificial, carry out various roles in different moments and at a different level. At the beginning they can select and decide which roles of the game will be played by each actor, also according to learning objectives and to previous results achieved. During the game, the OSA interacts with the narrative at the shell level and the game space level, while a BSA can help to maintain a high interaction level. At the end of the game, the tutor can build an individualized report regarding the overall interactions that occurred, record achievements and failures together with a preferred way to act, react and interact to build a detailed user profile. This report can also be useful to further customize the game/player interaction.

In the next section, we will introduce some examples of EG and present the design process that led to them by means of diagrams with formal notation.

4. SPA Applications to EG

In this section, we will report three relevant examples of how the SPA approach can be applied to EG design with particular attention to the formal representation of game elements. To this end we will use the following notation: The playground is represented as an empty rectangle, the circle represents OSA, and the square represents BSA. If the boundary is a full line, the agent is real, and if it is represented by a dashed line, the agent is artificial. The interactions are represented as lines: A continuous line represents a direct interaction, whereas the dashed line represents an indirect one, arcs represent feedback.

4.1. Block Magic

Block Magic [46,47] is an educational platform that exploits augmented reality based on RFID/NFC technology that allows building custom educational games with both physical and digital components. It consists of a set of magic blocks, a magic board/tablet device, and specific software (Figure 4). Magic blocks are an augmented version of traditional logic blocks, widespread structured materials, classically used in education. The technologies employed to augment are RFID/NFC sensors that allow to unite the manipulative approach, stimulated by logic blocks, and touch-screen technologies. An RFID system consists of an antenna and a transceiver, which can read the radio frequency and transfer the information to a device, and a small and low-cost tag, which is an integrated circuit containing the RF circuitry and information to be transmitted.

Figure 4. The Block Magic kit.

This configuration permits to a PC or a table, with BM software installed on, to connect with BM Magic Table, another relevant BM material. The Magic Table has a hidden antenna that recognizes each block, sends a signal to the PC/tablets, and produces feedback coherently with pupils learning path.

Each augmented magic block had an integrated/attached passive RFID sensor for wireless identification of every single block. A specially designed wireless RFID reader device, an active board, is used, which can read the RFID of a block and transmit the result to the BM software engine.

On the software side, the BM augmented blocks together with the Magic Table are complemented with software that includes a series of already-developed exercises and an authoring tool to build new ones.

The BM software engine is mainly formed by two parts: The first one is devoted to receiving input from the active board and generating an "action" (aural and visual). These actions implement the direct feedback the user can receive interacting with the system. This feedback is regulated by an embedded intelligent tutoring system [48,49] that ensures autonomous interaction between the user and the system, receiving active support, corrective indications, feedback, and positive reinforcement from the digital assistant on the outcome of the actions performed.

The second software component is devoted to customization, and it is dedicated to teachers, educators, etc., allowing them to build their exercises to be proposed to the child, focusing the attention on the skills the child needs to train more.

In BM, the narrative comprises the plot, the scenario, the characters, the setting and has the task of attracting the player and filling the game experience with meaning. The appropriate narrative allows to attract the child, so to immerse him/her in a completely different environment, that is relevant in every educational context. The narrative level exercises a framing effect on the core level.

The core is configured as an interaction between the player, the human OSA (or the players in a collective scenario), the teacher, another human OSA, and the artificial BSA. The interactions are mediated by physical materials: The Magic blocks.

Block Magic Representation in SPA Terms

From BM general description, we can move to BM description in terms of the SPA approach. As represented in Figure 5, in this case, we have two human OSA interacting: A learner and a teacher. Many important functions are played by the BSA, which is artificial. It provides feedback to the player (arc on the left) during the game, it affects the human OSA teacher proposing existing exercises and recording learner's interaction, it has an indirect effect on the learner OSA through the trainer OSA. The BSA is built according to adaptive tutoring systems theories [46,47].

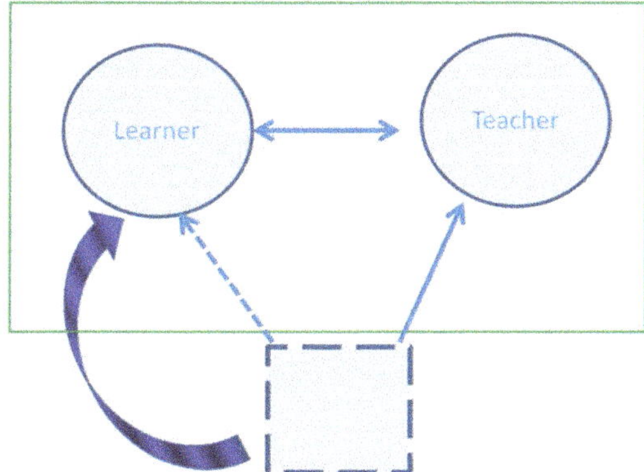

Figure 5. Block Magic represented in the SPA notation (the playground is represented as an empty rectangle, circles represent on-stage agents (OSA), and the square represents the BSA. Full-lined boundary indicates a real agent, dashed line indicates an artificial agent. A continuous line represents a direct interaction, a dashed line represents an indirect one, arcs represent feedback).

To validate the SPA approach, some data were collected with BM users. During the BM project (www.blockmagic.eu), trials were run in four schools in European Countries and involved about 250 students and 10 teachers of primary school and kindergarten. The teachers used pre-defined exercises but could also build their ones using the BM authoring tool, which is based on the SPA approach. After this process, researchers administered to teachers a structured questionnaire with 10 questions on a five-point Likert scale (5 indicating the most positive attitude) about the design process with SPA.

Results indicate that the design was facilitated by the SPA: In particular, it was appreciated for the possibility to quickly define interactions in the exercise (average = 4.30, st. dev. = 0,48; the point 5 in the scale corresponds to the most positive evaluation), to define the functions, especially the educational one, in the game in terms of agents (average = 4.10, st. dev. = 0,74) and to share in an easy and manageable way the idea with other professionals involved in the process (average = 4.40, st. dev. = 0.52). More details are reported in [46].

4.2. Enact: An EG to teach negotiation

Another example of SPA used to develop EGs is represented by Enact, implemented on a platform, based on recent psychological modeling through the application of current ICT research such as e-learning, mobility, internet, artificial intelligence [50,51].

The platform facilitates "learning by doing" experiences as the training scenarios that can be developed through EG can simulate real-life situations, and due to their verisimilitude, can enable the transfer of what has been learned to similar real-life contexts [52,53], developing the specific negotiation competence.

It is a single-player game designed to train users on effective communication and negotiation skills. A training scenario is populated by two 3D avatars, one controlled by the user and the other by the computer (the BOT), both able to express a range of communication aspects and elements by using verbal cues (e.g., vocal tone, shape of the speech bubble, and structure of the sentence), and non-verbal indicators (e.g., body posture, facial expression, eye contact, and gestures). These patterns of behavioral indicators have been identified in the communication model of assertiveness, passivity, and aggression [54].

On-stage agents within Enact are both the learner and the artificial agent with which the user interacts with during the game (see Section 4.2.1). OSAs perform their roles and interact with each other according to the theoretical principles of the five styles of handling interpersonal conflict proposed by Rahim and Bonoma [55] and Rahim [56] the psychological model adopted and underpinning the Enact game.

In other words, the main principles of the two theoretical psychological models of negotiation by Rahim and communication by Dryden and Constantinou underpinning the game, represent the rules defined in the core level that determine the OSAs' psychological and physical features. Rahim model differentiated five different styles of handling conflict on two basic dimensions: Concern for self and concern for others. The first dimension explains the degree (high or low) to which a person attempts to satisfy his or her own concern, while the second explains the degree (high or low) to which a person attempts to satisfy the concern of others. The combination of the two dimensions results in five styles of handling interpersonal conflict: Integrating, obliging, compromising, avoiding, dominating.

The five styles of handling interpersonal conflicts are described, as follows:

Avoiding (low concern for self and others) has been associated with withdrawal, buck-passing, or sidestepping situations.

Obliging (low concern for self and high concern for others) is associated with attempting to play down the differences and emphasize commonalities to satisfy the concern of the other party.

Dominating (high concern for self and low concern for others) has been identified with a win-lose orientation or with forcing behavior to win one's position.

Compromising (intermediate in concern for self and others) involves give-and-take where both parties give up something to make a mutually acceptable decision.

Integrating (high concern for self and others) involves openness, exchange of information and examination of differences to reach an effective solution acceptable to both parties.

Moreover the conflicting scenarios have been designed according to a series of variables which combination resulted in 25 different conflicting scenarios animated by 24 different characters, such as type of conflict (if based on divergence or convergence), gender (if player and agent have the same or opposite gender, so that the interactions can result as male-male (or female-female) and male-female (or female-male), and ethnic variables (to allow a user-avatar interaction covering different ethnic groups).

The user is introduced to the game with a scene explaining the conflicting situation, the role assigned to the user and her goal within the given scenario (shell level). Each exchange between the user and BOT is organized in a five-state scene (one for each of Rahim's styles of handling conflicts), which includes one turn of speech for each party. Each exchange is related to a gesture and/or facial expression that shows the way the sentence will be communicated to the BOT (core level).

After the user's answer, the BOT computes it according to the embedded psychological models, that is, for example, a dominating BOT will show predominantly aggressive and authoritative behaviors. Conversely, an obliging BOT will show an overall passive and submissive attitude towards the negotiation (Figure 6).

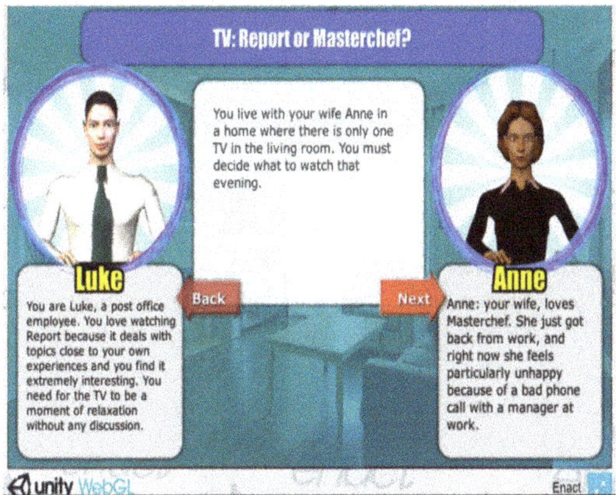

Figure 6. OSAs interacting at the shell level at the beginning of Enact. Introduction to the different OSAs.

The user starts the game by pressing the "play" button that brings the player on the game scene: The user's avatar is presented in a small window at the left upper corner of the screen, while the BOT represents the main character focused on by the camera.

The user's five possible choices are shown below the small avatar window, while the responses of the BOT are shown in the text bubble appearing over its head.

When the mouse is over one of the five user sentences (on the left-hand side of the screen), the animation (non-verbal behavior) related to that sentence is shown in the top-left window.

The innovative aspect of the Enact game is represented by its assessment feature that complements the training aspect. It implements soft skills measurements with an innovative rigorous psychometric approach, that offers the users the opportunity to assess her/his handling conflict styles, along with her negotiation and communication skills.

The assessment within Enact corresponds to the core of what we have defined as evaluation/tutoring level and represents the playful way through which the user can be assessed in a standardized manner according to the abovementioned Rahim's model.

The assessment of the player is based on the preferred negotiation styles used during a series of negotiation scenarios, given the description of the five styles provided by Rahim, and "pen and pencil" ROCI II instrument developed by the author. ROCI II is designed to measure the five independent dimensions of the styles of handling interpersonal conflict (integrating, obliging, dominating, avoiding, and compromising). The instrument contains three Forms A, B, and C to measure how a person handles her (his) conflict with her (his) supervisor, subordinates, and peers, respectively.

The Enact assessment is also fundamental for the automatic elaboration of a training strategy tailored to the specific development areas of the player, to create an effective learner-centered environment, where the user activity is focused on the areas of behavior that mostly require improvements.

Enact profiles resulted from user's game experiences are correlated with those obtained by the users through the administration of the ROCI ROCI-II (Rahim organizational conflict inventory-II). For this reason, the Enact tool has been designed to return a score directly comparable with the ROCI-II to produce scores for each of the five styles of handling conflict contemplated in Rahim's model: Collaborating, accommodating, dominating, avoiding, and compromising. In addition to the ROCI-II form C, the other four psychological tests have been administered: (a) A short version of BIG five personality inventory, (b) assertive efficacy test, (c) self-efficacy test, and (d) coping test. The aim was also to investigate possible relationships between high scores of self-efficacy and relevant personality traits with the styles adopted by the Enact users and related positive effects on negotiation processes observed within the game sessions.

All the test takers had to play Enact and fill in the electronic form containing the five psychological tools in a row, in random order so to avoid bias related to the order of presentation. The users were asked to negotiate with an avatar in 10 different scenarios. The assessment took about one hour.

The system collects the data about the user's behavior and choices and creates a model of the player that will then be used for generating tailored information to be used in the training session. The score and profile of the player's negotiation skills are actually calculated by summing the independent concern for self and concern for other variables gathered during interactions, which are represented within every sentence that the user can choose.

In the assessment session, the artificial agent's behavior is static, not adaptive, and reflects a specific negotiation style for each of the scenarios.

The tutoring system is available only after the assessment has been completed. Thus, it will intervene during the training scenarios and at the end of the game session in order to provide useful information to the user about his or her performance related to the BOT he or she is currently interacting with and to his or her general behavior when managing conflicting situations.

The user is given a profile based on the Rahim model related to the specific situations he or she played, together with advice about how to improve the efficacy of his or her communication and the changes achieved since the assessment profiling.

The profile emerges mainly through a comparison of the behavior of the user and the style of the artificial agent she interacted with.

Furthermore, we have highlighted the importance of offering the user with immediate feedback about his/her performances. An example of immediate feedback is provided in the Enact game session by the on-stage agent (Figure 7).

Figure 7. Examples of verbal and nonverbal indicators expressed by OSAs during the conflicting interaction.

The BOT, which the user interacts with, displays immediateness of the interaction with an aggressive, assertive, obligingness facial expressions and body posture (non-verbal communication) and gives verbal feedback through the text (Figure 7).

4.2.1. Enact Representation in SPA Terms

In the Enact game, as shown in Figure 8, there are two on-stage agents: The learner who plays the game and the artificial agent with which the user interacts in the scenario. The human OSA performs his/her role according to his/her negotiation style, whereas the artificial OSA acts according to the implementation of Rahim's principles. The OSAs interact directly with the other with questions and answers. In this case, there is a BSA, a tutor who is artificial and interacts directly with the artificial OSA and indirectly with the human OSA. It is outside the playground but affects directly the artificial OSA and indirectly the human OSA.

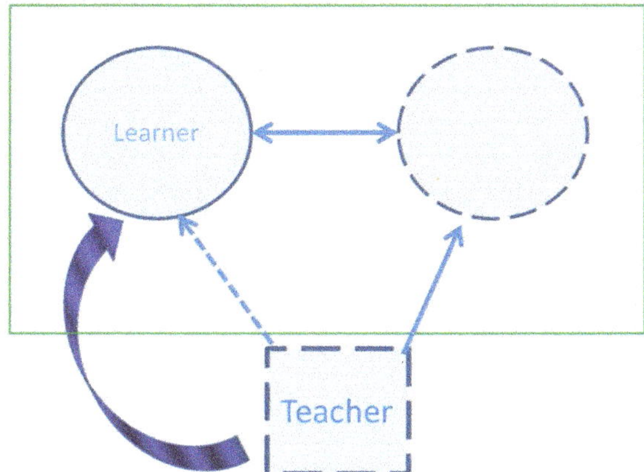

Figure 8. Enact game represented in the SPA approach (the playground is represented as an empty rectangle, circle represent OSA and square represent BSA. Full-lined boundary indicates a real agent, dashed line indicates an artificial agent. A continuous line represents a direct interaction, a dashed line represents an indirect one, arcs represent feedback).

At the end of the game, it also provides human OSA with relevant feedback.

During the Enact project, also the effectiveness of the SPA approach was investigated. Indeed, the Enact game was pre-validated in two iterations: The first one allowed to collect feedback by the means of a questionnaire on the quality of the interface and the BOT. The questionnaire was composed of eight questions on a five-point Likert scale. The complete results are reported in [2–53].

Data showed that the overall feedback was extremely positive. The second iteration involved the participants playing with different scenarios, and then a questionnaire of 13 questions on a Likert scale was administered. Additionally, in this case, the feedback provided by users was positive.

On the qualitative side, the people involved in the design and implementation of the Enact game, using the SPA was useful and allowed them to efficiently collaborate with other professionals involved in the game development.

4.3. *Eutopia*

4.3.1. Eutopia: EG to Train Soft Skills Based on Role-playing Mechanisms

Eutopia represents a specific application of SPA to develop EGs, as it is not just a game but rather a platform with which it is possible to create an unlimited number of role-playing games.

Eutopia platform can acknowledge many years of experience underpinning several European projects, such as Sisine, Sinapsi, Eutopia-Mt, Proactive, and S-cube project. Eutopia has been used and tested in different contexts and by different group targets (university, training institutions and agencies, MEs and SMEs, public administration, as well as non-governmental organizations and social enterprises) and for the development of various kinds of competencies (negotiation, international mediation, negotiation, communication, leadership, team building, time management, motivation, decision making, and problem solving).

Eutopia takes inspiration from the technology used in multiplayer games and embeds role-play methodology as a psycho-pedagogical approach.

The underpinning learning approach is based on open dynamics so that there is not an exclusive way to achieve the desired learning objectives.

The technological dimension allows a virtual extension of traditional face-to-face role-playing activity that is transposed it into a digital setting. This enhances the potential of the training experience in which learners are involved. Eutopia recreates a graphical word populated by virtual actors (avatar) controlled by real users.

While role-playing methodology that derives from psychodrama and sociodrama [33] has learning purposes, role-playing videogames are created for recreational purposes and take inspiration from pen-and-paper role-playing games. Indeed, role-play [30] has extensively been recognized as a powerful technique for enhancing the traditional training practice, boosting participants' learning experience, facilitating knowledge, and promoting skills, competencies, and group, as well as personal development, in face to face activities [57–60].

Since its origins, role-play technique has been variously adapted and applied to different settings and contexts, for different purposes and to many disciplines (e.g., psychology, organizational change, sociology, and pedagogy) for intensifying and accelerating learning and for developing new ways of understanding of concepts and knowledge.

Role-plays can be adopted to deal with personal (psychodrama) or collective (sociodrama) issues and used to exercise a variety of specific skills (learning simulations).

Moreover, role-play games can be considered as learning strategies that can be enhanced through technology by extending learning through added dimensions that may be impossible to conduct in face-to-face situations [61]. Among them, the so-called massive multiplayer online role-playing games (MMORPGs) and multi-user virtual environments (MUVEs) as, for example, Second Life (http://secondlife.com/education/) and Active Worlds (http://www.activeworlds.com/edu/).

MMORPGs derive from role-playing video games, which in turn take their origins from pen-and-paper role-playing games (e.g., Dungeons and Dragons) and use much of the same terminology, settings, and game mechanics.

Regarding the technological dimension, Eutopia, in addition to the functions normally provided in MMORPGs and MUVEs, offers specific features designed to facilitate its use in distance learning. In particular, it has been used to develop a variety of role-playing games for the development of different soft skills.

In summary, the platform is based on a client/server architecture, which comprises three different software pieces for users:

- Editor—for trainers, allowing the design of personalized storyboards and role-play learning scenarios
- Client—for both trainees and trainers, allowing them to interact with the 3D environments and with each other through text chat messages and non-verbal modalities
- Viewer—for visualizing recorded group interactions and sessions along with text-based exchanges.

Trainers through and within Eutopia assume potentially different roles. They can act as a playwright by writing storyboards, as a screenwriter by personalizing training scenarios, as a casting director by assigning roles to be played out, as a movie director by monitoring and guiding participants' actions and behaviors, a as director of photography by selecting relevant dynamics to be recorded, a as film critic by giving actors personalized feedback (debriefing phase).

Trainers by creating storyboards can define properties of training scenarios along with psychological and physical features of the different roles to be played by participants (Figure 9). They also act as a guide for using the learning platform features at their best to explore the learning potential of available tools.

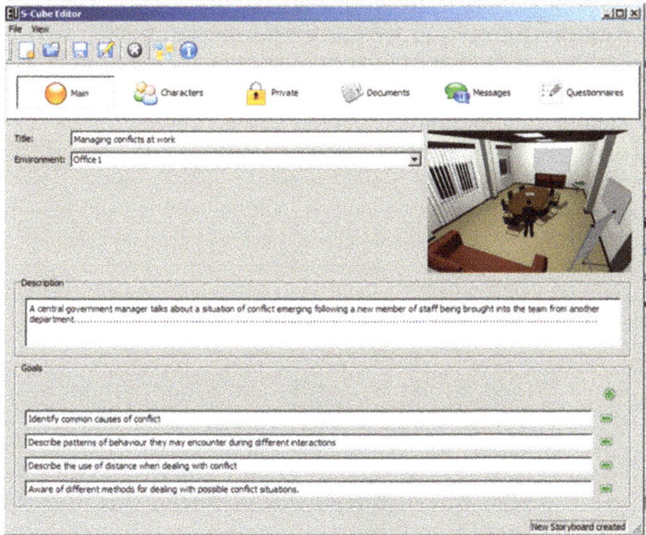

Figure 9. Script definition.

The use of feedback and debriefing systems allows the exploitation of all the potential of trainers' guide, facilitation, and support.

The Eutopia virtual environment provides an avatar-based system of communication, mediated by the artificial agents representing both human being trainer and the learners, respectively BSA and OSA.

By using the Eutopia Editor, trainers can write the storyboard for online multiplayer games. Its design requires an accurate definition of learning goals, narrative, and roles to be enacted and of the physical and psychological features of avatars (Figure 10).

Learners act out their roles interacting in a virtual, navigable environment provided by the system, through controlling virtual alter egos, the avatars.

These represent what we have defines as OSAs, as they directly act and interact in the virtual environment by influencing the dynamics of the game and impacting on its process.

Learners can communicate via short text messages, which appear in bubble cartoons over their avatars' heads.

They can also interact by using various forms of para- and non-verbal communication (expressed by emoticons and facial expressions that can be assumed by avatars).

For example, players can decide the loudness (shown by the font size of the text in the bubble) and emotional tone (shown by the shape and color of the bubble) of a message.

Players can control the gestures and body movements of avatars, for example, by making the avatar wave goodbye, point at someone, or hug someone.

They can "whisper" messages to each other, that is, send messages are that are visible only to players directly involved in the conversation and to the trainer.

Finally, they can communicate with the trainer and raise any questions to receive guidance or clarification. Trainers after scripting and starting the role-playing session can intervene during interaction among learners in two possible ways.

The first is to act as an invisible stage director that is to behave as a back-stage agent by using a variety of features to observe interactions among players. The second is by directly intervening in the game. For example, she can take the role of a character in the scenario and play the game like other players. However, they can also activate events to change the dynamics of actual interactions. These represent cases in which the role of BSAs coincides with that of OSAs.

When the game is concluded, they can provide players with personalized feedback assessing whether the group and individual goals have been achieved and to what extent, encourage group discussion and examine the most significant aspects and dynamics emerged, as well as the main strategies adopted by players.

Indeed, an embedded tutoring tool enables to record training sessions, and replay role-play session interactions for tutors to provide feedback to significant interactions between participants to encourage the communication process, mutual sharing, self-reflection, and self-discovery and help in identifying potential areas of personal development. Feedback can be provided immediately after role play or in a later feedback session

Figure 10. Avatar control as a way to explore an online session.

4.3.2. Eutopia Representation in SPA Terms

In the Eutopia platform, as shown in Figure 11, there are many on-stage agents that interact virtually: They are human OSA. They perform their role following the defined script and following the trainer (BSA indications). In this case, all the agents interact both directly and indirectly.

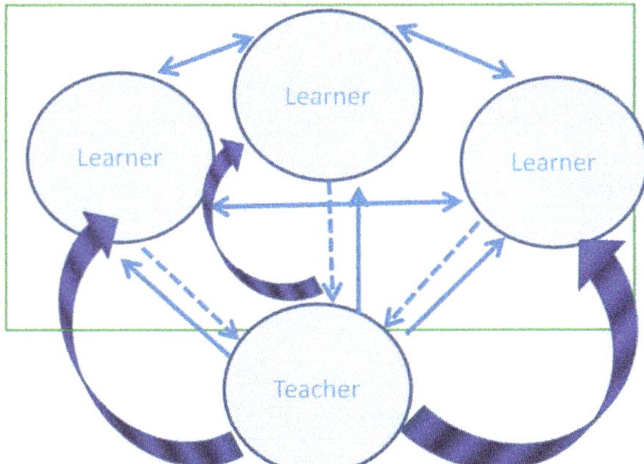

Figure 11. Eutopia platform represented with SPA notation (the playground is represented as an empty rectangle, circles represent OSA and squares represent BSA. Full-lined boundary indicates a real agent, dashed line represents an artificial agent. A continuous line represents a direct interaction, a dashed line represents an indirect one, arcs represent feedback).

At the end of the interaction, the BSA offers to OSA feedback and reflections on the different interactions.

Eutopia has been used and tested in different contexts and by different group targets (university, training institutions and agencies, MEs and SMEs, public administration, as well as non-governmental organizations) and for the development of various kinds of soft skills within different research projects, such as Sisine, Sinapsi, Eutopia-Mt, Proactive, S-cube (more information at www.nac.unina.it). In particular, to study the attitude towards the SPA approach, the perception of 18 experienced professionals (educators, trainers, psychologists, and educationalists adopting role-playing activities in traditional settings) on the use of role-playing games in educational and training contexts, with a specific focus on the Eutopia platform was investigated.

They completed a questionnaire on their perception of how online role play can encourage and foster meaningful learning experiences among participants. More details are reported in [2].

With regard to the methodological effectiveness of online role-play via SPA agents, we can affirm that it is generally considered as effective. A large consensus amongst the professionals was found on the role of the trainer, both virtual and real as conceived in the SPA approach.

5. Discussion and Conclusions

In this paper, we have introduced the SPA approach to developing EG. This approach presents various advantages. It opens the way to adopting automatic control systems and software based on artificial intelligence systems to model OSA and BSA behavior, as shown in the application section. By these means, it is possible to delegate both on-stage and backstage functionalities to intelligent and autonomous artificial agents, making it possible to run EG with mixed teams composed of human and artificial agents. It is, in fact, easier to build artificial agents to support the educational enterprise rather than model separately educational functions and features. Moreover, SPA allows to reproduce, model, and feed the dialogic interaction offering a formal representation of the people involved in the learning/teaching dynamic.

SPA offers an effective methodology to build up games moving on the shell and core level as well as the educational one. This means, that the same core level can be combined with different shell levels so as to be adapted to different contexts and allow to compare various populations (i.e., children and adolescents) and various areas of application (i.e., education, training or assessment).

Last but not least, it proposes a comprehensive framework that can be easily understood by specialists with different expertise. In EG design and development, education specialists, teachers, trainers are involved as well as computer scientists, software engineers, etc. These specialists can share their knowledge through this framework in a very effective way.

However, a possible shortcoming of this approach comes from the consideration that there are games, as well as educational software, for which there might be no need to define the rules of interaction in terms of psychological agents.

It is possible to summarize that the strongest point of the approach used within SPA to develop EGs is related mainly to the educational aspect allowing users to foster transversal skills through innovative approaches to teaching, learning, and assessment. The EGs proposed are based on two different educational approaches reflected in the implementation of the SPAs. Form a technological perspective is possible to distinguish EGs more cantered on allowing a virtual extension of traditional face-to-face psychodramatic mechanism and experiences (e.g., Eutopia), and those that instead reproduce "artificial" worlds based on computer-simulated, formal models about specific phenomena or theories to investigate (Enact and BM). From an educational and user-centered perspective, it is possible to identify two main categories. One category can represent the extent to which, while playing the game, the user has to express herself through behavioral acts that involve her body or other forms of interactions, such as an actor would do on stage. Those elements correspond to the traditional behavioral domain that plays a prominent role in psychodrama as we have highlighted for Eutopia and Enact, though with a different grade of involvement and immersion. Situations like BM in which the user is asked to perform abstract and strategic forms of decision-making are different from and yet complementary to these kinds of games. Here, the user's logical and reasoning aspects are prominently highlighted. The educational approach underpinning Eutopia is based on open dynamics. Therefore, there is no unique way to achieve the desired learning objectives. The technological dimension enhances the potential of the training experience because it makes a virtual extension of traditional face-to-face role-playing activity possible, transposing it into a digital setting. What emerges is that the figure of the trainer simultaneously represents a source of strength and weakness. On the one hand, it is undeniable that a real BSA trainer can enrich game performance by providing facilitation and adaptable performance feedback. On the other hand, the study presented shows that the need for fully skilled trainers may increase the cost and time of training.

Moreover, the dynamics resulting from the gameplay depend on learners, rather than on any form of artificial intelligence. This means that participants are offered a far richer, more open, learning experience than would have been possible if they had to interact with artificial OSAs and BSAs. However, the disadvantages of this method are represented by high cost and time consumption in organizing and managing the complexity of the virtual learning scenarios, as well as interactions among participants. Indeed, the critical element that emerged is related to the trainers/teachers' role in managing the online role-plays, and their need to be skilled in mastering different competencies at once.

Those limits have induced the authors to consider the advantages of introducing game technologies less dependent on the supervision of real BSAs, such as Enact.

In this case, although the system allows users to dramatize and enact role-plays, the complexity of the dynamics between OSAs is limited by the rule of the game to a certain number of actions, and the responsibility of the BSA is certainly reduced. Therefore, the assessment and observation of learning experience is less subjected to the influence and interpretation of many other potential interfering variables.

While Eutopia and Enact allow users to experience direct involvement with the learning objectives through a personal dramatization by acting out roles, BM points instead more on the logical and reasoning aspects involved in the gameplay. In this case, a set of formal rules and interactions embedded in the game needs to be followed for learners to achieve the relevant learning objectives.

This brings us to another aspect of our experience that is the appropriateness of the use of EGs. The decision on which game to use depend largely on the skills to be developed, as well as the resources and the time allocated for achieving the learning objectives. For instance, if a learning objective regards training from the cognitive domain, and the priority is making players learn and assess specific skills or behaviors (e.g., problem-solving requiring a quick response), the ideal methodology is more likely to be based on more structured games, as BM. Indeed, the educational resources and learning path that learners have to follow is easily accessible from learners at any time from anywhere. However, the set of formal rules and interactions to be followed to achieve the relevant learning objectives are embedded in the software and do not require a constant presence of experienced real external guidance as BSA. BM and Enact can drive the player to a stable training outcome more rapidly than in open dynamic situations, like Eutopia. Therefore, the advantage of this method lies in the fact that it is very low cost, as after an initial phase to familiarize users with the system, and it can be used without the guidance of a trainer, as the system is self-regulated and enables learners to achieve objectives rapidly. Conversely, if the competencies that are meant to develop are more related to aspects of emotional awareness, self-assessment, and self-confidence, we think that a situation methodologically such Eutopia, closer to the traditional role-play technique, might be the most appropriated. For all the EGs presented, we can acknowledge that the strength of providing the software with authoring systems has been a valued an extremely beneficial aspect as it allows trainers to rapidly develop their scenarios, personalizing their work for specific target populations with specific learning needs. In this light, there are many possible and potential areas of application.

The strongest points of the approach used within EGs are related mainly to the central role assigned to the player in the training or assessment processes developed within the software. The users can enhance their attitudes towards different skills, improve their capabilities, understanding, and practice with the support of the tutoring system provided, and following customized training sessions.

The experiences of the EU projects confirmed the value of using information technology as a tool placed in the hands of a trainer for the development of controlled ad hoc learning exercises, rather than being considered a simple replacement for trainers and learners.

The SPA approach presents a novel element of flexibility, both in delivery and practice of different skills and competencies training, where users can broaden the practice of different skills outside the traditional classroom approach by leveraging Internet technologies. However, what is even more interesting, professionals can be in total control of the model implemented, the training and the assessment processes. Furthermore, every skill or competence that requires the exploitations of people's interactions could benefit from such realization of an SPA to develop EGs.

The reason is that, whatever skills need to be transferred in the digital role-play, the educational technological level represented within the software enables modification both of the narratives and the educational models underlining the training requirements.

Author Contributions: Conceptualization, M.P., E.D., D.M., O.M.; methodology, M.P., E.D., D.M., O.M.; writing—original draft preparation, M.P.; writing—review and editing, M.P., E.D., supervision, D.M.; project administration, O.M.

Funding: BM Game was developed within the LLP lifelong Learning programme; the Enact Game was developed within the framework of the ENACT Project, funded by EACEA under Erasmus+ KA3 measure; EUTOPIA has been developed within the framework of projects funded by Lifelong Learning Programme (EACEA).

Acknowledgments: The authors would like to thank Onofrio Gigliotta for the graphical abstract design and drawing.

Conflicts of Interest: The authors declare no conflict of interest. The funders had no role in the design of the study; in the collection, analyses, or interpretation of data; in the writing of the manuscript, or in the decision to publish the results.

References

1. Bacigalupo, M.; Kampylis, P.; Punie, Y.; Van den Brande, G. *EntreComp: The Entrepreneurship Competence Framework*; Publication Office of the European Union: Luxembourg, 2016; p. 14.
2. Dell'Aquila, E.; Marocco, D.; Ponticorvo, M.; di Ferdinando, A.; Schembri, M.; Miglino, O. *Educational Games for Soft-Skills Training in Digital Environments: New Perspectives*; Springer: Berlin, Germany, 2016.
3. Johnson, R.T.; Johnson, D.W. Active learning: Cooperation in the classroom. *Ann. Rep. Educ. Psychol. Jpn.* **2008**, *47*, 29–30. [CrossRef]
4. Dewey, J. Experience and education. *Educ. Forum* **1986**, *50*, 241–252. [CrossRef]
5. Michael, D.R.; Chen, S.L. *Serious Games: Games That Educate, Train, and Inform*; Muska & Lipman/Premier-Trade: New York, NY, USA, 2005.
6. De Freitas, S. Are games effective learning tools? A review of educational games. *J. Educ. Technol. Soc.* **2018**, *21*, 74–84.
7. Xun, G.; Ifenthaler, D. Designing engaging educational games and assessing engagement in game-based learning. In *Gamification in Education: Breakthroughs in Research and Practice*; IGI Global: Hershey, PA, USA, 2018; pp. 1–19.
8. Shute, V.; Seyedahmad, R.; Xi, L. *Supporting Learning in Educational Games: Promises and Challenges. Learning in a Digital World*; Springer: Berlin, Germany, 2019; pp. 59–81.
9. Prensky, M. Digital game-based learning. *Comput. Entertain.* **2003**, *1*, 21. [CrossRef]
10. Kapp, K.M. *The gamification of learning and instruction*; Wiley: Hoboken, NJ, USA, 2012.
11. Abt, C.C. *Serious Games*; University Press of America: Lanham, MD, USA, 1987.
12. Connolly, T.M.; Boyle, E.A.; MacArthur, E.; Hainey, T.; Boyle, J.M. A systematic literature review of empirical evidence on computer games and serious games. *Comput. Educ.* **2012**, *59*, 661–686. [CrossRef]
13. Boyle, E.A.; Hainey, T.; Connolly, T.M.; Gray, G.; Earp, J.; Ott, M.; Lim, T.; Ninaus, M.; Ribeiro, C.; Pereira, J. An update to the systematic literature review of empirical evidence of the impacts and outcomes of computer games and serious games. *Comput. Educ.* **2016**, *94*, 178–192. [CrossRef]
14. Arnab, S.; Lim, T.; Carvalho, M.B.; Bellotti, F.; De Freitas, S.; Louchart, S.; Suttie, N.; Berta, R.; De Gloria, A. Mapping learning and game mechanics for serious games analysis. *Br. J. Educ. Technol.* **2015**, *46*, 391–411. [CrossRef]
15. Carvalho, M.B.; Bellotti, F.; Berta, R.; De Gloria, A.; Sedano, C.I.; Hauge, J.B.; Hu, J.; Rauterberg, M. An activity theory-based model for serious games analysis and conceptual design. *Comput. Educ.* **2015**, *87*, 166–181. [CrossRef]
16. Anderson, L.W.; Krathwohl, D.R. *A Taxonomy for Learning, Teaching and Assessing: A Revision of Bloom's Taxonomy*; Longman Publishing: Harlow, UK, 2001.
17. Bloom, B.S. *Taxonomy of Educational Objectives. Vol. 1: Cognitive Domain*; Edwards Bros.: Ann Arbor, MI, USA, 1956; pp. 20–24.
18. Danto, A.C.; Goehr, L.; Ankersmit, F. *Narration and Knowledge*; Columbia University Press: New York, NY, USA, 2007.
19. Neitzel, B. *Narrativity in Computer Games. Handbook of Computer Game Studies*; MIT Press Cambridge: Cambridge, MA, USA, 2005; pp. 227–245.
20. Gregory, J. *Game Engine Architecture*; CRC Press: Boca Raton, FL, USA, 2017.
21. Freire, M.; Serrano-Laguna, Á.; Iglesias, B.M.; Martínez-Ortiz, I.; Moreno-Ger, P.; Fernández-Manjón, B. *Game Learning Analytics: Learning Analytics for Serious Games. Learning, Design, and Technology: An International Compendium of Theory, Research, Practice, and Policy*; Springer: Berlin, Germany, 2016; pp. 1–29.
22. Serrano-Laguna, Á.; Torrente, J.; Moreno-Ger, P.; Fernández-Manjón, B. Application of learning analytics in educational videogames. *Entertain. Comput.* **2014**, *5*, 313–322. [CrossRef]
23. Seaton, J.; Graf, S.; Chang, M.; Farhmand, A. Incorporating learning analytics in an educational game to provide players with information about how to improve their performance. In Proceedings of the 2018 IEEE 18th International Conference on Advanced Learning Technologies (ICALT), Mumbai, India, 9–13 July 2018.
24. Siemens, G.; Long, P. Penetrating the fog: Analytics in learning and education. *Educ. Rev.* **2011**, *46*, 30.
25. Hodkinson, P.; Biesta, G.; James, D. Understanding learning cultures. *Educ. Rev.* **2007**, *59*, 415–427. [CrossRef]
26. Griffin, J. Learning to learn in informal science settings. *Res. Sci. Educ.* **1994**, *24*, 121–128. [CrossRef]

27. Halverson, E.R.; Sheridan, K. The maker movement in education. *Harv. Educ. Rev.* **2014**, *84*, 495–504. [CrossRef]
28. Adomssent, M.; Godemann, J.; Michelsen, G.; Barth, M.; Rieckmann, M.; Stoltenberg, U. Developing key competencies for sustainable development in higher education. *Int. J. Sustain. High. Educ.* **2007**, *8*, 416–430.
29. Ponticorvo, M.; Di Fuccio, R.; Di Ferdinando, A.; Miglino, O. An agent-based modelling approach to build up educational digital games for kindergarten and primary schools. *Expert Syst.* **2017**, *34*, e12196. [CrossRef]
30. Ponticorvo, M.; Di Fuccio, R.; Ferrara, F.; Rega, A.; Miglino, O. Multisensory educational materials: Five senses to learn. In Proceedings of the International Conference in Methodologies and intelligent Systems for Techhnology Enhanced Learning, Toledo, Spain, 20–22 Jun 2018; pp. 45–52.
31. Ponticorvo, M.; Rega, A.; Di Ferdinando, A.; Marocco, D.; Miglino, O. Approaches to embed bio-inspired computational algorithms in educational and serious games. In Proceedings of the 1st International Workshop on Cognition and Artificial Intelligence for Human-Centred Design 2017, Melbourne, Australia, 19 August 20107; pp. 8–14.
32. Montessori, M. *The Montessori Method*; Transaction Publishers: Piscataway, NJ, USA, 2013.
33. Moreno, J.L. *Who Shall Survive?: A New Approach to the Problem of Human Interrelations*; Nervous and Mental Disease Publishing Co.: Washington, DC, USA, 1934; pp. 2–20.
34. Moizer, J.; Lean, J.; Dell'Aquila, E.; Walsh, P.; Keary, A. An approach to evaluating the user experience of Serious Games. *Comput. Educ.* **2019**, *136*, 141–151. [CrossRef]
35. Ponticorvo, M.; Miglino, O. Hyper activity books for children: How technology can open books to multisensory learning, narration and assessment. *QWERTY Open Interdiscip. J. Technol. Cult. Educ.* **2018**, *13*, 46–61.
36. Dell'Aquila, E.; Vallone, F.; Zurlo, M.C.; Marocco, D. Creating digital environments for interethnic conflict management. In *International Conference in Methodologies and Intelligent Systems for Techhnology Enhanced Learning*; Springer: Berlin, Germany, 2019.
37. Jeuring, J.; Grosfeld, F.; Heeren, B.; Hulsbergen, M.; IJntema, R.; Jonker, V.; Mastenbroek, N.; van der Smagt, M.; Wijmans, F.; Wolters, M.; et al. Communicate!—A serious game for communication skills. In *Design for Teaching and Learning in a Networked World*; Springer: Berlin, Germany, 2015; pp. 513–517.
38. Johnson, W.L. Using virtual role-play to prepare for cross-cultural communication. In Proceedings of the 5th International Conference on Applied Human Factors and Ergonomics AHFE 2014, Krakow, Poland, 19–23 July 2014.
39. Core, M.; Traum, D.; Lane, H.C.; Swartout, W.; Gratch, J.; Van Lent, M.; Marsella, S. Teaching negotiation skills through practice and refl ection with virtual humans. *Simulation* **2006**, *82*, 685–701. [CrossRef]
40. Ebner, N.; Kovach, K.K. Simulation 2.0: The resurrection. In *Re-Thinking Negotiation Teaching Series, Vol. 2: Venturing Beyond the Classroom*; DRI Press: Saint Paul, MN, USA, 2011; p. 245.
41. Haferkamp, N.; Kraemer, N.; Linehan, C.; Schembri, M. Training disaster communication by means of serious games in virtual environments. *Entertain. Comput.* **2011**, 2, 81–88. [CrossRef]
42. Kolb, D.A. *Experiental Learning*; Englewood Cliffs/Prentice Hall: Upper Saddle River, NJ, USA, 1984.
43. Coppard, L.C.; Goodman, F.L. Urban gaming/simulation: A handbook for educators and trainers. In *School of Education*; University of Michigan: Ann Arbor, MI, USA, 1979.
44. Thatcher, D.C. Promoting learning through games and simulations. *Simul. Gaming* **1990**, *21*, 262–273. [CrossRef]
45. Petranek, C.F.; Corey, S.; Black, R. Three levels of learning in simulations: Participating, debriefing, and journal writing. *Simul. Gaming* **1992**, *23*, 174–185. [CrossRef]
46. Di Ferdinando, A.; di Fuccio, R.; Ponticorvo, M.; Miglino, O. Block magic: A prototype bridging digital and physical educational materials to support children learning processes. In *Smart Education and Smart e-Learning*; Springer: Berlin, Germany, 2015; pp. 171–180.
47. Di Fuccio, R.; Ponticorvo, M.; Di Ferdinando, A.; Miglino, O. Towards hyper activity books for children. Connecting activity books and Montessori-like educational materials. In *Design for Teaching and Learning in a Networked World*; Springer: Berlin, Germany, 2015; pp. 401–406.
48. Nkambou, R.; Mizoguchi, R.; Bourdeau, J. *Advances in Intelligent Tutoring Systems*; Springer Science & Business Media: Berlin, Germany, 2010.
49. Freedman, R.; Ali, S.S.; McRoy, S. What is an intelligent tutoring system? *Intelligence* **2000**, *11*, 15–16. [CrossRef]

50. Pacella, D.; Dell'Aquila, E.; Marocco, D.; Furnell, S. Toward an automatic classification of negotiation styles using natural language processing. In Proceedings of the International Conference on Intelligent Virtual Agents, Stockholm, Sweden, 27–30 August 2017; pp. 339–342.
51. Ponticorvo, M.; Di Ferdinando, A.; Marocco, D.; Miglino, O. Bio-inspired computational algorithms in educational and serious games: Some examples. In Proceedings of the European Conference on Technology Enhanced Learning, Lyon, France, 13–16 September 2016; pp. 636–639.
52. Ferrara, F.; Ponticorvo, M.; Di Ferdinando, A.; Miglino, O. Tangible interfaces for cognitive assessment and training in children: LogicART. In *Smart Education and e-Learning*; Springer: Berlin, Germany, 2015; pp. 329–338.
53. Marocco, D.; Pacella, D.; Dell'Aquila, E.; Di Ferdinando, A. Grounding serious game design on scientific findings: The case of ENACT on soft skills training and assessment. In *Design for Teaching and Learning in a Networked World*; Conole, G., Klobučar, T., Rensing, C., Konert, J., Lavoué, É., Eds.; Springer: Berlin, Germany, 2015; pp. 441–446.
54. Dryden, W.; Constantinou, D. *Assertiveness Step by Step*; Sheldon Press: London, UK, 2004.
55. Rahim, M.A. A measure of styles of handling interpersonal conflict. *Acad. Manag. J.* **1983**, *26*, 368–376.
56. Rahim, M.A.; Bonoma, T.V. Managing organizational conflict: A model for diagnosis and intervention. *Psychol. Rep.* **1979**, *44*, 1323–1344. [CrossRef]
57. McGill, I.; Beaty, L. *Action learning: A Guide for Professional, Management & Educational Development*; Psychology Press: London, UK, 2001.
58. Shaw, M.E.; Corsini, R.J.; Blake, R.R.; Mouton, J.S. *Role Playing: A Practical Manual for Group Moderators*; University Associates: San Diego, CA, USA, 1980.
59. Turner, D.A. *Roleplays: A Sourcebook of Activities for Trainers*; Kogan Page: London, UK, 1992.
60. Van Ments, M. *The Effective Use of Role-Play: Practical Techniques for Improving Learning*; Kogan Page: London, UK, 1999.
61. Garrison, D.R. *E-learning in the 21st Century: A Framework for Research and Practice*; Routledge: Abingdon, UK, 2011.

Article

Towards Agent Organizations Interoperability: A Model Driven Engineering Approach

Luciano R. Coutinho [1], Anarosa A. F. Brandão [2], Olivier Boissier [3] and Jaime S. Sichman [2,*]

[1] Departamento de Informática (DEINF), Universidade Federal do Maranhão (UFMA), Av. dos Portugueses, 1966, Bacanga, São Luís 65080-805, Brazil; luciano.rc@ufma.br

[2] Laboratório de Técnicas Inteligentes (LTI), Escola Politécnica (EP), Universidade de São Paulo (USP), Av. Prof. Luciano Gualberto, 158, travessa 3, São Paulo 05508-010, Brazil; anarosa.brandao@usp.br

[3] Department of Computer Science and Intelligent Systems, ENS Mines Saint-Etienne, 158, cours Fauriel, CEDEX 2, 42023 Saint-Étienne, France; Olivier.Boissier@emse.fr

* Correspondence: jaime.sichman@usp.br; Tel.: +55-11-3091-0625

Received: 4 May 2019; Accepted: 7 June 2019; Published: 13 June 2019

Abstract: In the research and development of *multiagent systems* (MAS), one of the central issues is how to conciliate the autonomy of the agents with a desirable and stable behavior of the MAS as a whole. *Agent organizations* have been proposed as a suitable metaphor for engineering social order in MAS. However, this emphasis has led to several proposals of *organizational models* for MAS design, thus creating an *organizational interoperability problem*: How to ensure that agents, possibly designed to work with different organizational models, could interact and collectively solve problems? In this paper, we have adopted techniques from *Model Driven Engineering* to handle this problem. In particular, we propose an abstract and integrated view of the main concepts that have been used to specify agent organizations, based on several organizational models present in the literature. We apply this integrated view to design MAORI, a model-based architecture for organizational interoperability. We present a MAORI application example that has shown that our approach is computationally feasible, enabling agents endowed with heterogeneous organizational models to cooperatively solve a problem.

Keywords: interoperability; multiagent systems; organizational models

---

1. Introduction

In the research and development of *multiagent systems* (MAS), one of the central issues is how to conciliate the autonomy of the agents with a desirable and stable behavior of the MAS as a whole. Borrowing ideas from the Social Sciences, some authors have named this issue the *problem of social order*: "How to obtain from local design and programming, and from local actions, interests, and views, some desirable and relatively predictable/stable emergent result" [1]. A closely related issue is the *problem of social consensus* often characterized as how to reach agreement with regard to some aspect or quantity of interest in a network of agents by combining the local preferences or states of individual agents [2,3]. Both social order and social consensus are fundamental problems in the design of MAS. While social order stresses the idea that agent behaviors must be coherent with the MAS global purpose, social consensus highlights the need of agreement among agents working together for a global purpose.

Faced with these problems, especially in the context of open MAS (i.e., systems formed by a dynamical population of agents provided by different developers), several researchers have argued in favour of using the *human organizations* as a proper metaphor for engineering MAS [4–7]. Human organizations, whose typical examples are firms, clubs, corporations, etc., are *collectivities* pursuing *specific goals* and exhibiting *formalized social structures* [8]. Goals are specific to the extent that they are explicitly and clearly defined. Social structures are formalized in such a way that patterns of structuring and behaving (such as *roles*, *role relations*, *procedures*, *protocols*, *norms*, etc.) are precisely specified regardless of personal traits and relations of any individual part of the organization. Thus, by conceiving a MAS as an organization—or more generally as a bigger system formed by several organizations, hereafter called *agent organizations*—the basic idea is to promote social order and consensus in a top-down fashion. The idea is to have the agents' actions and interactions governed by formalized "social structures", defined above the agents level, in order to enable the MAS (seen as a collective entity) to achieve definite global goals.

This emphasis on organizations as a suitable metaphor for engineering social order has led to several proposals of *organizational models* for MAS design [4]. From the perspective of software development, most of these organizational models can be characterized as *domain specific modeling languages* [9]. That is, they provide a specialized *conceptual structure* (metamodel) embodied in some *concrete syntax* (notation) by means of which the designer can write formal representations of the social structure of agent organizations. Such representations, called *organizational specifications*, are then used as specification artifacts driving the development of agents and MAS.

On the one hand, the existence of many organizational models favours the organizational design of MAS since, with various proposals, experience and best practices are accumulated. On the other hand, a great variety of organizational models introduces heterogeneity in the development of MAS. As a direct consequence of this heterogeneity, mainly in the case of open MAS, a new and important interoperability issue arises: If to enter and fully work in an organization the agents should be designed to "understand" and comply with an organizational specification of a given kind (i.e., conforming to some organizational model), then, in addition to the communication language and the domain ontology, the organizational model is something that the agents are supposed to share in order to properly work together. In other words, how can we provide means for a set of agents, immersed in a common environment, to evolve, reason, decide and interact with each other based on organizational concepts, since their organizational models may differ? In this paper, we call this issue the *organizational interoperability problem*.

We can think about four basic approaches to solve the organizational interoperability problem: *Standardization*, *universal agents*, *delegation* and *adaptation*. Standardization consists in providing interoperability by eliminating the root of the problem, the diversity, by means of a *standard model* that has to be accepted and used by all developers [10]. The universal agents approach implies the creation of agents which are able to deal with several different organizational models [11]. Delegation means creating specialized services in middleware layers (like proxies [12] and governors [13]) to whom agents may delegate reasoning and decision mechanisms related to organizational issues. Adaptation, by its turn, is a solution based on the possibility of defining mappings between models [14]. From these mappings, an *adapter* is created, a component that converts specifications from a model to another model [15,16].

Each of these approaches have their pros and cons. Standardization fully eliminates the problem but it is politically and economically difficult to achieve and lets aside legacy systems. Universal agents must be updated every time a new model is created or changed. Delegation practically vanishes the agents organizational autonomy. Adaptation deals with legacy systems but it is technically difficult to achieve, if not impossible, when there are no meaningful mappings between the models.

Motivated by the organizational interoperability problem and the basic approaches to it, all of which presuppose an integrated knowledge of the organizational models used to engineer agent organizations, the objective of this paper is to analyze the conceptual structures of several organizational models present in the literature and, based on this analysis, to propose an abstract and integrated view of the main concepts that have been used to specify agent organizations. We believe that the abstract view of organizational models we put forward can be used both as basis for defining essential mappings and for future standardization efforts of organizational models.

This work is based on a previous work [17] in which we did a review of several prominent organizational models to answer the questions: How are the conceptual structures of the models related? Are there basic similarities? What are they? The answers we have given to these questions, in terms of *modeling dimensions*, are summarized in Section 2. The idea of modeling dimensions describes the organizational models basic similarities in broad terms. It characterizes the macrostructure of organizational models.

In this paper we deepen our analysis by further exploring the conceptual structure within each modeling dimension. In this sense, we seek to characterize the common concepts and their relations found in existing organizational models along the modeling dimensions. The specific questions we propose to answer in this paper are: Inside each modeling dimension, what are the recurring modeling concepts? Is it possible to combine these recurring concepts into a coherent whole? How this can be used in a solution to the organizational interoperability problem? In approaching these questions, we have used techniques from Model Driven Engineering (MDE) [18]. Specifically, we think of organizational models as domain specific modeling languages whose conceptual structures are represented by means of metamodels. Thus, to address the questions systematically, we propose an iterative integration method, described in Section 3, aiming at building an integrated metamodel out of particular metamodels.

A central step in the integration method is the identification of correspondences between the conceptual structure of organizational models represented by metamodels. To assist this identification, in Section 4, we analyze the recurring concepts of existing organizational models along the modeling dimensions. The result is an *abstract conceptual structure* formed by the union of *conceptual patterns* found by comparing the organizational models. Relying on this abstract conceptual structure and using the proposed integration method, in Section 5 we show how to effectively integrate (part of) three existing organizational models. To put into perspective the integration of organizational models, we then discuss in Section 6 a solution based on adaptation for the organizational interoperability problem. In this solution, named MAORI (*Model-based Architecture for ORganizational Interoperability*) [19], the mappings between organizational models are defined indirectly by using the integration we have proposed.

In Section 7, we compare our proposal to related work in the literature. To the best of our knowledge, the systematic integrated analysis of organizational models we propose is novel, constituting the main contribution of the paper. Its importance, as already hinted, lies on serving as a common ground for aligning organizational models and as a starting point towards standardization. The MDE approach we apply is also a contribution and advancement in the state of the art. Looking at the literature, we found that few organizational models are defined in terms of explicit metamodels. Then, our representation of existing organizational models and their integration by means of formal metamodels helps in further the understanding of their features and limitations. Finally, in Section 8, we present our conclusions and future work.

2. Organizational Models for MAS

This section presents organizational models for multiagent systems by classifying their content in modeling dimensions that were adopted to define a method for the integration of organizational models. As a result of using the method, we build an *abstract conceptual structure* to deal with the organizational interoperability problem within MAS.

2.1. Modeling Dimensions

After a detailed analysis of a significant part of the existing organizational models, we have noted a lot of similarities and complementary issues regarding their conceptual structures. The common points identified were classified into some recurring themes we have called *modeling dimensions* for agent organizations [17]. In what follows, we discuss the modeling dimensions identified. Then we move to a short overview of the various organizational models analyzed. Ending the section, we show a comparative table summarizing the models analyzed along the modeling dimensions identified.

2.1.1. Fundamental Aspects of Systems

In general, designed systems exhibit some fundamental aspects that, from an engineering standpoint, are natural candidates for modeling. Firstly, there is the *functional behavior* of the system—the input (stimulus) to output (response) relations that couple the system to external elements composing the environment in which the system is situated. In modeling this aspect, the system is commonly depicted as a black box whose internal constitution, at first, does not matter. What really matters is that the environment imposes functional requirements, such as operations or tasks, that the system as a whole is supposed to perform. Further, these functional requirements may be subdivided in a recursive way until reaching atomic operations or tasks arranged in a given ordering (dependency graph). Later, when the innards are determined, the actual execution of the atomic operations or tasks can be associated with specific components of the system and their interactions. Modeling the functional behavior is a common practice both in developing computer systems and in representing organizational processes. Modeling techniques such as DFD (Data Flow Diagram) [20], and the activity diagrams of UML (Unified Modeling Language) [21] are typical examples of that.

Another fundamental aspect is the *internal structure* of the system. In contrast, to model the internal structure of a system means to represent it as a transparent box. It means to represent the break down of the system in its constituent parts (components and subsystems) and the relations interconnecting these parts. Like functional modeling, modeling the internal structure of a system is a recurring theme in system design. In software development, for example, the class diagrams and the component diagrams of UML serve to this purpose. In the case of human organizations, a traditional form of structural modeling is the creation of organograms describing the divisions, roles and hierarchical relationships existing inside an organization.

A third candidate for modeling is the *structural behavior* of the system. Roughly, it consists of the "movement" of the internal structure of a system towards the realization of some desired functional behavior. Thus, when modeling the structural behavior, we also see the system as a transparent box. What we try to represent is the ordering of interactions occurring over time among the constituent parts of a system. These interactions make the system work, i.e., perform some expected task or operation in its environment. Examples of this type of modeling are the sequence and collaboration diagrams of UML.

2.1.2. Primary Modeling Dimensions

From the premise that designed systems in general, not only agent organizations, exhibit these three fundamental aspects as natural modeling concerns, we have used them as a first classification scheme for separating the modeling concepts of organizational models into cohesive categories. Consequently, we define:

- The *functional dimension*, in which we place the modeling concepts used to represent the functional behavior of agent organizations;
- the *structural dimension*, composed by modeling concepts used to represent the internal structure of agent organizations; and
- the *interactive* or *dialogical dimension*, grouping the modeling concepts relative to the representation of the structural behavior of agent organizations.

2.1.3. Social Systems and the Normative Dimension

While the functional, structural and interactive dimensions can be justified by analysing the modeling of systems in general, they are not sufficient to classify all modeling concepts appearing in organizational models.

According to [22], three basic types of systems and corresponding models can be identified: *Deterministic*, i.e., systems and models in which neither the parts nor the whole are purposeful; *animated*, i.e., systems and models in which the whole is purposeful but the parts are not; and *social*, i.e., systems and models in which both the whole and the parts are purposeful. A fourth type is also considered, *ecological*, i.e., systems and models in which the parts are purposeful but the whole is not. Given this classification, we can say that traditional software systems are deterministic, autonomous agents are animated, and agent organizations are social. Bigger and more encompassing MAS aggregating agents and agent organizations form ecological systems.

Being deterministic, traditional software systems tend to have an architecture in which the functional behavior, the internal structure and the structural behavior are foreseen in detail. Their components are not conceived as purposeful entities with autonomous behaviors. On the contrary, they are designed to obey rigidly what is fixed in the architectural specification of the system.

Regarding agent organizations characterized as social systems, the idea of agents obeying rigidly the prescriptions of functional, structural and interactive specifications is not realistic. Agents are conceived as self interested components, especially in open MAS. Therefore, neither the functional, structural and interactive specifications can be very detailed to the point of precisely determining the minutiae of the joint structuring and behaving of the agents, nor one can assume benevolence from the agents with respect to the organizational goals.

In this context our analysis is that the specification of *norms* (permissions, prohibitions, obligations, etc.), as occurs in human organizations design, are also expected to show up in organizational models. They will work as a complementary mechanism helping to couple more flexibly the agents to the organization. On the one hand, norms provide explicit means to capture interdependencies among the functional, structural and interactive aspects (e.g., agent playing a given role in the internal structure is obliged to behave functionally or structurally in a given way). On the other hand, norms can be used to explicitly regulate sanctions or penalties to deviant behavior.

Accordingly we define a fourth and last category for the analysis of organizational modeling concepts:

- The *normative dimension*, characterized by modeling concepts to further restrict, regulate and interrelate elements from the other modeling dimensions, given the expected autonomous behavior of the agents.

2.2. Models Review

Now we pass to a quick description of concrete organizational models taking into account how they cover the four dimensions of modeling identified. We describe six models—TAEMS [23], AGR [5], STEAM [24], MOISE+ [25], ISLANDER [26] and OPERA [27]. We think of these as good exemplars showing how models have evolved towards a full coverage of the organizational modeling dimensions. Other models are mentioned at the end of the review.

2.2.1. TAEMS

In TAEMS (Task Analysis, Environment Modeling, and Simulation) the basic modeling concept is the notion of *task*. In essence, by using TAEMS we can specify *tasks structures* composed by the definition of *tasks*, *resources*, *tasks relationships*, and *task groups*. A task group is an independent collection of interrelated tasks. There are two kinds of task relationships: *Subtask* and *non-local effects* relationships. The subtask relationship links a parent task to child task explicitly defining a task decomposition tree. Individual tasks that do not have child tasks are called *methods*. Methods are primitive tasks that agents should be able to perform. Non-local effects are task relationships that have positive or negative effects in the quality, costs or duration of the related tasks. Examples of possible non-local effects are: *Facilitates*, *enables*, *hinders*, *limits*, etc.

TAEMS is a model specialized exclusively in the specification of the functional behavior of agent organizations. A TEAMS specification, the task structure, only represents what should be done by the agents alone (method definitions) or in groups (task groups). It tells nothing about the internal structuring or explicit interactions to realize the specified tasks.

2.2.2. ARG

AGR (Agent, Group, Role) is the evolution of the AALAADIN model [28]. In AGR *agent*, *group* and *role* are the primitive modeling concepts. An agent is an active, communicating entity playing one or more roles within one or more groups. No constraints are placed upon the architecture of an agent or about its mental capabilities. A group is a set of agents sharing some common characteristics. A group is the context for a pattern of activities and is used for partitioning organizations. An agent can participate at the same time in one or more groups. Agents may communicate if and only if they belong to the same group. A role is the abstract representation of a functional position of an agent in a group. Roles are local to groups, and a role must be requested by an agent.

AGR is a model providing a minimalist structural view of organizations. There is no concepts for modeling functional behavior. The specification of an organization, called *organizational structure*, is in essence the depiction of the internal structure of the organization in terms of roles, roles constraints and group structures. AGR also says that agents can have their joint behavior orchestrated by interaction protocols, but the nature and the primitives to describe such protocols are left open.

2.2.3. STEAM

STEAM (a Shell for TEAMwork) is a model whose focus is teamwork. In STEAM an agent organization is conceived as an *agent team*. Two separate hierarchies are used to specify the internal structure and functional behavior of a team: A *subteam* and *roles* hierarchy (or *organization hierarchy*), and a hierarchy of joint activities (or *operator hierarchy*). The subteam and roles hierarchy is a tree in which the root represents a team, the internal nodes the possible subteams and the leaves the individual agent roles. The joint activity hierarchy is also a tree whose nodes are called *operators*. Leaf operators represent atomic activities. Internal operators represent a *reactive plan*, i.e., the decomposition of an activity into interrelated subactivities. For each individual role or subteam, it is assigned one or more operators from the activity hierarchy.

With STEAM we see a first model that combines the structural (subteams and role hierarchy) and functional modeling dimensions (operator hierarchy).

2.2.4. MOISE+

MOISE+ (Model of Organization for multI-agent SystEms) is a model that explicitly divides the specification of an agent organization in three parts: The *structural*, the *functional* and the *deontic* specifications. The structural specification defines the internal structuring of agents through the notions of *roles*, *roles relations* and *group specifications*. A role defines a set of constraints the agent has to accept to enter in a group. Role relations are *links* (*communication*, *acquaintance* and *authority*) and *compatibilities* from a source role to a target role. A group specification consists in role definitions, subgroup definitions (group decomposition), links and compatibilities definitions, role cardinalities and subgroup cardinalities. The functional specification describes how an agent organization usually achieves its *global goals*, i.e., how these goals are decomposed (by *plans*) and distributed to the agents (by *missions*). Global goals, plans and missions are specified by means of a *social scheme*. A social scheme can be seen as a goal decomposition tree where the root is a global goal, the internal nodes are *plan operators* (sequence, choice, parallel) to decompose goals into subgoals, and the leaves are atomic goals that can be achieved by an individual agent. Missions are coherent sets of goals; hence, an agent that is committed to a mission is responsible for the satisfaction of all its component goals. Finally, the deontic specification associates roles to missions by means of *permissions* and *obligations*.

Like STEAM, MOISE+ addresses both the functional and structural dimensions of modeling. However, MOISE+ goes further and provides concepts for modeling normative aspects (deontic specification). The deontic concepts allow a flexible coupling between the functional and structural specifications that is not seen in STEAM.

2.2.5. ISLANDER

ISLANDER is a declarative language for specifying *electronic institutions*. According to ([26] p. 348), "Institutions establish how interactions of a certain sort will and must be structured within an organization". In ISLANDER, an electronic institution is composed of four basic elements: A *dialogic framework*, *scenes* definitions, a *performative structure*, and *norms* definitions. In the dialogic framework it is defined the participating *roles* and their *relationships*. Each role defines a pattern of behavior within the institution and any agent within an institution is required to adopt some of them. A scene is a collection of agents playing different roles in interaction with each other in order to realize a given activity. Every scene follows a well-defined *communication protocol*. The performative structure establishes relationships among scenes. The idea is to specify a network of scenes that characterizes more complex activities. The norms component of an electronic institution defines the *commitments*, *obligations* and *rights* of participating roles.

With ISLANDER we perceive a change of focus from functional to structural behavior. Unlike the previous models, there is no concepts for explicitly modeling goals and plans (goal decompositions). All behavior is specified by means of direct interactions between roles (dialogs) and regulated by the definitions of norms.

2.2.6. OPERA

In OperA (Organizations per Agents) an agent organization is specified in terms of four structures: The social, the interaction, the normative and the communicative structures. In the *social structure* are defined *roles*, *objectives*, *groups* and *role dependencies*. Roles identify activities and services necessary to achieve social objectives. Groups provide means to collectively refer to a set of roles. Role dependencies describe how the roles are related in terms of objective realization. The *interaction structure* defines how the main activity of an agent organization is supposed to happen. This definition is done in terms of *scenes*, *scene scripts*, *scene transitions* and *role evolution relations*. Scenes are representations of specific interactions. A scene script is described by its players (roles or groups), scene norms (expected behavior of actors in a scene) and a desired interaction pattern. Scene transitions are used to coordinate scenes by defining the ordering and synchronization of the scenes. Role evolution relations specify the constraints that hold for the role-enacting agents as they move from scene to scene respecting the defined transitions. The normative structure gathers all the norms that are defined during the specification of roles, groups, and scene scripts. Norms are specified as formal logical expressions. Finally, the communicative structure describes the set of performatives and the domain concepts used in the interaction structure by the role enacting agents.

OPERA is a model that addresses all the identified modeling dimensions. Nevertheless, we note that the functional and structural modeling of OPERA is less developed than the others models, the interactive modeling is comparable to what is found in ISLANDER, and the normative modeling is the most elaborated of all models analysed. In OPERA norms are expressed in a formalism called LCR (Logic for Contract Representation).

2.2.7. Other Models

The literature on organizational models is vast. For reasons of space and scope, we briefly mention other models below:

- ODML—*Organizational Design Modeling Language* [29]—a minimalist organizational model that provides elements to model and evaluate structural aspects of organizations.
- AGRE—*Agent, Group, Role, Environment* [30]—an extension of AGR that takes into account physical and social environments. The main idea is that agents are situated in domains called *spaces*. The spaces can be physical (*areas*) or social (*groups*).

- MOISE^{Inst} [31]—an extension of MOISE+ that allows for *Contextual Specifications* (*contexts* and transitions between contexts) and *Normative Specifications* (*norms* set) in the modeling of an MAS organization.
- OMNI—*Organizational Model for Normative Institutions* [32]—an unification of two other models: the OperA and the HarmonIA framework [33].
- MAS-ML—*MAS Modeling Language* [34]—a modeling language that extends UML with elements of the TAO conceptual framework [35]. Regarding organizational modeling, one of the distinguishing features of MAS-ML is that organization technically are an extension of an *AgentClass* classifier. This means that organizations are conceived as kinds of agents.
- MACODO [36] is an organizational model for context-driven dynamic agent organizations where organization, agent, role and context are abstract concepts related to the system structure; capabilities, role positions and role contracts are related to the system functioning and there are laws for governing inter-organizational (merge law) or intra-organizational (join law) interactions. MACODO also provide a middleware that allows the implementation and execution of systems modeled following the MACODO model.

In addition to these, we also find in the literature on agent oriented software engineering (AOSE) methodologies a strong concern about organization modeling during the analysis and design of MAS. The Gaia [37] and Tropos [38] are some examples of AOSE methodologies that incorporate the concept of organization in their metamodels.

2.3. Models Comparison

More than large differences, we perceive several similarities and complementarity among the conceptual structures of the organizational models analysed, as we show in Table 1. The commonalities occur in two levels. In a first macro level, they occur as the dimensions of organizational modeling that were identified. For example, all models except TAEMS present concepts to represent the internal structure of organizations (structural dimension). All models except AGR and ISLANDER promote the functional behavior modeling (functional dimension). ISLANDER and OPERA present very similar concepts for representing the structural behavior of organizations (dialogical dimension). Normative concepts appear in MOISE+, ISLANDER and OPERA (normative dimension).

Table 1. Organizational models comparison.

Model	Functional	Structural	Dialogical	Normative
TAEMS	method, task, subtask relation, non-local effect	none	none	none
AGR	none	role, group, role relation	interaction protocol	none
STEAM	operator, plan, dependency	team, individual role	none	none
MOISE+	goal, plan, mission	role, group, role relations	none	deontic relations
ISLANDER	none	roles, role relations	scene, transition, interaction protocols	obligation
OPERA	objective, subobjective	role, group, dependency	scene, transition, interaction patterns	obligation

On a more detailed level of analyses, one can still identify various modeling patterns within each dimension. In the next section, we propose an iterative integration method that rely on these patterns, whose formal description is presented in Section 4.

3. Method for the Integration of Organizational Models

We advocate that modeling dimensions for agent organizations are useful not only to analyse and compare organizational models, but also they serve as a starting point for the *conceptual integration* of organizational models. As we have mentioned in the Introduction, when the objective is to provide organizational interoperability, a consistent conceptual integration of organizational models is a fundamental and necessary element. In order to systematically perform such integration (having in mind the problem of organizational interoperability), we have defined an iterative integration method that is discussed in this section.

3.1. General Process

Let OM_1, OM_2, ..., OM_n be n organizational models to be integrated. In broad lines, we understand by a *conceptual integration* of $OM_1, \ldots, OM_n$ the process of representing, correlating and joining the *conceptual structure* (i.e., the modeling concepts and their interrelationships) of each OM_i obtaining as the final result an *integrated metamodel* MM^{int} whose conceptual structure subsumes the structure of all OM_i. This idea of conceptual integration, as an iterative process, is depicted in Figure 1.

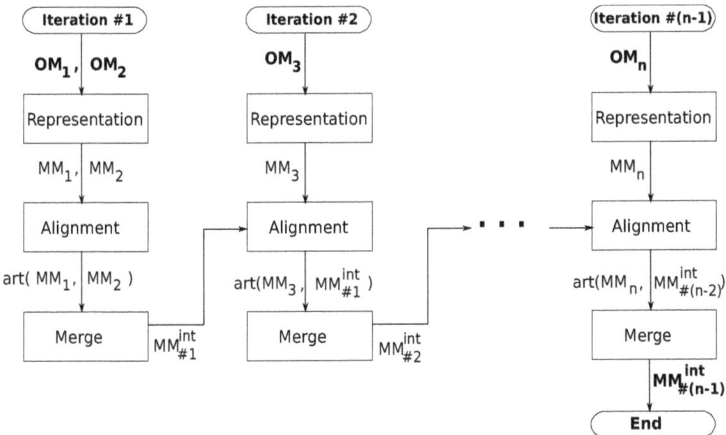

Figure 1. General conceptual integration method.

More specifically, for n models we have $n-1$ iterations. In the first one, three sequencial steps are performed for OM_1 and OM_2:

1. Analysis and representation of the conceptual structure of OM_1 and OM_2 as *metamodels* MM_1 and MM_2, respectively. In the representation it is used a common *metamodeling language*.
2. Comparison of MM_1 and MM_2, and identification of *correspondences* between the conceptual structure expressed in both metamodels. Such correspondences represent semantic overlapping areas between MM_1 and MM_2, i.e., modeling concepts and concept relationships that are assumed to have equivalent or similar interpretations. In general, the identified correspondences are explicitly expressed as *articulations* $art(MM_1, MM_2)$ between MM_1 and MM_2, as described next.
3. From MM_1, MM_2 and $art(MM_1, MM_2)$, we produce an integrated metamodel $MM^{int}_{\#1}$ that (*i*) is as expressive as both MM_1 and MM_2 (in the sense of retaining all concepts, relations and restrictions found in both MM_1 and MM_2); and (*ii*) avoids unnecessary replication of elements which were declared equivalent or similar by means of the correspondences between MM_1 and MM_2.

From the second iteration onward, the same three steps are performed for each $OM_{3 \leq i \leq n}$ with the following differences: (1) Instead of MM_1 and MM_2 we have MM_i and $MM^{int}_{\#(i-2)}$, respectively; MM_i

is the metamodel representing OM_i and $MM^{int}_{\#(i-2)}$ is the integrated metamodel from the previous iteration; (2) in the place of $art(MM_1, MM_2)$ we have $art(MM_i, MM^{int}_{\#(i-2)})$ which is the articulation between MM_i and $MM^{int}_{\#(i-2)}$; and (3) instead of $MM^{int}_{\#1}$ we produce $MM^{int}_{\#(i-1)}$, i.e., the integrated metamodel resulting from joining MM_i to $MM^{int}_{\#(i-2)}$.

3.2. Metamodel Representation

In the first step, we assume a metamodel based representation of the organizational models. In this sense, our method adopts the way by which special purpose modeling languages are defined in the area of Model-Driven Engineering. Given this assumption, there are several metamodeling languages available for expressing the conceptual structure of the organizational models. Some of these languages are KM3 [39], MOF/OCL [40,41], XMF [42] and Ecore [43]; this last is used in this work, as illustrated in Sections 4 and 5.

3.3. Metamodel Alignment

In the second step, the definition of correspondences between modeling concepts is an inherently heuristic process. One possible heuristic is to use the modeling dimensions identified in Section 2. The basic idea is to divide the work along the modeling dimensions. For each model, we start by classifying its modeling constructs in one or more dimensions. Then, for each dimension covered by the models, we identify the corresponding modeling constructs. In this way, the functional modeling concepts of one model are put into correspondence with the functional concepts of the other model, the structural concepts of one model with the structural concepts of the other, and so on.

Another heuristic we put forward for aligning the conceptual structure of organizational models is to take into account some basic conceptual patterns found in the models. These patterns are described in Section 4 in the form of an abstract organizational model.

3.4. Metamodel Merging

Unlike the alignment, the merging of metamodels is a more deterministic process that can be fully automated by using several algorithms reported in the literature [44–46]. In general, these proposals for (meta)model merging can be described as merging based on *graphs* and *morphisms*. In this case, the metamodels are abstractly conceived as graphs and the correspondences between two metamodels assume the form of an articulation between graphs. If MM_1 and MM_2 are two metamodels viewed as graphs, then an articulation $art(MM_1, MM_2)$ between them is a triple composed of a graph G^{art} and two morphisms $m_1 : G^{art} \rightarrow MM_1$ and $m_2 : G^{art} \rightarrow MM_2$:

$$art(MM_1, MM_2) = < G^{art}, m_1, m_2 >$$

Intuitively, the idea is that G^{art} is a representation of the common concepts and relations found in M_1 and M_2, and the morphisms m_1 and m_2 are the links mapping this common concepts and relations to their counterpart in both MM_1 and MM_2.

Once characterized as graphs, the merging of two metamodels MM_1 and MM_2 is in essence an *amalgamated sum* (or *pushout*) of MM_1 and MM_2, modulo $art(MM_1, MM_2)$:

$$merge(MM_1, MM_2) = MM_1 \oplus_{art(MM_1, MM_2)} MM_2 = < MM^{int}, m'_1, m'_2 >$$

where MM^{int} is the resulting integrated metamodel at the end of the iteration and m'_1 and m'_2 are morphisms $m'_1 : MM_1 \rightarrow G^{int}$, $m'_2 : MM_2 \rightarrow G^{int}$. The integrated metamodel MM^{int} consists of the union of the nodes (concepts) and edges (concept relationships) of MM_1 and MM_2, where correspondent elements as described via $art(MM_1, MM_2)$ are treated as only one element [44,46]. In this way MM^{int} retains all non-duplicate information in MM_1 and MM_2 collapsing the elements that $art(MM_1, MM_2)$ declares redundant. The morphisms $m'_1 : MM_1 \rightarrow G^{int}$ and $m'_2 : MM_2 \rightarrow G^{int}$

describe how translating from the particular to the integrated metamodel. The inverse, translating from the integrated to the particular metamodels, is performed via G^{art}, m_1 and m_2.

4. Abstract Conceptual Structure of Organizational Models

In this section we compare in detail the conceptual structure of the organization models discussed in Section 2. With this comparison, we intend to explicitly show two main findings: (*i*) We can identify patterns in the conceptual structure of organizational models inside each modeling dimension, if we homogenize the terminology used and abstract some particularities of each model; and (*ii*) the patterns identified can be consistently combined into a single conceptual structure (metamodel) that represents in an essential and integrated way the conceptual structures of organizational models. In the next subsections, we detail the basic patterns that emerge when one look more closely to the conceptual structures of the organizational models proposed in the literature. Each subsection focus on a modeling dimension previously discussed in Section 2. At the end, we combine the conceptual patterns obtaining in this way the abstract organizational metamodel.

4.1. Functional Dimension

In the functional dimension, we found concepts for the specification of the functional behavior of an agent organization, i.e., the collective behavior of agents when the internal structure of their organization is not taken into account. Looking at Table 1, we can see that this dimension occurs in TAEMS, STEAM, MOISE+ and OPERA. In these models, the functional specifications follow a general pattern which is illustrated in Figure 2.

4.1.1. Graphs of Hierarchical Plans and Goal Relationships

In essence, the general pattern can be characterized as directed graphs where:

- The nodes correspond to *goals* (in MOISE+), *operators* (in STEAM), *objectives* (in OPERA) or *tasks* (in TAEMS), to be achieved or done by the agents in an organization;
- the edges represent:
    - Either the acyclic decomposition of a *goal* (*operator*, *objective* or *task*) into *subgoals* (*suboperators*, *subobjectives* or *subtasks*), giving rise to the notion of *hierarchical plans*,
    - or binary relationships between *goals* (*operators*, *objectives* or *tasks*), like the *depends* relation in STEAM or the *non-local effects* in TAEMS.

Further, in each graph there is one *root node* that corresponds to a primary goal whose planning and future achievement is prescribed by the structure of the graph. Such graphs of hierarchical plans and goal relationships receive the names of "task group" in TAEMS, "operator hierarchy" in STEAM, "social scheme" in MOISE+ and "role objective definition" in OPERA.

In Figure 2, the conceptual pattern identified in the functional dimension is represented by means of an Ecore metamodel. The classes and references composing the metamodel are described in Table 2. Additional contextual constraints are presented in Table 3. These are written in OCL and formalize the static semantics of the metamodel (conceptual pattern).

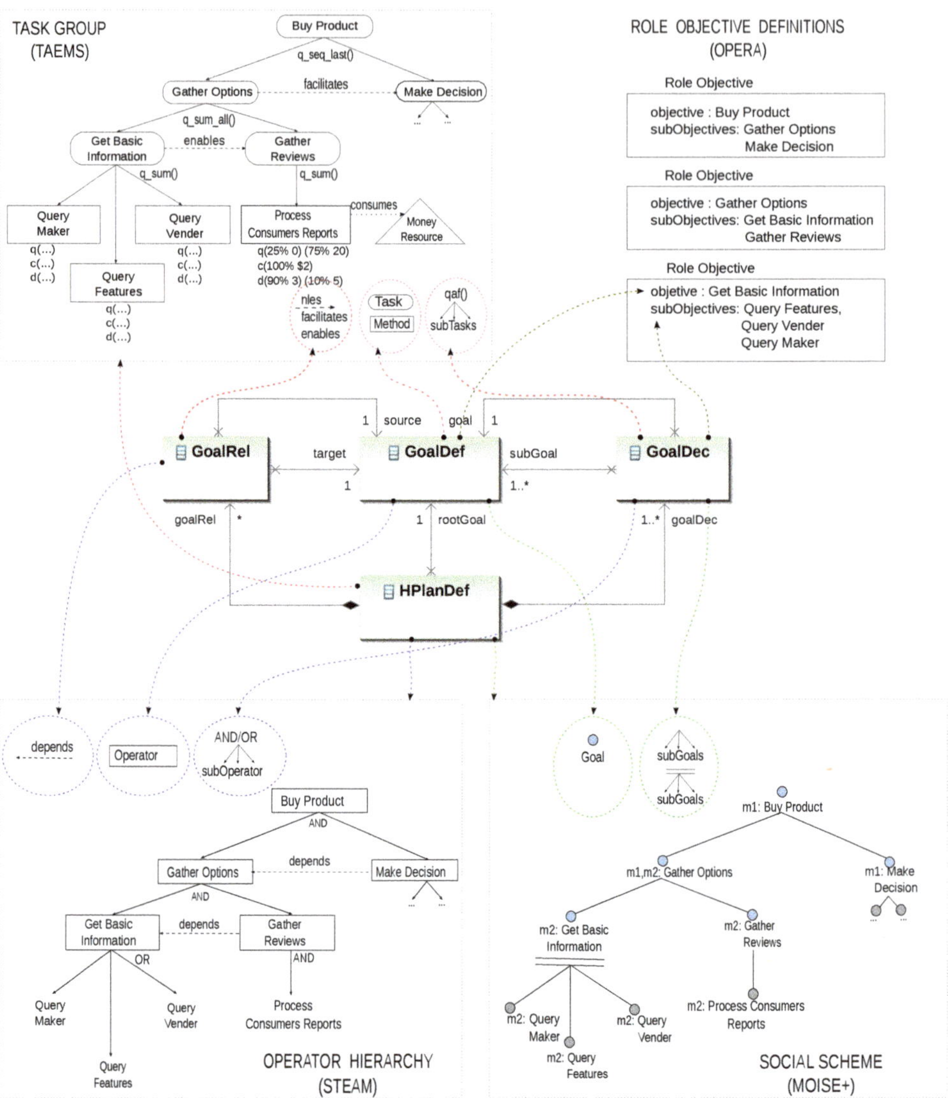

Figure 2. Similar functional specifications written in TAEMS, OPERA, STEAM and MOISE+, which prescribe how an agent is supposed to proceed for buying a product in an electronic market (example taken from [47]). The class diagram (Ecore metamodel) in the center of the figure represents the conceptual pattern identifiable in the various approaches of functional modeling. The dotted arrows detail what concepts are captured by what classes of the metamodel.

Table 2. Functional specification pattern: Classes description.

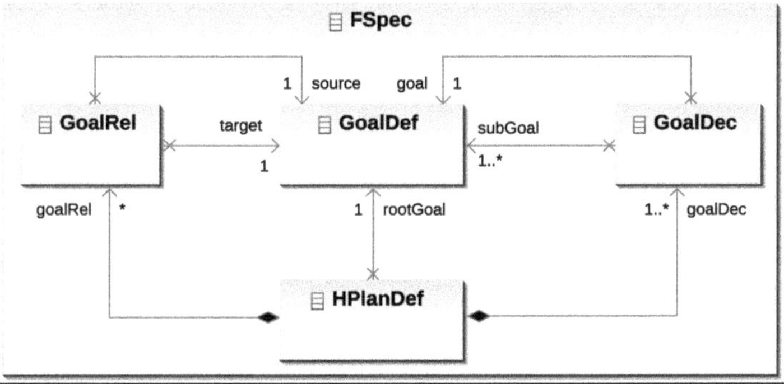

Class	Description
FSpec	Represents the concept of *functional specification*, i.e., an organizational specification restricted to the functional dimension.
GoalDef	Part of a functional specification that represents the definition of a *goal*. Abstracts the concepts of *task* in TAEMS, *operator* in STEAM, *goal* in MOISE+, or *objective* in OPERA.
HPlanDef	Part of a functional specification that represents the definition of a graph of *hierarchical plans* and *goal relationships*. HPlanDef is characterized by a unique *root goal* (the GoalDef referenced by HPlanDef::rootGoal), one or more *goal decompositions* (referenced by HPlanDef::goalDec) and zero or more *goal relationships* (referenced by HPlanDef::goalRel). Abstracts the concepts of *task group* in TAEMS, *operator hierarchy* in STEAM, *social scheme* in MOISE+ and *role objective definition* in OPERA.
GoalDec	Part of a HPlanDef graph that represents the general concept of *goal decomposition*, i.e., the decomposition of one major *goal*(GoalDec::goal reference) into one or more minor direct *subgoals*(GoalDec::subGoal reference). Abstracts the relationship of *subtasks* or *local effects* in TAEMS, the concept of *plan* in STEAM and MOISE+, and the *subobjective definitions* in OPERA.
GoalRel	Part of a HPlanDef graph that represents *goal relationships* directed from one *source goal*) GoalRel::source reference) to one *target goal* (GoalRel::target reference. Abstracts the concept of *non-local effects* (*facilitates*, *enables*, etc.) in TAEMS and the *depends* relation in STEAM.

Table 3. Functional specification pattern: Contextual constraints.

Class	Auxiliary Definitions

`HPlanDef`	`::getAllSubGoal(goal:GoalDef) : Set(GoalDef)`, a query that returns the set of all subgoals in which a given goal is direct or indirectly decomposed in the context of a `HPlanDef` graph, via goal decomposition referenced by `HPlanDef::goalDec`
	
	In OCL:
	context `HPlanDef`
	def: `getAllSubGoal(goal : GoalDef) : Set(GoalDef)`
	`= getSubGoal(goal) -> union(`
	`getSubGoal(goal) -> collect (sg | getAllSubGoal(sg)));`
	def: `getSubGoal(goal : GoalDef) : Set(GoalDef)`
	`= getGoalDec(goal) -> collect (gd | gd.subGoal);`
	def: `getGoalDec(goal : GoalDef) : Set(GoalDec)`
	`= goalDec(goal) -> serlect (gd | gd.goal = goal);`

	Contextual Constraints

	(1) In the context of a `HPlanDef` graph, all decompositions of a goal into subgoals must be reachable from the root goal. In other words, each goal decomposed in subgoals in a `HPlanDef` graph is either the root goal itself or a direct or indirect subgoal of the root goal.
	In OCL:
	context `HPlanDef`
	inv: `goalDec -> forAll (gd |`
	`root.Goal = gd.goal or`
	`getAllSubGoal(rootGoal) -> includes(gd.goal) )`

	(2) In the context of a `HPlanDef` graph, the source and target goals of all goal relationships must pertain to the collection of subgoals of the root goal of the graph. In other words, each goal relationship must connect only subgoals of the root goal of a `HPlanDef` graph. In OCL: **context** `HPlanDef` **inv:** `getAllSubGoal(rootGoal) -> includesAll(` `goalRel -> collect (gr | Set(gr.source, gr.target)))`

	(3) Restricted to goal decomposition, all `HPlanDef` graphs must be acyclic. In other words, in the context of a `HPlanDef` graph, no goal can be, direct or indirectly, a subgoal of itself.
	In OCL:
	context `HPlanDef`
	inv: `not getAllSubGoal(rootGoal) -> includes(rootGoal)`
	`and getAllSubGoal(rootGoal) -> forallAll( sg |`
	`not getAllSubGoal(sg) -> includes(sg) )`

4.1.2. Particularities

Besides the common aspects represented in the functional specification pattern (Tables 2 and 3), there are some particular aspects in the organizational models that should be highlighted.

In TAEMS, there are the concepts of *resources* and *non-local effects between tasks and resources*. These concepts were not considered as part of the functional specification pattern presented because they occur only in TAEMS.

In MOISE+, there is no notion of *binary relationships between goals*. There is, however, the particular notions of *mission* and *preferences among missions*, which do not occur in the other organization models analysed, but only in MOISE+. Thus, like the concepts of *resources* and *resource non-local effects* of TAEMS, the notions of *mission* and *mission preferences* also do not appear in the functional specification pattern presented.

In OPERA, the functional modeling is done implicitly as part of a *role definition* (structural modeling). In this way, we have a functional modeling with little resources when compared to what can be found in TAEMS, STEAM and MOISE+. Quoting [27], a *role definition* is done specifying one or more *role objectives* γ and

> "[each] role objective γ can be further described by specifying a set of subobjectives that must hold in order to achieve objective γ. Subobjectives give an indication of how an objective should be achieved, that is, describe the states that must be part of any plan that an agent enacting the role will specify to achieve that objective. However, subobjectives abstract from any temporal issues that must be present in a plan, and as such must not be equated with plans." (pp. 60–61)

From this passage, we conclude that there is no explicit notion of *hierarchical plans* in OPERA. Even so, the general notion of *goal decomposition* can be identified in OPERA, as we have done in the previous subsection, by making it to correspond to the notion of *subobjectives specification* in OPERA (see Figure 2).

Lastly, we note that the abstract concepts of *goal decomposition* and *goal relationships* admit concrete subtypes of various kinds in the organizational models analysed. For instance, in TAEMS, the kinds of goal decomposition are denominated *quality accumulation functions* (*qafs*) and the kinds of goal relationships are named *non-local effects* (*nles*): Examples of *qafs* are `q_seq_last` (all subtasks must be completed in order, and overall quality is the quality of last task), `q_sum_all` (all subtasks must be completed in no specific order, and overall quality is the aggregate quality of all subtasks) and `q_exactly_one` (only one subtask may be performed, and overall quality is the quality of the single task performed); regarding specific *nles*, two of them are *facilitates* (when information from one task reduces or changes the search space making some other task easier to solve) and *enables* (when information from one task is a prerequisite for doing another task). In STEAM, there are two basic concrete subtypes of goal decomposition: AND (when all suboperators must be done to realize a given operator), and OR (when at least one suboperator must be done to realize a given operator), which are complemented by the *depends* relationship (that establishes a partial order for doing operators). Finally, in MOISE+, there is not the concept of *goal relationship*, only three subtypes of goal decomposition: *Sequence* (when subgoals must be achieved in some order to achieve a given goal), *choice* (when only one subgoal must be achieve to fulfill a given goal), and *parallelism* (when all subgoals can be pursued at the same time in order to achieve a given goal).

4.2. Structural Dimension

Forming the structural dimension, we have modeling concepts used to specify the internal structure of an agent organization in which the agents must engage to become an active member of the organization. From Table 1, five organizational models provide concepts for creating structural specifications. They are: AGR, TAEMS, MOISE+, ISLANDER and OPERA. Looking carefully at the structural specifications one is able to produce using these models, like the ones shown in Figure 3, we realize that the structural dimension of organizational modeling can also be characterized by an abstract conceptual pattern.

4.2.1. Graphs of Roles and Groups

In the structural dimension, the organizational models present three fundamental concepts: Role, group and role relationships. The instantiation of these three interrelated concepts forms the specification of the internal structure of agent organizations.

In essence, structural specifications can be characterized directed graphs where:

- The nodes correspond
  - either to the definition of *groups* (in AGR, MOISE+ and OPERA) or *teams* (in STEAMS),
  - or to the definition of *roles* (in all models);
- the edges represent
  - either the *decomposition of a group in subgroups* (or a team in subteams), forming a *group (team) hierarchy*,
  - or binary *relationships between roles*,

- or *links from a role to a group* (or *subteam*) in which the role can be played by agents.

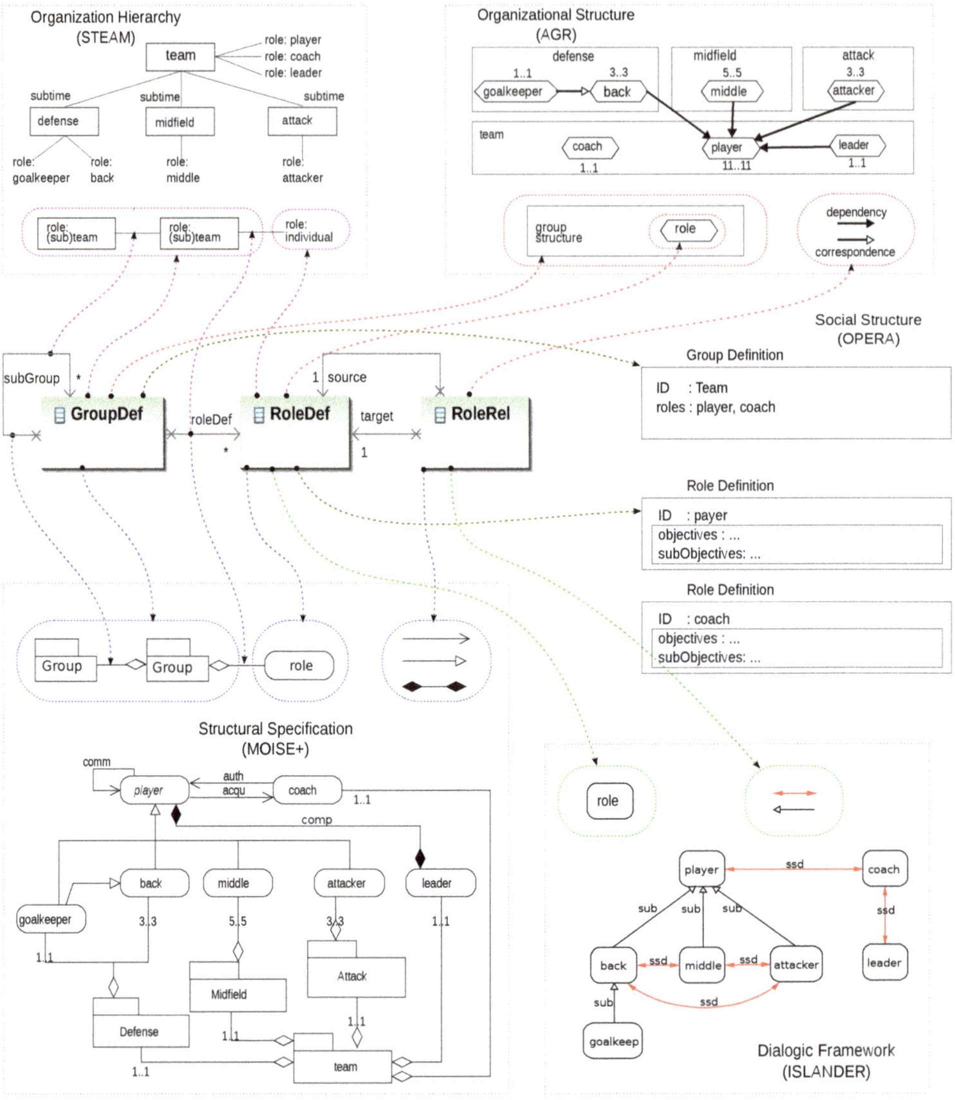

Figure 3. Similar structural specifications written in STEAM, AGR, OPERA, MOISE+ and ISLANDER, for a team of agents part of a simulated soccer game (example taken from [24,25]). The team is composed of eleven *players* (one *goalkeeper*, three *backs*, five *mid-filders* and three *attackers*) and one *coach*, and is divided in three groups: *Defense*, *midfield* and *attack*. The class diagram (Ecore metamodel) in the center of the figure represents the conceptual pattern identifiable in the various approaches of structural modeling. The dotted arrows detail what concepts are captured by what classes of the metamodel.

With regard to the *group* (*team*) *hierarchy*, there are two situations present in the models. On the one hand, we have a unique *root group* that represents the organization as a whole (the hierarchy is a rooted tree), what occurs in STEAM, MOISE+ and OPERA. On the other hand, there is no explicit *root group*, or even the decomposition of groups in subgroups, what is the case of AGR. Regarding the *binary relationships between roles*, in general they are directed and, in the models AGR, MOISE+ and ISLANDER they are subdivided in various kinds with diverse interpretations.

Such graphs of roles and groups receive the names of "organization hierarchy" in STEAM, "organizational structure" in AGR, "structural specification" in MOISE+. In ISLANDER, they are part of the definition of a "dialogic framework". In OPERA, they form the "social structure".

In Figure 3, the conceptual pattern identified in the structural dimension is represented by means of an Ecore metamodel. This metamodel is presented in more detail in Table 4.

Table 4. Structural specification pattern.

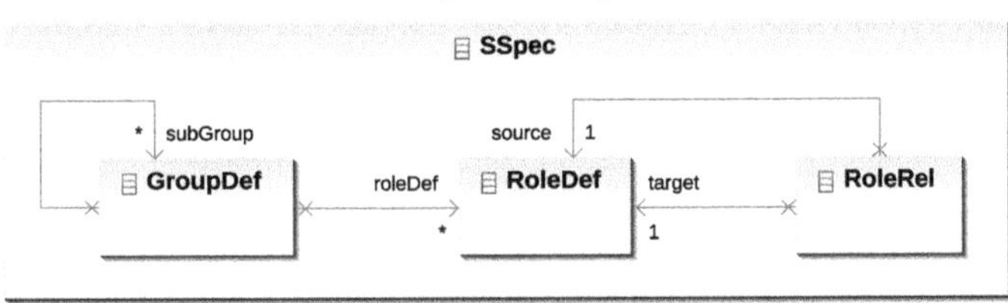

Class	Description
`SSpec`	Represents the concept of *structural specification*, i.e., organizational specifications restricted to the structural dimension.
`GroupDef`	Part of a structural specification that represents a *group definition*. Each group definition is characterized by the declaration of the roles that agents can play in the group (the `RoleDef` referenced by `GroupDef::rootDef`), and by possible subgroup definitions (referenced by `GroupDef::subGroup`). Abstracts the concepts of *group structure* in AGR, *subteam role* in STEAM, *group specification* in MOISE+, and *group definition* in OPERA.
`RoleDef`	Part of a structural specification that represents a *role definition*. Abstracts the concepts of *role* in AGR, MOISE+, ISLANDER and OPERA, and *individual role* in STAEMS.
`RoleRel`	Part of a structural specification that represents a *direct relationship* between two roles: a *source role* referenced by `RoleRel::source`, and a *target role* referenced by `RoleRel::target`. Abstracts the concept of *role constraints* in AGR, *role relations* and *inheritance* in MOISE+, *static separation of duties* (*ssd*) and *subroles* in ISLANDER.
	Auxiliary definitions
`GroupDef`	`::getAllSubGroup() : Set(GroupDef)`, a query that returns the set of all group definitions that, direct or indirectly, are subgroups of a given `GroupDef` via the `GroupDef::subGroup` reference. In OCL:

```
context  GroupDef
def:     getAllSubGroup( ) :  Set(GroupDef)
         = subGroup(goal) -> union(
                   subGroup -> collect (sg | sg.getAllSubGroup()));
```

Contextual Constraints

(1) All group definition shall reference at least one role definition or one subgroup definition, or both. In OCL:

```
context  GroupDef
inv:     roleDef -> notEmpty() or subGroup -> notEmpty()
```

(2) No group definition can be, direct or indirectly, a subgroup of itself, i.e., the group definitions shall form an acyclic directed graph.
In OCL:

```
context  GroupDef
inv:     not self.getAllSubGroup() -> includes(self)
```

4.2.2. Particularities

Besides the similarities that give rise to the structural specification pattern, the organizational models also differ in some particular points. Four of these points deserve mention.

The first one is related to the definition of subgroups. In STEAM and MOISE+, the definition of subgroups forms a real hierarchy, i.e., an non cyclic graph. On the other hand, in AGR and OPERA, there is no explicit subgroup relationships between group definitions. From structural specification pattern perspective, this fact can be expressed in the following way: In AGR and OPERA, for each group definition g (instance of `GroupDef`), the collection of its subgroups is empty, i.e., `g.subGroup = {}`. On the other hand, in STEAM and MOISE+, there exist group definitions g such that `g.subGroup` $\neq$ `{}`.

The second point is the notion of *cardinality* that is found in AGR and MOISE+ but not in the other models. Cardinalities can be defined for roles or subgroups. In the case of roles, cardinalities indicate a maximum and a minimum number of agents allowed per role in the context of a group. Regarding cardinalities of groups, they determine how many subgroups of a given type can be created in the context of a group. In AGR, cardinalities are attributes of role and group. In MOISE+, they are attributes of the association between role and group, or group and subgroup. For this reason, and observing that they are not an explicitly feature of the majority of the models, we have chosen not to explicitly represent cardinalities in the structural specification pattern.

As third point, we observe that the abstract notion of *structural relationships between roles* (class `RoleRel`, Table 4) admits diverse concrete subtypes, analogously to what happens with the notion of *goal relationships* (class `RoleRel`, Table 2). In AGR, for instance, there exist two subtypes: *Correspondence* (which states that agents playing one role will automatically play another one) and *dependency* (which rules out the possibility of an agent to play one role if it is not playing another role). In MOISE+, three subtypes: *Links* (which declare the possible relationships of *communication*, *authority* and *acquaintance* between roles), *compatibility* (which determines that two roles can be played at the same time by the same agent) and *inheritance* (which states that one role, besides its own features, also has all the features, like links and compatibilities, of another role). In ISLANDER, two subtypes: The concept of *subroles* (which is similar to the concept of inheritance in MOISE+), and the concept of *static separation of duties* (which means the opposite of the concept of *compatibility* in MOISE+).

In OPERA, there is only one type of binary directed relationship between roles: The *dependency*. Nevertheless, differently from the other organizational models, the concept of *dependency between roles* in OPERA is not properly a structural but rather a functional relationship. In other words, in OPERA, the dependency relationship reflects directly the decomposition of a goal into subgoals, elements of the functional dimension. When one of the subgoals defined in the scope of a role is a goal of another role, then there exist the dependency relationship between the two roles in OPERA. Such idea is different from the structural dependency present in AGR, which indicates that the fact of playing a given role is a prerequisite for playing another role.

The last point that should be mentioned concerns the nature of the group definitions. In all analysed models, except MOISE+, agents playing any roles in the same group may, in principle, exchange messages. Further, in the absence of explicit constraints, such as incompatibilities or dependencies, the agents are free to play the roles they wish in a given group. In MOISE+, we have the opposite situation. If there is no explicitly stated communications links or compatibility relations between roles, the agents are not allowed to exchange messages or play more than one role in the same group. Moreover, in MOISE+, communication links or compatibility relations are not limited to a single group, but can be specified between roles defined in different groups leading to possible inter-group collaborations. In other models, such as AGR, this inter-group collaboration can also be achieved by means explicitly defined correspondence links between two roles in different groups. In this way, an agent playing one of the roles automatically plays the other role and can participate in more than one group at the same time.

4.3. Dialogical Dimension

The dialogical dimension is characterized by concepts to prescribe (or describe) the direct interaction by means of message exchanging that occurs between role playing agents in order to achieve organizational goals. Among the organizational models considered in this work, only ISLANDER and OPERA offer explicit concepts for dialogical modeling (see Table 1). In these two models, the dialogical specifications are written according to approximate conceptual structures, as can be seen in Figure 4.

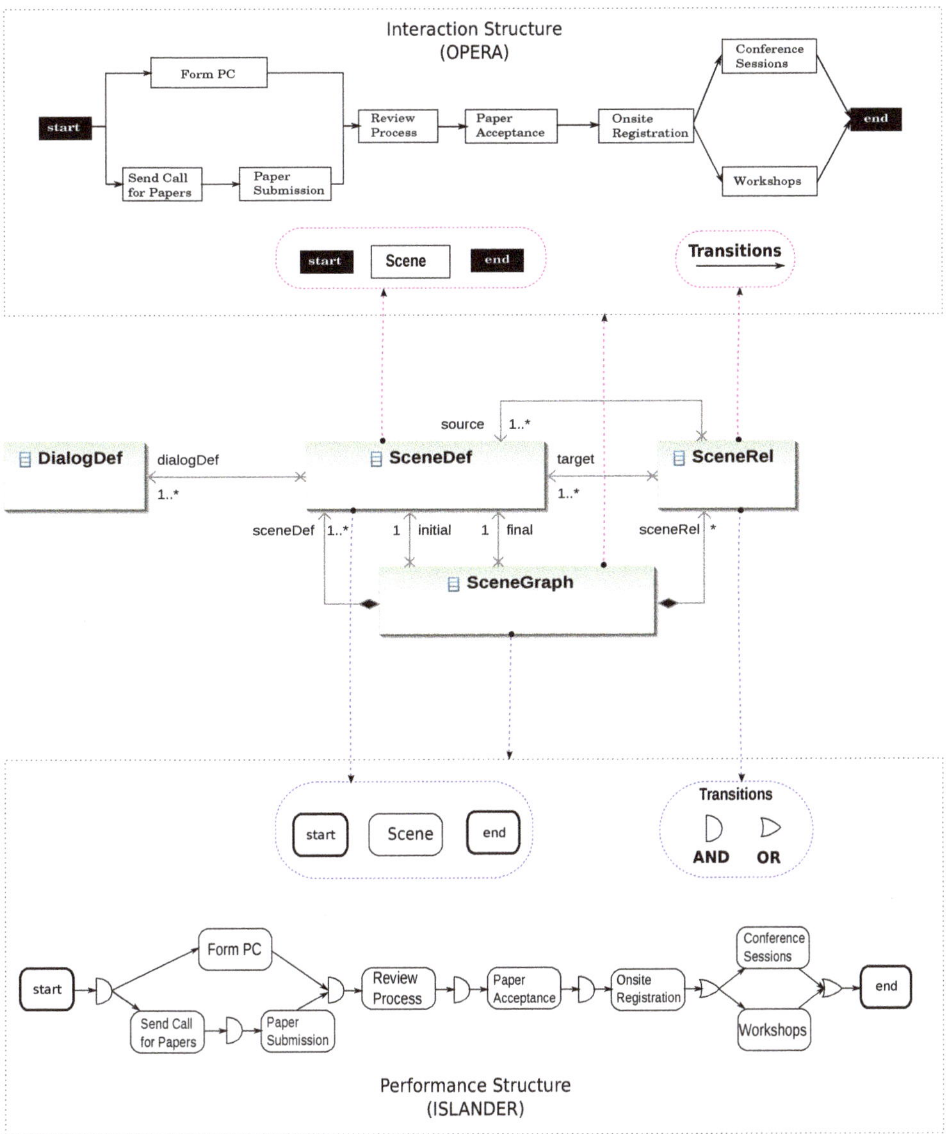

Figure 4. Similar dialogical specifications written in OPERA and ISLANDER for an agent based conference management system (example taken from [27] chapter 3). The class diagram (Ecore metamodel) in the center represents the conceptual pattern identifiable in the two approaches of dialogical modeling. The dotted arrows detail what concepts are captured by what classes of the metamodel.

4.3.1. Hypergraphs of Scenes

On a macro level, both in ISLANDER and in OPERA, the direct interactions by message exchanging are partitioned into *scenes*. Scenes are structured and coordinated by means of directed *hypergraphs* (directed graphs where some edges, called *hyperedges*, can connect any number of sources and target nodes)in which:

- Nodes correspond to the definitions of *scenes*;
- hiperedges represent partial ordering and/or synchronization relationships from many source scenes to many target scenes, giving rise to the concept of *scene transitions*.

In a well formed hypergraph of scenes, there is:

- An *initial scene*, i.e., the node from which agents playing roles have access to the other scenes of the hypergraph. Starting in the initial scene and following the transitions, all the scenes that make up a hypergraph should be achievable;
- a *final scene*, i.e., the node in which the dialogic participation of agents within an organization ends successfully. As the dual of the initial scene, the final scene has to be achievable from any scene in a hypergraph, otherwise the hypergraph of scenes is not well formed.

In ISLANDER, the hypergraphs of scenes are named "performance structures", and in OPERA they are called "interaction structures".

On a micro level, the interactions within each scene are governed by one or more predefined *dialogue scripts*. These scripts correspond to the concepts of *scene protocol* in ISLANDER and *interaction pattern* in OPERA. Dialogue scripts are not detailed in Figure 4. The reason is that the intra-scene (micro level) dialogical specifications have distinct natures both in ISLANDER and in OPERA, as will be discussed in the sequel.

In Figure 4, the conceptual pattern identified in the dialogical specifications of ISLANDER and OPERA is captured by means of an Ecore metamodel. This metamodel is described in details in Tables 5 and 6.

4.3.2. Particularities

In both ISLANDER and OPERA, the dialogical specification consists in a network of scenes in which all possible or desirable episodes of direct interaction within an organization are planned and orchestrated. As mentioned earlier, this common structure takes place at the macro level. This means that the joint activity characteristic of agent organization, under a broad point of view, is ruled by the presented hypergraphs of scenes.

The main difference between ISLANDER and OPERA occurs at the micro level. In other words, restricting the point of view to each particular scene, instead of the network of scenes, the models analyzed have different ways to specify how agents can or should interact.

On the one hand, in ISLANDER, there is the notion of *scene protocol*. In a scene protocol, one represents in detail a communication protocol in which are specified all the involved roles, and the sequencing of all possible message exchanges (*illocution schemes*), in On the other hand, in OPERA, there is the notion of interaction pattern. Unlike scene protocols, an interaction pattern does not determine in detail the exchange of messages in a given scene. Instead, it delimits a partial order between scene states (*landmarks*) towards achieving the objectives related to the scene. Any detailed communication protocol used in a scene should respect the established interaction pattern.

4.4. Normative Dimension

The normative modeling dimension is characterized by the general concept of *norm* (*permissions*, *obligations*, etc.). Norms occur in organizational specifications as a mechanism that interrelates and complements the functional, structural and dialogical specifications. Three organizational models we have analyzed present concepts to create normative specifications. They are: MOISE+, ISLANDER and OPERA (see Table 1).

4.4.1. The Concept of Norm

In the organizational models analyzed, unlike the other dimensions, the normative specifications are not created as graph like structures. Instead, they assume the form of *textual normative expressions*. This general pattern is expressed in Table 7.

Table 5. Dialogical specification pattern: Classes description.

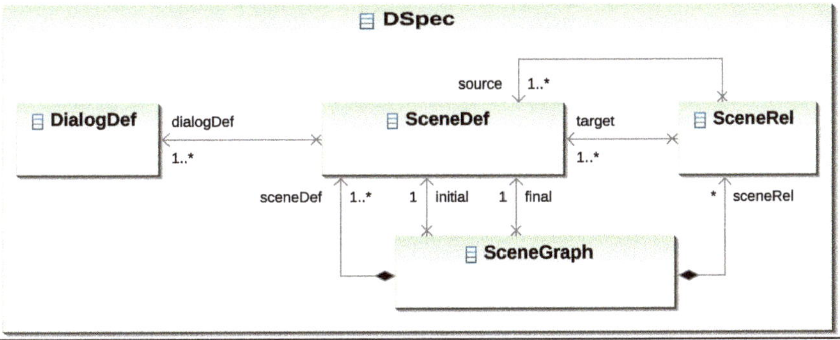

Class	Description
`DSpec`	Represents the concept of *dialogical specification*, i.e., an organizational specification restricted to the dialogical dimension.
`SceneGraph`	Part of a dialogical specification that represents the definition of a *hypergraph of scenes*. `SceneGraph` is characterized by one or more *scene definitions* (referenced by `SceneGraph::sceneDef`), and several *scene relationships* (referenced by `SceneGraph::sceneRel`). Among the scenes, there are one *initial scene* (referenced by `SceneGraph::initial`) and one *final scene* (referenced by `SceneGraph:: final`). Abstracts the concepts of *performance structure* in ISLANDER and *interaction structure* in OPERA.
`SceneDef`	Part of a hypergraph of scenes that represents the definition of a *scene*. Each scene definition is characterized by the declaration of a *dialogue script*, via reference `SceneDef::dialogDef`. Abstracts the concepts of *scene definition* in ISLANDER and in OPERA.
`SceneRel`	Part of a hypergraph scene that represents a directed relationship from one or more source scenes (referenced by `SceneRel::source`), to one or more target scenes (referenced by `SceneRel::target`). Abstracts the concepts of *scene transitions* in OPERA and in ISLANDER.
`DialogDef`	Part of of a dialogical specification that represents the definition of a *dialogue script*. Abstracts the concepts of *scene protocol* in ISLANDER, and *interaction pattern* in OPERA.

Table 6. Dialogical specification pattern: Contextual constraints.

Class	Auxiliary Definitions
`SceneGraph`	OCL queries that return immediate predecessors and successors of a scene definition in the context of a hypergraph of scenes.

```
context  SceneGraph
def:     getSourceSceneDef(sceneDef : SceneDef): Set(SceneDef)
         = getSceneRelHavingTarget(sceneDef) -> collect(sr | sr.source);

def:     getTargetSceneDef(sceneDef : SceneDef): Set(SceneDef)
         = getSceneRelHavingSource(sceneDef) -> collect(sr | sr.target);

def:     getSceneRelHavingTarget(sceneDef : SceneDef): Set(SceneRel)
         = sceneRel) -> select(sr | sr.target -> includes(sceneDef));

def:     getSceneRelHavingSource(sceneDef : SceneDef): Set(SceneRel)
         = sceneRel) -> select(sr | sr.source -> includes(sceneDef));
```

Contextual Constraints

(1) The initial and final scenes are scenes defined in the context of the same `SceneGraph`. In OCL:

```
context  SceneGraph
inv:     sceneDef-> includes(initial) and
inv:     sceneDef-> includes(final)
```

(2) In a `SceneGraph`, there should be no relationships arriving at the initial scene or departing from the final scene. In OCL:

```
inv:  getSceneRelHavingTarget(initial)-> isEmpty() and
inv:  getSceneRelHavingSource(final)-> isEmpty()
```

(3) In a `SceneGraph`, all scene definitions must directly or indirectly be reachable from the initial scene as well as lead to the final scene, via the scene relationships. In OCL:

```
context  SceneGraph
inv:     sceneDef-> forAll(sd |
                   getTargetClosure(Set{initial} -> includes(sd) and
                   getSourceClosure(Set{final} -> includes(sd))

def:     getTargetClosure(sceneSet : Set(SceneDef)): Set(SceneDef)
         = let newSceneSet = sceneSet-> collect(scene |
                   getTargetSceneDef(scene) -> including(scene)
                   ) in
         if sceneSet -> includesAll(newSceneSet)
              sceneSet
         else
         getTargetClosure(newSceneSet)
         endif;
def:     getSourceClosure( sceneSet : Set(SceneDef)): Set(SceneDef)
         = let newSceneSet = sceneSet-> collect(scene |
                   getSourceSceneDef(scene) -> including(scene)
                   ) in
         if sceneSet -> includesAll(newSceneSet)
              sceneSet
         else
         getSourceClosure(newSceneSet)
         endif;
```

(4) Every scene relationship must only involve scenes defined within the context of the same `SceneGraph`. In OCL:

```
context  SceneGraph
inv:     sceneRel-> forAll(sr |
                   sceneDef -> includesAll(sr.source) and
                   sceneDef -> includesAll(sr.target))
```

4.4.2. Particularities

In ISLANDER, norms are written as logical expressions in accordance with the format:

$$(s_1, \gamma_1) \wedge ... \wedge (s_m, \gamma_m) \;\wedge\; e_1 \wedge ... \wedge e_n \;\wedge\; \neg((s_{m+1}, \gamma_{m+1}) \wedge ... \wedge (s_{m+n}, \gamma_{m+n})) \rightarrow obl_1 \wedge ... \wedge obl_p$$

where (s_1, γ_1), ..., (s_{m+n}, γ_{m+n}) are pairs of *scenes* and *illocution schemes*, e_1, ..., e_n are *boolean expressions* over illocution scheme variables, $\neg$ is a *defeasible negation*, and obl_1, ..., obl_p are *obligations*. "The meaning of these rules is that if the illocutions (s_1, γ_1), ..., (s_{m+n}, γ_{m+n}) have been uttered, the expressions e_1, ..., e_n are satisfied and the illocutions (s_{m+1}, γ_{m+1}), ..., (s_{m+n}, γ_{m+n}) have *not* been uttered, the obligations obl_1, ..., obl_p hold" ([48] p. 38).

In OPERA, norms are specified using logical expressions written in a formalism called LCR (*Logic for Contract Representation*). There are three types of norms: *Obligations, permissions* and *prohibitions*. The following excerpt from ([27] p. 149) summarizes the syntax for writing these modalities.

```
<Norm>::= OBLIGED(<id>,<Norm-Form>)|PERMITTED(<id>,<Norm-Form>)|
FORBIDDEN(<id>,<Norm-Form>)
```

`OBLIGED(<id >,<Norm-Form>)` represents an obligation of the agent playing the role referenced by `<id>` in achieving the state `<Norm-Form>` described as an LCR formula ([27] chapter 4). Based on the notion of *obligation*, the concepts of *permission* and *prohibition* are defined. A permission `PERMITTED<id>,<Norm-Form>)` is an abbreviation for `¬OBLIGED(<id>,¬<Norm-Form>)`. In turn, a prohibition `FORBIDDEN(<id>,<Norm-Form>)` means the same as `OBLIGED(<id>, ¬<Norm-Form>)`.

Table 7. Normative specification pattern: Class description.

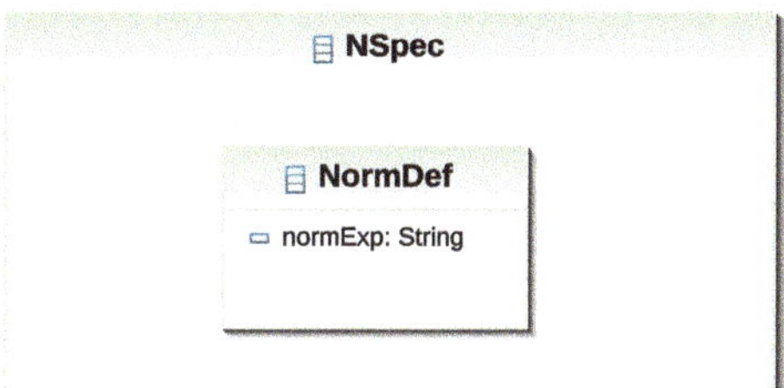

Class	Description
NSpec	Represents the concept of *normative specification*, i.e., organizational specifications restricted to the normative dimension.
NormDef	Central part of a normative specification. Represents the definition of a *norm*. In general, norm definitions are characterized by a *normative expression* (attribute `NormDef::normExp`) that refers to elements found in the structural, functional and dialogical dimensions. The form and meaning of the normative expression vary considerably in the organizational models analyzed. Abstracts the concepts of *deontic relation* found in MOISE +, and the particular concepts of *norm* found in ISLANDER and OPERA.

Finally, in MOISE+, the general concept of *norm* is translated into the notion of *deontic relations* that link *roles* to *missions*. There are two types of deontic relations, *permissions* and *obligations*:

> "A permission *per*(ρ, *m*, *tc*) states that an agent within the role ρ may be committed to the mission *m*. Temporal constraints (*tc*) are established for the permission, that is, they determine a set of time periods when the permission is valid ... An obligation *obl*(ρ, *m*, *tc*) states that an agent within the role ρ is required to commit to the mission *m* in the time periods determined by *tc*." ([49] pp. 46–47)

In this case, the normative expressions *per*(ρ, *m*, *tc*) and *obl*(ρ, *m*, *tc*) are less comprehensive than what is found in OPERA and ISLANDER.

4.5. Abstract Organizational Metamodel

All the patterns identified in the organizational modeling dimensions, and previously discussed in Sections 4.1–4.4, can be combined to form an *abstract organizational metamodel*. This abstract metamodel, as shown in Figure 5, characterizes the common conceptual structure of the organizational models analyzed.

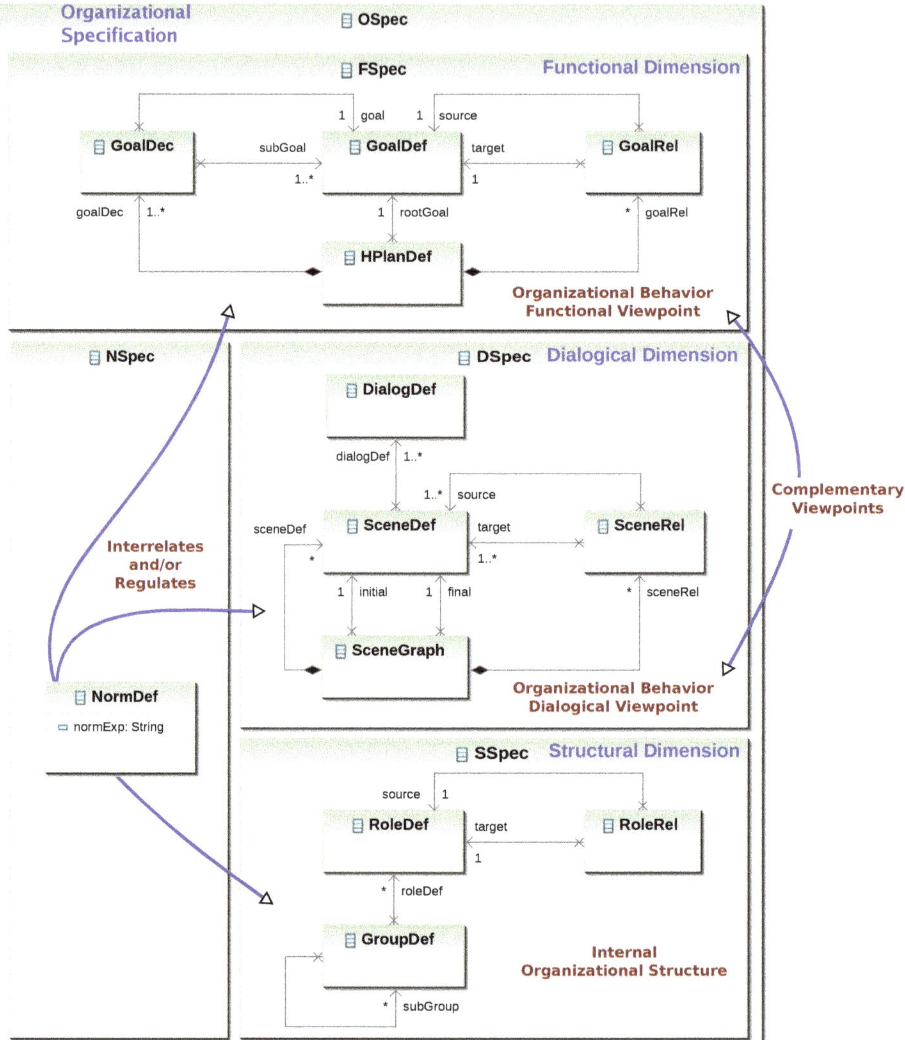

Figure 5. Abstract organizational metamodel.

By means of this abstract organization metamodel, we can see that the normative dimension works as a glue among the three others. It interrelates and/or regulates the organization behavior (be it functional or dialogical) and the organizational internal structure (in the sense of allowing or forcing the association of certain functional and/or dialogical elements with certain structural elements). Last, but not least, it makes it clear that structuring of organizational modeling dimensions greatly helps in making the notions independent and self-contained, while linked via normative bonds.

5. Integration Method Application

In this section we present an application of the integration method described in Section 3, guided by conceptual patterns identified in Section 4. We show how to apply the method to integrate the AGR, STEAM and MOISE+ models. Since a complete description of the integration process involve many details, we will restrict the discussion to the structural dimension.

As shown in Figure 1, we need to perform two iterations of the method to integrate three models. First we merge AGR and STEAM. Then, we merge MOISE+ to the result of the previous iteration.

5.1. First Iteration

The representation of AGR and STEAM as Ecore/OCL metamodels is shown in Figure 6. Below, on the left side, we have the AGR metamodel; on the right side, the STEAM metamodel (both restricted to the structural dimension). Above, mediating the alignment of the metamodels, we see the conceptual pattern of Section 4.2.

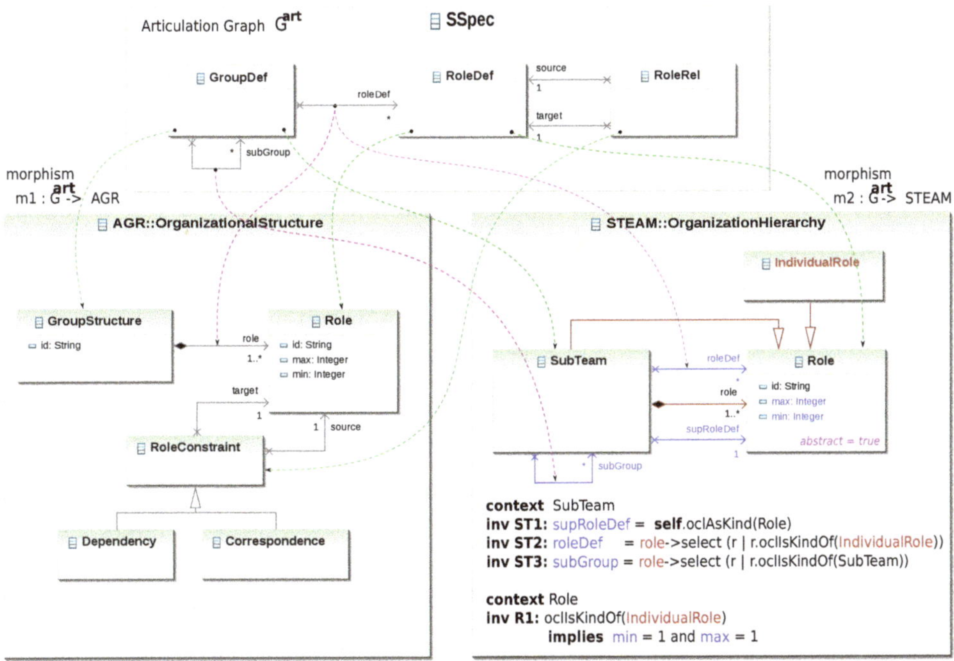

Figure 6. Alignment between AGR and STEAM.

In the alignment, the *organizational structure* of AGR and the *organization hierarchy* of STEAM are identified as similar specifications. The concepts of *group* (AGR) and *subteam* (STEAM) are declared similar concepts, both identified with the general concept of *group definition*. The same happens with the concepts of *role* (AGR) and *role* (STEAM), both identified with the general concept of *role definition*.

Intuitively, when we take into account only the terms used, these basic correspondences between AGR and STEAM are reasonable. However, when we look more closely to the specific relationships among the concepts, it is possible to see that there is a stronger coupling between *subteam* and *role* in STEAM than the one that exists between the correspondent concepts of *group* and *role* in AGR. In STEAM, the notion of *role* is abstract, being materialized both in the specification of activities for groups of agents as a whole and in the specification of activities for individual agents (*individual role*). This notion is represented in the metamodel as an abstract class `STEAM::Role` with two concrete subtypes `STEAM::SubTeam` and `STEAM::IndividualRole`. As a consequence, every instance of `STEAM::SubTeam` besides corresponding to a *group definition* is also a kind of *role definition*.

This coupling between the concepts of *group* and *role definitions* is absent in AGR and is not foreseen in the structural pattern of Section 4.2. To easy the merging of these different views regarding the nature of the concepts, one possibility is to interpret the generalization relation between `STEAM::SubTeam` and `STEAM::Role` as a composition relationship, similar to the application of the "replace inheritance with delegation" refactoring, as proposed in [50]. When this is done we posit a derived reference from `STEAM::SubTeam` to `STEAM::Role` named `supRoleDef`. This derived reference when used in the place of the original generalization decouples the concepts of *group definition* and *role definition* while permitting to represent the same information in a slightly different way.

In Figure 6 there are two other derived references: `roleDef` and `subGroup`; both extracted from the original reference `role` between `STEAM::SubTeam` and `STEAM::Role`. The rationale for these is to make explicit that, in reality, not only *role definitions* but also *group definitions* can be associated with a *subteam* via the reference `role`. Once these derived references become explicit, we can do a more fine grained matching between the STEAM metamodel and the corresponding abstract concepts of *group* and *role definition*.

Ending the comparison between the metamodels, we note that in AGR the *group definitions* cannot be decomposed into *sub groups* and there is the concept of *role relation* materialized as *role constraints* (*dependencies* and *correspondences*). In STEAM, there is no explicit *role relations* and no explicit *role cardinality*. In the case of *individual role* there is an implicit cardinality of *min=1* and *max=1*.

Finally, concluding the iteration, we merge the metamodels taking into account the correspondences identified. The resultant integrated metamodel $MM^{int}_{\#1}$ is shown in Figure 7, where we retain the terminology of the abstract structural pattern of Section 4.2. See online version for colors. The elements added to the structural pattern from the specific metamodels are depicted in blue. The elements marked in red in the STEAM metamodel (Figure 6) are not included in the integrated metamodel, being replaced by derived references aforementioned. Essentially, $MM^{int}_{\#1}$ is an *almagamated sum* of the AGR and STEAM metamodels (viewed as graphs) modulo the alignment (articulation) between AGR and STEAM, as describe in Section 3.4. For simplicity, the morphisms from $MM^{int}_{\#1}$ to the AGR and STEAM metamodels are omitted.

5.2. Second Iteration

In the second iteration, we integrate the MOISE+ metamodel (structural dimension) to the resulting metamodel $MM^{int}_{\#1}$ obtained in the previous iteration. The MOISE+ metamodel is shown on the left side of Figure 8. On the right side, we have $MM^{int}_{\#1}$ (from Figure 7) augmented with derived classes and relationships. On the middle, there is the articulation graph between MOISE+ and $MM^{int}_{\#1}$ metamodels. Since the articulation graph preserves the class and reference names from $MM^{int}_{\#1}$, for simplicity we have omitted the morphism $m_2 : G^{art} \rightarrow MM^{int}_{\#1}$.

In $MM^{int}_{\#1}$ there are three classes `GroupDef`, `RoleDef` and `RoleRel` which represent the main concepts for the structural specification of agent organizations. In MOISE+, the correspondent classes are `GroupSpecification`, `Role` and `RoleRelation`, respectively. Similar to class $MM^{int}_{\#1}$`::GroupDef`, class `MOISE+::GroupSpecification` represents the definition of a *group* in which it is possible to specify *roles* and *subgroups*. Like $MM^{int}_{\#1}$`::RoleDef`, the class `MOISE+::Role` denotes a *role definition* associated with *group definitions*. Both $MM^{int}_{\#1}$`::RoleRel` and `MOISE+::RoleRelation` characterize *role relationships* from a target to a source *role definition*.

Apart from this basic agreement, there are some particularities regarding how these concepts occur in MOISE+ that leads to an extension of $MM^{int}_{\#1}$. One first particularity is the way in which *group definitions* are linked to *role definitions* and *subgroups*. In the integration of AGR and STEAM, *group definitions* are linked to *role definitions* and to *subgroups* by means of the `roleDef` and `subGroup` references, respectively. In the MOISE+ metamodel the correspondent links are not represented by references but by the classes `GroupRole` and `SubGroup`, respectively.

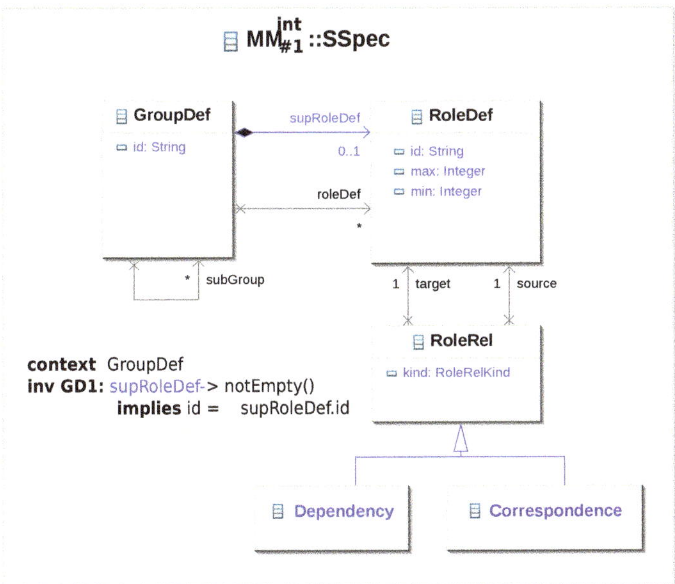

Figure 7. Integrated metamodel $MM^{int}_{\#1}$.

By means of MOISE+::GroupRole and MOISE+::SubGroup, the same *role definition* or *subgroup* can have different cardinalities, one for each *group definition* in which the definition is referenced. The cardinalities are represented by the attributes max (for the maximum number of agents per role, or subgroups in a group) and min (for the minimum number).

In $MM^{int}_{\#1}$ this flexibility is not possible as long as the cardinalities are declared directly as attributes of the *role definition*, and not as attributes of the relation between a *group definition* and *role* or *subgroup definition*. In this way, we note that the information about role and group cardinalities of MOISE+ can not always be expressed in the current integration of ARG and STEAM. However, the converse is always possible, as it can be shown by means of the derived classes $MM^{int}_{\#1}$::RoleRef and $MM^{int}_{\#1}$::GroupRef in the upper right of Figure 8.

The derived class $MM^{int}_{\#1}$::RoleRef is the implict correspondent of MOISE+::GroupRole. Similar to class MOISE+::GroupRole, class $MM^{int}_{\#1}$::RoleRef has attributes max and min and makes reference a single *role definition*. In the context of $MM^{int}_{\#1}$::GroupDef the derivation of $MM^{int}_{\#1}$::RoleRef is specified by the invariant GD2 shown in the bottom right of Figure 8. This invariant establishes that for each instance rd of $MM^{int}_{\#1}$::RoleDef (referenced by roleDef), there must exist (be created) an instance rr of $MM^{int}_{\#1}$::RoleRef that points to rd and has the attributes rr.min = rd.min and rr.max = rd.max.

By its turn, the class $MM^{int}_{\#1}$::GroupRef is the derived correspondent of MOISE+::SubGroup. In the context of $MM^{int}_{\#1}$::GroupDef the derivation of $MM^{int}_{\#1}$::GroupRef is specified by the invariant GD3 shown in the bottom right of Figure 8. The invariant establishes that for each instance sg of $MM^{int}_{\#1}$::GroupDef (referenced by subGroup), there must exist (be created) an instance gr of $MM^{int}_{\#1}$::GroupRef pointing at sg and having the attributes gr.min = sg.supRoleDef.min and gr.max = sg.supRoleDef.max.

As long as they are more expressive, the classes RoleRef and GroupRef are used in the articulation graph replacing the references roleDef and subGroup present in $MM^{int}_{\#1}$. Therefore the replaced references are marked to be left out during the merge step at the end of the iteration. In addition, the attributes max and min in the class $MM^{int}_{\#1}$::RoleDef are marked once the same information is now represented as attributes of RoleRef and GroupRef in the articulation graph.

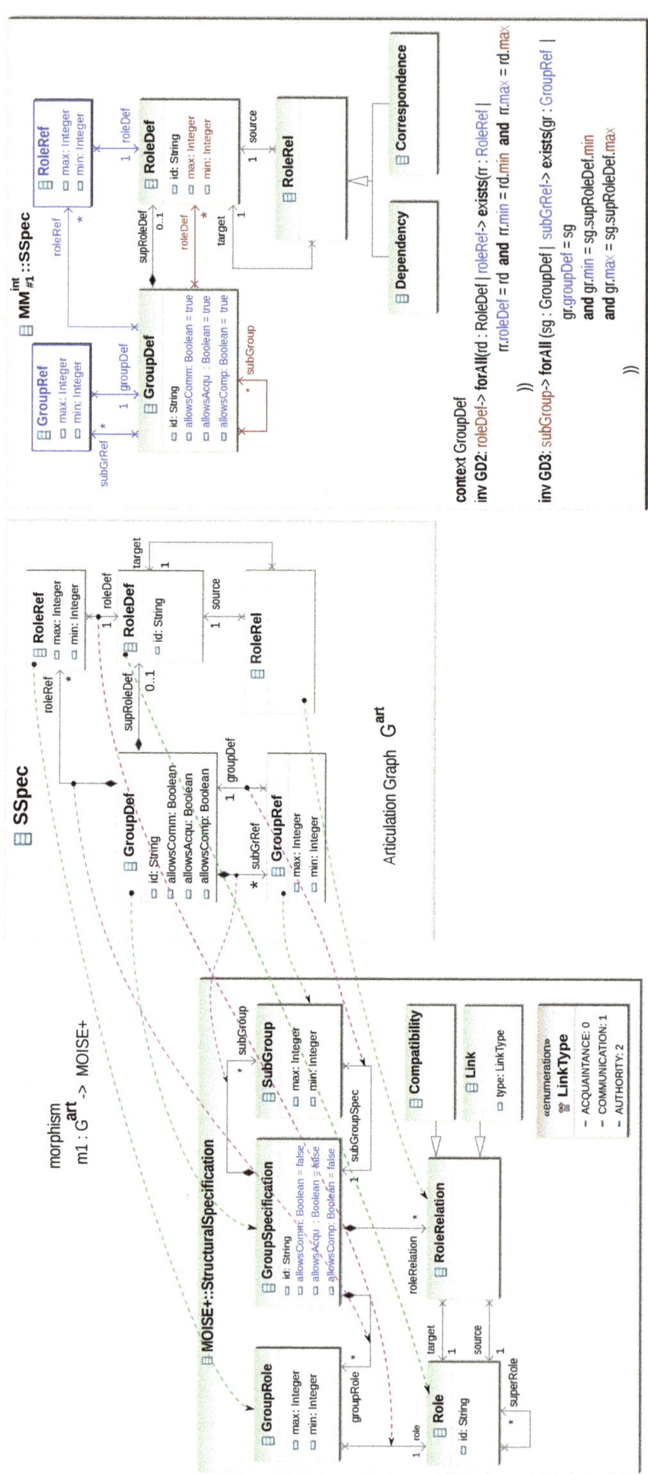

Figure 8. Alignment between MOISE+ and $MM^{int}_{\#1}$.

A second peculiarity of MOISE+ concerns the possibility of defining *links* and *compatibilities* between *roles* as part of a *group specification*. In this regard, there are two observation to be made. Firstly, *links* and *compatibilities* are new kinds of *role relation*, not present in $MM^{int}_{\#1}$. In fact, we observe that the *link* and *compatibility* concepts, in essence, differ from the *dependency* and *correspondence* concepts found in AGR. On the one hand, in MOISE+, a *link* enable the *acquaintance*, *communication*, or *authority* between roles; and a *compatibility* indicates that an agent playing a role can also play another role. On the other hand, in AGR, a *dependency* indicates that an agent only can play a role if it previously commits itself to another one; and a *correspondence* means that to play a role automatically implies to play another role.

The second observation is about the nature of the *group definition* concept behind the notions of *links* and *compatibilities*. By default, a *group definition* in MOISE+ does not enable the *compatibility* or any *link* between roles. In other words, if not stated explicitly, an agent playing a role do not have permission to play another role, or even to interact with agents playing another role, neither in the same nor in other groups. If *compatibilities* and *links* are needed, this must be explicitly specified in the *group definition*.

Conversely, in AGR and STEAM a *group definition* does not implies *a priori* any restriction regarding *compatibility* and *link* among the roles. With the exception of explicit *dependencies* relationships and cardinality restrictions, in AGR and STEAM specifications the agents are free to play the roles they want and are not blocked with respect to interacting with any agent playing some other role.

In the articulation presented in Figure 8, these observations are made explicit by means of the derived attributes `allowsComm`, `allowsAcqu` and `allowsComp` in the context *group definitions*. For `MOISE+::GroupSpecification`, these attributes have the value `false`. This represents respectively the *communication*, *acquaintance* and *compatibility* restrictions existing in MOISE+. On the other side, for $MM^{int}_{\#1}$`::GroupDef` the three attributes assume the value `true` indicating the absence of the respective restrictions in AGR and STEAM.

Despite their opposite nature, we note that AGR and STEAM *group definitions* can be expressed in MOISE+. To this end, one has to explicitly define *communication*, *acquaintance*, and *compatibilities* relationships between all roles in a *group definition*. However, the converse is not always possible without losing information.

Ending the comparison, in MOISE+ there is a third form of *role relationship*: the *inheritance*. In MOISE+ metamodel this relationship is represented by the reference `superRole` involving instances of the class `MOISE+::Role`. In the articulation between MOISE+ and $MM^{int}_{\#1}$, an alternative form of representing inheritance could be as a subclass of `RoleRel`. As the inheritance relation does not have a direct effect on the behavior of the agents as the other role relations, being only a way of simplifying role definitions in MOISE+, we have opted to preserve the representation of this relation as the reference `superRole` rather than defining a new subclass for `RoleRel`.

Finally, concluding the iteration, we merge the metamodels taking into account the identified correspondences. The result is show in Figure 9.

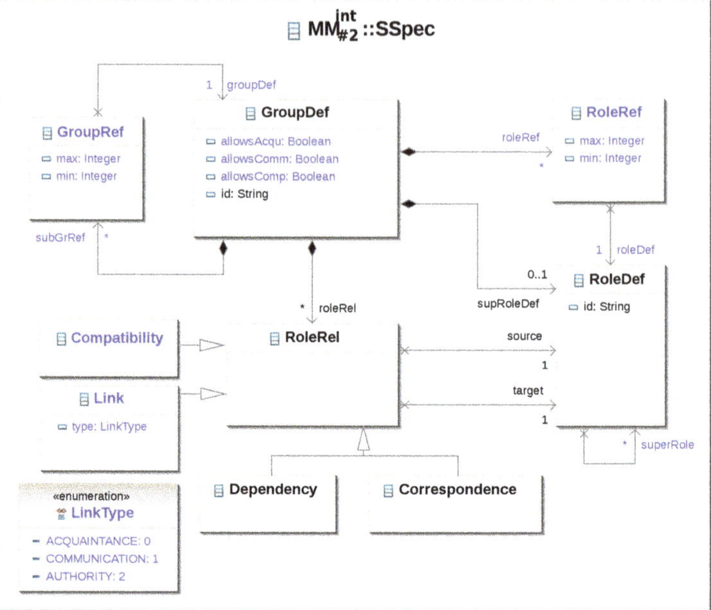

Figure 9. Integrated metamodel $MM_{\#2}^{int}$.

6. Organizational Interoperability Approach

Adopting an organization-centered perspective [4], the engineering of MAS can be described as a process that starts with the creation of an *organizational specification* written in conformance to an *organizational model*. This specification is a prescription of the desired patterns of joint activity that should occur inside the MAS towards some desired purpose. Once the organizational specification is done, it is used as the input to an *organizational infrastructure*. In general, what we mean by organizational infrastructure is some kind of middleware supposed to interpret the specification and reify the *organization* of the MAS outside the agents. In this respect, it should maintain an internal *organizational state* and offer to the agents an interface for accessing and modifying this state. The information maintained in the organizational state contains a list of the members of the organization, what roles they are playing, what groups are active in the organization, among others. Finally, with the organizational infrastructure materializing the desired agent organization, it is time to develop application domain agents (not necessarily by the same designer of the organization specification) that can enter and interact inside it by accessing the available organizational interface.

Regarding *organizational infrastructures* there are several approaches for the engineering of (open) agent organizations [12,13,36,47,51–54]. On the one hand, the availability of a wide range of diverse models and infrastructures has made the development of agent organizations feasible. On the other hand, such a diversity introduced an important new interoperability challenge for agent designers: How to deal with heterogeneous organizational models and infrastructures? Whenever an agent is build to enter some MAS it has to be able to interact with the other participants using a particular agent communication language as well as to understand received messages against a given domain ontology. Besides this, if the MAS was designed as an agent organization, the entering agent has also to be able to access a particular organizational infrastructure and to interpret its underlying organizational model. In this way, the agent design can become tailored to a particular organizational approach.

For instance, suppose that several e-business applications designed as open agent organizations are available on the Internet. In addition, assume that these applications are heterogeneous regarding the organizational technology applied to build them. To put it in more concrete terms, let us suppose two agents organizations: One built upon the S-MOISE+ [53] organizational middleware, based on the MOISE+ model, and the other by using MADKIT [54] platform, based on ARG model. In this setup (and assuming a shared common agent communication language and domain ontology), the agent designers face the following problem: The native S-MOISE+ agents do not interoperate with the MADKIT platform, and vice-versa. Thus, it is not directly possible, for instance, to write an agent code that enter both e-business agent organizations in the search of products and/or services on behalf of its users. Such fact limits the range of applicability of S-MOISE+ and MADKIT agents which, in turn, limits the idea of open MAS.

As mentioned in the Introduction, four basic approaches can be envisioned for this organizational interoperability problem. One of them is to bridge the interface between the external agents and the agent organization by means of model mapping (Adaptation). By using such mappings it is possible to provide adapted copies of the specification and state of a given organizational model/infrastructure "understood" by the external agents.

In what follows we describe MAORI—a *Model-based Architecture for ORganizational Interoperability* [19]. MAORI is an experimental framework for providing organizational interoperability following the line of adaptation. Its main objective is to show how the integration of organizational models presented in this paper can possibly be used in a solution for the problem of organizational interoperability.

6.1. MAORI Overview

MAORI is structured along three layers, as it may be seen in Figure 10:

- In the bottom, there is the *Model Integration* (M2M) layer—the purpose of this layer is to provide an integrated view and transformations between the organizational models represented as metamodels;
- in the middle, there is the *Organizational Interoperability* (ORI) layer—this layer is formed by components that use the M2M layer to translate and adapt the specification and state of agent organizations from one source organizational infrastructure to one or more target organizational infrastructures;
- in the top, there is the *Organizational Infrastructure* (MAS) layer—this layer corresponds to the available infrastructures for implementing organization-centered MAS.

6.2. Model Integration Layer

M2M layer is composed of *metamodels* and *transformations*. For each organizational model OM_i, there is a corresponding metamodel MM_i. The metamodel MM^{int} is the conceptual integration of all MM_i, as described in Section 5.

The transformations are functions that implement the morphisms between the integrated metamodel MM^{int} and the particular metamodels MM_i (Section 3.4). There are two types of transformations. One type is **transf**$(from : MM_i) : MM^{int}$, which converts from MM_i to MM^{int}. The other type is **transf**$(from : MM^{int}) : MM_i$, which converts from MM^{int} to MM_i. In this way, M2M main functionality is to provide transformations that can be combined to translate specifications and states between organizational models/infrastructures.

6.3. Organizational Interoperability Layer

ORI layer works as an extension of organizational infrastructures. In order to enable heterogeneous agents in the same organization, ORI adds two basic components to the organizational infrastructures: *providers* and *adapters*.

Providers are responsible for exporting the organizational specification/state of agent organizations. In this case, to export means to use **transf**$(from : MM_i) : MM^{int}$ to convert the specification/state from from a source MM_i to the integrated metamodel MM^{int}. Adapters are responsible for importing the organizational specification/state that was exported by a provider. The import is done by using **transf**$(from : MM^{int}) : MM_i$.

Imagine a scenario where an agent functions on a given organizational infrastructure and consider an agent organization implemented on a different organizational infrastructure. If the agent wants to participate in the organization, an adapter has to be instantiated in the organizational infrastructure of the entering agent. Initially, the responsibility of the adapter is to locate the appropriate provider, establish a connection with it, ask for the organizational specification/state and finally translate this specification/state to the target organizational infrastructure of the entering agent. In this way, for each heterogeneous agent organization there will be an organizational provider. Connected to this provider, there will be several organizational adapters, one for each organizational infrastructure in which there could be external heterogeneous agents.

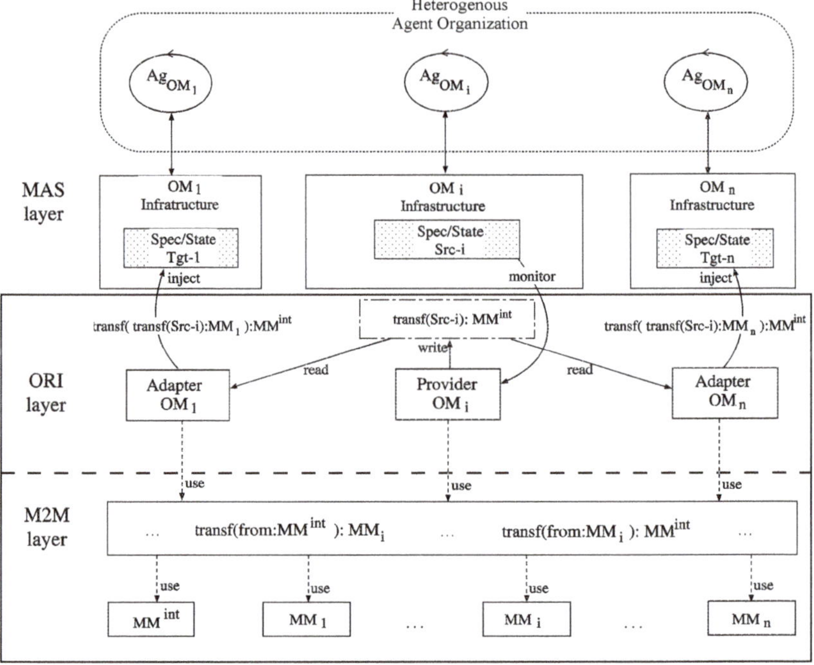

Figure 10. MAORI framework (redesigned from [19]).

6.4. MAORI Implementation

MAORI was implemented in the Java programming language. The metamodels in M2M layer were coded using the Eclipse Modeling Framework (EMF) [55]. Regarding the transformations, there were first prototyped in the Atlas Transformation Language (ATL) [56] and then ported to Java for performance reasons. An implementation was developed as a proof of concept of the ORI layer considering the MADKIT and S-MOISE+ organizational infrastructures.

To evaluate MAORI, some agent organizations were developed. One is the example of a group of agents that wants to write a paper and use for this purpose an explicit organization to help them to collaborate. The organization consists in a group composed of: One agent in the role of *coordinator* (who controls the process and writes the introduction and conclusion of the paper), one to five agents in the role of *collaborator* (who writes the paper sections) and one agent in the role of *librarian* (who compiles the bibliography). Taking this simple example, some experiments were performed. One of them considered an organization composed of five agents—one coordinator (Eric), three collaborators (Greg, Joel and Mark) and one librarian (Carol). Initially the organization was started in the MADKIT platform. In addition, in MADKIT, the agents Eric and Carol were started and, after that, in the S-MOISE+ platform an organizational adapter was started to import the organization state. The three remaining agents (Greg, Joel and Carol) are started in S-MOISE+. They perceived and enter the organization by requesting the role of collaborator. At this point, the interaction begins: The agents in S-MOISE+ are now members of an organization running in MADKIT. More details about MAORI may be found in [19].

7. Related Work

The proposition of abstract structures, such as patterns, and integrated metamodel to enable interoperability among organization-centered multiagent systems is somehow new. Therefore, related work must be considered in several areas from business to services, including multiagent systems. In the following we contextualize our work within this broad scenario.

In the multiagent systems area, Pechoucek and Marik [57] adopted a model-driven approach to propose a general metamodel for developing multiagent systems. The metamodel is defined as a Platform Independent Model (PIM) considering the MDA abstraction levels definition [18]. In order to identify a unified metamodel to support the development of agent-based systems, they considered seven views: Multiagent view; Agent view; Behavioural view; Organization view; Role view; Interaction view and Environment view. They adopted the top down approach to develop the metamodels that represent each of the views aforementioned after analysing some existing agent-oriented modeling languages, methodologies and programming languages. Moreover, it was conceived to be used independently of the agent-oriented methodology, modeling and programming languages. Nevertheless, their main purpose was to support the development of MAS using a model-driven approach than providing means for interoperability among MAS. Our work presents some similarities with theirs since the proposed integrated metamodel could be used independently of the Organization Model adopted to design and implement an organization-centered MAS. In addition, the abstract structures where defined based on organization-centered multiagent systems dimensions, in a similar approach of theirs when stating their metamodel concerning some views. Nevertheless, although using a model driven approach, we adopted it to define the way integration occurs using the bottom-up approach to define the integrated metamodel, based on existing Agent Organization Models and their underlying metamodels. By doing that we foster interoperability to organization-centered MAS during design time, by providing means of transforming the design of a MAS with an underlying organization model into another one, or during execution time, as presented in Section 6.

Muramatsu and colleagues [58] provided organizational interoperability by using organizational artifacts within the environment where the MAS is situated. They adopted a normative language to describe the organizational structure in artifacts. In this sense, their work is similar to ours while adopting a common language (in our case a common metamodel) to describe several organizational models.

Isern and colleagues [59] classified organizational structures according to organizational paradigms, such as (i) hierarchy, (ii) holarchy, (iii) coalition, (iv) team, (v) congregation, (vi) society, (vii) federation and (viii) market, to support the design of MAS using existing agent-oriented methodologies and organizational models. The main purpose of their work is to provide information for MAS developers that would like to adopt an organization approach to develop MAS and did not know what Organization Model or agent-oriented methodology to choose. Their work is related to ours in the sense they adopted metamodels' characteristics of existing organizational models and existing agent-oriented methodologies to classify such organizational structures as patterns.

Karaenke et al. [60] proposed an inter-organizational interoperability architecture based on multiagent systems, web services and semantic web technologies. In their work, the MAS did not present an underlying organizational model and agents adopt the "head body" paradigm to include web services technologies to provide interoperability among enterprise information systems. Therefore, interoperability is focused on system-to-system communication using web services technologies, which limits the kind of systems that may participate in such communication. Our proposition is broader in the sense that it provides a solution for interoperability among open organization-centered MAS independently of their underlying organizational Model.

A template description for agent-oriented patterns was given by Oluyomi and colleagues [61]. Based on a classification scheme, they organized agent technology concepts into categories and then identified agent-oriented pattern description templates for each category. Eight agent-oriented pattern templates were described to support the modeling of multiagent systems. Examples of the conformity between the proposed templates and their adoption during the design phase of existing agent-oriented methodologies were provided. Comparing with our proposition, their work adopts a similar approach when considering categories (in our case we adopted dimensions) to guide the patterns definition. In addition, their work is situated in the model level instead of the metamodel level as ours, since their objective is to improve communication among AOSE developers. Nevertheless, the adoption of an integrated metamodel obtained via model driven transformations combined with organizational dimensions give to our patterns high level of formality whenever compared with theirs.

Chella et al. [62] defined agent-oriented patterns to develop multiagent system to support robot programming. The proposed patterns were created based on an existing layered architecture for programming robots [63]. Patterns are described considering three aspects: The problem description; the definition of the solution in terms of MAS models and the description of the solution in terms of implementation. They just define some patterns for an specific domain based on the templates proposed by [61]. The only relation between their work and our is that pattern definition is based on some criteria that can be classified as dimension or category.

Organizational interoperability and integration issues are not new concerns for the administrative practice and research, specially after the wide acceptance and use of Information and Commmunication Technologies in their business models [64]. Several frameworks to provide organization interoperability were defined and even in this domain some dimensions were considered to define such frameworks.

8. Conclusions and Future Work

The research reported in this work has consisted in the use Model Driven Engineering techniques to address the *organizational interoperability problem*: How can we provide means for a set of agents, immersed in a common environment, to evolve, reason, decide and interact with each other based on organizational concepts, since their organizational models may differ? In order to achieve this goal, we have proposed an abstract and integrated view of the main concepts that have been used to specify agent organizations, based on the analysis of several organizational models present in the literature. In this model, we captured the recurring modeling concepts, that were coherently combined into an abstract conceptual structure. We have then presented an adaptation-based solution for the organizational interoperability problem, when we have defined the mappings between different organizational models, by using this abstract conceptual structure. We have built our abstract conceptual structure based on six organizational models (STEAM, MOISE+, AGR, OPERA, TEAMS, ISLANDER), presented in Section 2. For brevity, in Section 5 we illustrated the application of our integration method using three of these models (MOISE+, AGR, STEAM), and concerning exclusively the structural dimension.

A first extension of this work would be to build an integrated metamodel that could cope with OPERA and TEAMS. Moreover, we could evaluate how the other organizational models mentioned in Section 2.2.7 would affect our abstract conceptual structure. Concerning the MAORI framework, described in Section 6, we have tested its use by interoperating two organizational infrastructures, S-MOISE+ and MADKIT. A second extension of this work would be to test the franework with other organizational infrastructures, like AMELI [13] and ORA4MAS [52]. Finally, we would like to test our model driven approach to solve other MAS interoperability problems, like the ones mentioned in Section 1.

Author Contributions: Conceptualization, L.R.C, J.S.S. and O.B.; Formalization, implementation and experiments, L.R.C and A.A.F.B.; Supervision, J.S.S and O.B.; Writing and proofreading the paper, all authors.

Funding: "This research was partially funded by FAPEMA grant number 127/04 and CAPES, Brazil, Grant 1511/06-8". Jaime Sichman was partially supported by CNPq and FAPESP, Brazil. Anarosa Alves Franco Brandão was partially supported by grant #010/20620-5 and #014/03297-7, São Paulo Research Foundation (FAPESP), Brazil.

Conflicts of Interest: The authors declare no conflicts of interest.

References

1. Castelfranchi, C. Engineering social order. In Proceedings of the Engineering Societies in the Agents World: First International Workshop, ESAW 2000, Berlin, Germany, 21 August 2000; Revised Papers; Omicini, A., Tolksdorf, R., Zambonelli, F., Eds.; Springer: Berlin/Heidelberg, Germany, 2000; Volume 1972, pp. 1–18.
2. Olfati-Saber, R.; Fax, J.A.; Murray, R.M. Consensus and Cooperation in Networked Multi-Agent Systems. *Proc. IEEE* **2007**, *95*, 215–233. [CrossRef]
3. Shang, Y. Hybrid consensus for averager–copier–voter networks with non-rational agents. *Chaos Solitons Fract.* **2018**, *110*, 244–251. [CrossRef]
4. Boissier, O.; Hübner, J.F.; Sichman, J.S. Organisational oriented programming from closed to open organizations. In Proceedings of the 7th International Workshop, Engineering Societies in the Agents World VII, ESAW 2006, Dublin, Ireland, 6–8 September 2006; pp. 86–105.
5. Ferber, J.; Gutknecht, O.; Michel, F. From agents to organizations: An organizational view of multi-agent systems. In Proceedings of the Agent-Oriented Software Engineering IV: 4th International Workshop, AOSE 2003, Melbourne, Australia, 15 July 2003.
6. Gasser, L. Perspectives on organizations in multi-agent systems. In *Multi-Agent Systems and Applications: 9th ECCAI Advanced Course, ACAI 2001, and Agent Link's 3rd European Agent Systems Summer School, EASSS 2001, Prague, Czech Republic, 2–13 July 2001; Selected Tutorial Papers*; Springer: Berlin/Heidelberg, Germany, 2001; Volume 2086, pp. 1–16.
7. Zambonelli, F.; Jennings, N.R.; Wooldridge, M. Organisational abstractions for the analysis and design of multi-agent systems. In Proceedings of the Agent-Oriented Software Engineering: First International Workshop, AOSE 2000, Limerick, Ireland, 10 June 2000; Revised Papers; Ciancarini, P., Wooldridge, M., Eds.; Springer: Berlin/Heidelberg, Germany, 2001; Volume 1957, pp. 235–251.
8. Scott, W.R. *Organizations: Rational, Natural and Open Systems*, 4th ed.; Prentice Hall: Upper Saddle River, NJ, USA, 1998.
9. Kleppe, A. *Software Language Engineering: Creating Domain-Specific Languages Using Metamodels*; Addison Wesley: Reading, UK, 2008.
10. Poslad, S.; Charlton, P. *Standardizing Agent Interoperability: The FIPA Approach*; Springer: Berlin, Germany, 2001; Number 2086 in LNAI, pp. 98–117.
11. Magnin, L.; Pham, V.T.; Dury, A.; Besson, N.; Thiefaine, A. Our guest agents are welcome to your agent platforms. In Proceedings of the ACM Symposium on Applied Computing 2002, Madrid, Spain, 11–14 March 2002; pp. 107–114.
12. Tambe, M.; Pynadath, D.V. Towards heterogeneous agent teams. In *Multi-Agent Systems and Applications, Proceedings of the 9th ECCAI Advanced Course, ACAI 2001, and Agent Link's 3rd European Agent Systems Summer School, EASSS 2001, Prague, Czech Republic, 2–13 July 2001; Selected Tutorial Papers*; Springer: Berlin/Heidelberg, Germany, 2001; Volume 2086, pp. 187–210.
13. Esteva, M.; Rosell, B.; Rodríguez-Aguilar, J.A.; Arcos, J.L. AMELI: An agent-based middleware for electronic institutions. In Proceedings of the Third International Joint Conference on Autonomous Agents and Multiagent Systems (AAMAS'04), New York, NY, USA, 19–23 July 2004; Volume I, pp. 236–243.
14. Nardin, L.G.; Brandão, A.A.F.; Sichman, J.S. Experiments on semantic interoperability of agent reputation models using the SOARI architecture. *Eng. Appl. Artif. Intell.* **2011**, *24*, 1461–1471. [CrossRef]
15. Giampapa, J.A.; Paolucci, M.; Sycara, K. Agent Interoperation Across Multiagent System Boundaries. In Proceedings of the Fourth International Conference on Autonomous Agents (Agents 2000), Barcelona, Spain, 3–7 June 2000; pp. 179–186.
16. Wegner, P. Interoperability. *ACM Comput. Surv.* **1996**, *28*, 285–287. [CrossRef]

17. Coutinho, L.; Sichman, J.; Boissier, O. Modelling Dimensions for Agent Organizations. In *Handbook of Research on Multi-Agent Systems: Semantics and Dynamics of Organizational Models*; Dignum, V., Ed.; IGI Global: Hershey, PA, USA, 2009; Chapter II, pp. 18–50.
18. Schmidt, D.C. Model-driven engineering. *IEEE Comput.* **2006**, *36*, 25–31. [CrossRef]
19. Coutinho, L.R.; Brandão, A.A.F.; Sichman, J.S.; Hübner, J.F.; Boissier, O. A Model-Based Architecture for Organizational Interoperability in Open Multiagent Systems. In *Coordination, Organizations, Institutions and Norms in Agent Systems V*; Padget, J., Artikis, A., Vasconcelos, W., Stathis, K., Silva, V., Matson, E., Polleres, A., Eds.; Springer: Heidelberg, Germany, 2010; Volume 6069, pp. 102–113, doi:10.1007/978-3-642-14962-7_7.
20. Yourdon, E. *Análise Estruturada Moderna*; Campus: Rio de Janeiro, Brazil, 1992.
21. Booch, G.; Rumbaugh, J.; Jacobson, I. *The Unified Modeling Language*; Addison Wesley: Reading, UK, 1999.
22. Ackoff, R.L. *Ackoff's Best, His Classic Writings on Management*; John Wiley & Sons: New York, NY, USA, 1999.
23. Decker, K.; Lesser, V. Task environment centered design of organizations. In *AAAI Spring Symposium on Computational Organization Design*; AAAI: Menlo Park, CA, USA, 1994; pp. 32–38.
24. Tambe, M.; Adibi, J.; Al-Onaizan, Y.; Erdem, A.; Kaminka, G.A.; Marsella, S.C.; Muslea, I. Building agent teams using an explicit teamwork model and learning. *Artif. Intell.* **1999**, *110*, 215–239. [CrossRef]
25. Hübner, J.F.; Sichman, J.S.; Boissier, O. A Model for the Structural, Functional, and Deontic Specification of Organizations in Multiagent Systems. In Proceedings of the Advances in Artificial Intelligence: 16th Brazilian Symposium on Artificial Intelligence, SBIA 2002 Porto de Galinhas/Recife, Brazil, 11–14 November 2002; Bittencourt, G., Ramalho, G.L., Eds.; Springer: Berlin/Heidelberg, Germany, 2002; Volume 2507, pp. 118–128.
26. Esteva, M.; Padget, J.; Sierra, C. Formalizing a language for institutions and norms. In Proceedings of the Intelligent Agents VIII: 8th International Workshop, ATAL 2001, Seattle, WA, USA, 1–3 August 2002; pp. 348–366.
27. Dignum, V. A Model for Organizational Interaction: Based on Agents, Founded in Logic. Ph.D. Thesis, Utrecht University, Utrecht, The Netherlands, 2004.
28. Ferber, J.; Gutknecht, O. A meta-model for the analysis and design of organizations in multi-agent systems. In Proceedings of the Third International Conference on Multi Agent Systems, Paris, France, 3–7 July 1998; pp. 128–135.
29. Horling, B.; Lesser, V. Quantitative organizational models for large-scale agent systems. In Proceedings of the International Workshop on Massively Multi-Agent Systems, Kyoto, Japan, 10–11 December 2004; pp. 297–312.
30. Ferber, J.; Michel, F.; Baez, J. AGRE: Integrating environments with organizations. In Proceedings of the Environments for Multi-Agent Systems: First International Workshop, E4MAS 2004, New York, NY, USA, 19 July 2004; Revised Selected Papers; Weyns, D., Parunak, H.V.D., Michel, F., Eds.; Springer: Berlin, Germany, 2005; Volume 3374, pp. 48–56. [CrossRef]
31. Gâteau, B.; Boissier, O.; Khadraoui, D.; Dubois, E. *MoiseInst: An Organizational Model for Specifying Rights and Duties of Autonomous Agents;* EUMAS, Gleizes, M.P., Kaminka, G.A., Nowé, A., Ossowski, S., Tuyls, K., Verbeeck, K., Eds.; Koninklijke Vlaamse Academie van Belie voor Wetenschappen en Kunsten: Brussels, Belgium, 2005; pp. 484–485.
32. Dignum, V.; Vázquez-Salceda, J.; Dignum, F. OMNI: Introducing social structure, norms and ontologies into agent organizations. In Proceedings of the Programming Multi-Agent Systems: Second International Workshop ProMAS 2004, New York, NY, USA, 20 July 2004; Selected Revised and Invited Papers; Bordini, R.H., Dastani, M., Dix, J., Seghrouchni, A.E.F., Eds.; Springer: Berlin/Heidelberg, Germany, 2005; Volume 3346, pp. 181–198.
33. Vázquez-Salceda, J.; Dignum, F. Modelling electronic organizations. In Proceedings of the Multi-Agent Systems and Applications III: 3rd International Central and Eastern European Conference on Multi-Agent Systems, CEEMAS 2003, Prague, Czech Republic, 16–18 June 2003; Marík, V., Müller, J., Pechoucek, M., Eds.; Springer: Berlin/Heidelberg, Germany, 2003; Volume 2691, pp. 584–593.
34. Silva, V.; Choren, R.; Lucena, C. A UML based approach for modeling and implementing multi-agent systems. In Proceedings of the Third International Joint Conference on Autonomous Agents and Multiagent Systems (AAMAS'04), New York, NY, USA, 19–23 July 2004; pp. 914–921.

35. Silva, V.; Garcia, A.; Brandão, A.; Chavez, C.; Lucena, C.; Alencar, P. Taming agents and objects in software engineering. In *Software Engineering for Large-Scale Multi-Agent Systems: Research Issues and Practical Applications*; Garcia, A., Lucena, C., Zambonelli, F., Omicini, A., Castro, J., Eds.; Springer: Berlin/Heidelberg, Germany, 2003; Volume 2603, pp. 1–26.
36. Weyns, D.; Haesevoets, R.; Helleboogh, A. The MACODO Organization Model for Context-driven Dynamic Agent Organizations. *ACM Trans. Auton. Adapt. Syst.* **2010**, *5*, 16:1–16:29. [CrossRef]
37. Zambonelli, F.; Jennings, N.R.; Wooldridge, M. Developing multiagent systems: The Gaia methodology. *ACM Trans. Softw. Eng. Methodol.* **2003**, *12*, 317–370. [CrossRef]
38. Bresciani, P.; Giorgini, P.; Giunchiglia, F.; Mylopoulos, J.; Perini, A. Tropos: An agent-oriented software development methodology. *J. AAMAS* **2004**, *8*, 203–236. [CrossRef]
39. Jouault, F.; Bézivin, J. KM3: A DSL for Metamodel Specification. In *Formal Methods for Open Object-Based Distributed Systems*; Gorrieri, R., Wehrheim, H., Eds.; Springer: Berlin/Heidelberg, Germany, 2006; Volume 4037, pp. 171–185, doi:10.1007/11768869_14.
40. OMG. *Meta-Object Facility (2.0) Core Specification*; OMG Document Formal/2006-01-01. Available online: https://www.omg.org/spec/MOF/2.0/About-MOF/ (accessed on 10 June 2019).
41. OMG. *Object Constraint Language*, Version 2.0, 2006. Available online: https://www.omg.org/spec/OCL/2.0/About-OCL/ (accessed on 10 June 2019).
42. Clark, T.; Sammut, P.; Willans, J. *Applied Metamodelling—A Foundation for Language Driven Development*; CETEVA: London, UK, 2008.
43. Steinberg, D.; Budinsky, F.; Paternostro, M.; Merks, E. *EMF: Eclipse Modeling Framework*; Addison-Wesley: Reading, UK, 2008.
44. Pottinger, R.; Bernstein, P.A. Merging models based on given correspondences. In Proceedings of the 29th International Conference on Very Large Data Bases, VLDB 2003, Berlin, Germany, 9–12 September 2003; Freytag, J.C., Lockemann, P.C., Abiteboul, S., Carey, M.J., Selinger, P.G., Heuer, A., Eds.; Morgan Kaufmann: San Francisco, CA, USA, 2003; pp. 826–873.
45. Sabetzadeh, M.; Easterbrook, S. View Merging in the Presence of Incompleteness and Inconsistency. *Requir. Eng. J.* **2006**, *11*, 174–193. [CrossRef]
46. Diskin, Z.; Dingel, J. A metamodel-independent framework for model transformation: Towards generic model management patterns in reverse engineering. In Proceedings of the 3rd International Workshop on Metamodels, Schemas, Grammars, and Ontologies, Genoa, Italy, 1–6 October 2006; Favre, J.M., Gasevic, D., Laemmel, R., Winter, A., Eds.; Springer: Berlin, Germany, 2007; Volume 4364, pp. 52–55.
47. Lesser, V.; Decker, K.; Wagner, T.; Carver, N.; Garvey, A.; Horling, B.; Neiman, D.; Podorozhny, P.; Prasad, N.N.; Raja, A.; et al. Evolution of the GPGP/TAEMS domain-independent coordination framework. *Auton. Agents Multi-Agent Syst.* **2004**, *9*, 87–143. [CrossRef]
48. Esteva, M. Electronic Institutions: From Specification to Development. Ph.D. Thesis, Institut d'Investigació en Intel.ligència Artificial, Bellaterra, Catalonia, Spain, 2003.
49. Hübner, J.F. Um modelo de reorganização de sistemas multiagentes. Tese (doutorado), Escola Politécnica da Universidade de São Paulo, 2003. Available online: http://www.teses.usp.br/teses/disponiveis/3/3141/tde-17052004-151854/ (accessed on 10 June 2019).
50. Fowler, M. *Refactoring: Improving the Design of Existing Code*; Addison Wesley: Reading, UK, 1999.
51. Boissier, O.; Bordini, R.H.; Hübner, J.F.; Ricci, A.; Santi, A. Multi-agent oriented programming with JaCaMo. *Sci. Comput. Program.* **2013**, *78*, 747–761. [CrossRef]
52. Kitio, R.; Boissier, O.; Hubner, J.F.; Ricci, A. Organisational artifacts and agents for open multi-agent organisations: Giving the power back to the agents. In Proceedings of the International Workshops COIN@AAMAS 2007, Durham, UK, 3–4 September 2007; Springer: Berlin, Germany, 2008; Volume 4870, pp. 171–186.
53. Hübner, J.F.; Sichman, J.S.; Boissier, O. S-Moise+: A middleware for developing organised multi-agent systems. In *Coordination, Organizations, Institutions, and Norms in Multi-Agent Systems*; Boissier, O., Padget, J., Dignum, V., Lindemann, G., Matson, E., Ossowski, S., Sichman, J.S., Vázquez-Salceda, J., Eds.; Springer: Heidelberg, 2006; LNCS (LNAI), Volume 3913, pp. 64–77.
54. Gutknecht, O.; Ferber, J. The MADKIT agent platform architecture. In *International Workshop on Infrastructure for Multi-Agent Systems: Infrastructure for Agents, Multi-Agent Systems, and Scalable Multi-Agent Systems*; Springer: London, UK, 2000; Volume 1887, pp. 48–55.

55. Eclipse Team. Eclipse Modeling Framework (EMF). Available online: http://www.eclipse.org/emf/ (accessed on 10 June 2019).
56. Jouault, F.; Kurtev, I. Transforming models with ATL. In Proceedings of the Satellite Events at the MoDELS 2005 Conference, Montego Bay, Jamaica, 2–7 October 2005; Bruel, J.M., Ed.; Springer: Berlin, Germany, 2006; Volume 3844, pp. 128–138.
57. Hahn, C.; Madrigal-Mora, C.; Fischer, K. A platform-independent metamodel for multiagent systems. *Auton. Agents Multi-Agent Syst.* **2009**, *18*, 239–266. [CrossRef]
58. Muramatsu, F.T.; Vitorello, T.M.; Brandão, A.A.F. Towards Organizational Interoperability through the Environment. In *Agent Environments for Multi-Agent Systems IV*; Weyns, D., Michel, F., Eds.; Springer International Publishing: Cham, Switzerland, 2015; pp. 209–231.
59. Isern, D.; Sanchez, D.; Moreno, A. Organizational structures supported by agent-oriented methodologies. *J. Syst. Softw.* **2011**, *84*, 169–184. [CrossRef]
60. Karaenke, P.; Schuele, M.; Micsik, A.; Kipp, A. Inter-organizational Interoperability through Integration of Multiagent, Web Service, and Semantic Web Technologies. In *Agent-Based Technologies and Applications for Enterprise Interoperability, (ATOP 2009, ATOP 2010)*; Fischer K., Müller J.P., Levy R., Eds; Lecture Notes in Business Information Processing, Volume 98; Springer: Berlin/Heidelberg, Germany, 2012, pp. 55–75.
61. Oluyomi, A.; Karunasekera, S.; Sterling, L. Description templates for agent-oriented patterns. *J. Syst. Softw.* **2008**, *81*, 20–36. [CrossRef]
62. Chella, A.; Cossentino, M.; Gaglio, S.; Sabatucci, L.; Seidita, V. Agent-oriented software patterns for rapid and affordable robot programming. *J. Syst. Softw.* **2010**, *83*, 557–573. [CrossRef]
63. Chella, A.; Cossentino, M.; Gaglio, S.; Sabatucci, L.; Seidita, V. An architecture for autonomous agents exploiting conceptual representations. *Robot. Auton. Syst.* **1998**, *25*, 231–240. [CrossRef]
64. Kubicek, H.; Cimander, R. Three dimensions of organizational interoperability. *Eur. J. Epractice* **2009**, *6*, 3–14.

Article

ARPS: A Framework for Development, Simulation, Evaluation, and Deployment of Multi-Agent Systems

Thiago Coelho Prado and Michael Bauer *

Department of Computer Science, Western University, London, ON N6A 3K7, Canada; tprado2@uwo.ca
* Correspondence: bauer@uwo.ca; Tel.: +1-519-661-3562

Received: 18 September 2019; Accepted: 18 October 2019; Published: 23 October 2019

Abstract: Multi-Agent Systems (MASs) are often used to optimize the use of the resources available in an environment. A flaw during the modelling phase or an unanticipated scenario during their execution, however, can make the agents behave not as planned. As a consequence, the resources can be poorly utilized and operate sub-optimized, but it can also bring the resources into an unexpected state. Such problems can be mitigated if there is a controlled environment to test the agents' behaviour before deployment. To this end, a simulated environment provides not only a way to test the agents' behaviour under different common scenarios but test them as well in adverse and rare state conditions. With this in mind, we have developed ARPS, an open-source framework that can be used to design computational agents, evaluate them in a simulated environment modelled after a real one, and then deploy and manage them seamlessly in the actual environment when the results of their evaluation are satisfactory.

Keywords: multi-agent systems; discrete event simulator; interoperability; agent and multi-agent applications

---

1. Introduction

In environments where resource management is critical, software agents can be employed to optimize those resources. When the task to accomplish this is too complex to be carried out by a single agent, multiple interacting agents can be used to achieve effectiveness. The Multi-Agent System (MAS) approach has been successfully applied in many domains, including: helping controllers in Air Traffic Control for making decisions based on the aircrafts' fuel availability, flight plan, weather, and any other relevant data [1]; optimizing distributed generators in smart grids for energy production, storage, and distribution [2]; or finding better arrangements for the components in manufacturing plants to increase the throughput and optimize material usage [3].

Many frameworks and toolkits enable the implementation of a MAS. They are not designed to provide seamless means of evaluation of the outcome of the agents in the system under certain scenarios before deployment. A flaw during the design and implementation of the agents, or an unpredicted state of the environment where they are acting, can not only interfere with the achievement of their goals, but can also bring the environment into a unexpected situation with unforeseen consequences.

Some work with MASs has made use of simulations to aid deployed agents to update their plans according to the current state of the environment. This means that the simulation is used as a planner tool, meant to help the agent to make a decision under a specified near future scenario rather than using the simulation in the process of designing the agent behaviour before deployment. This strategy has been applied to the field of multiple Unmanned Aerial Vehicles (UAV) [4], where communication is required for coordination and failures related to it can make the entities unreachable. The system simulates possible scenarios where communication is unavailable and an action by the UAV is expected. Another example is the simulation component used to detect conflicts and inconsistencies of resource allocation during the high-level planning in a manufacturing plant [5].

There is no general purpose MAS framework, to the best of our knowledge, that integrates the process of validation of the agents in a simulated environment before their deployment. To address this, we have developed ARPS, an open-source framework available at https://gitlab.com/arps/arps/ under MIT LICENSE [6], to seamlessly design, implement, assess the agents, and deploy the MAS after the results meet established criteria. ARPS stands for some of the core properties of agents: autonomy, reactivity, proactivity, and social ability. We use the management of resources of a data centre to illustrate our approach and how it can be used in other domains.

In Section 2 we cover related work. Following this, Section 3 describes the background and architecture of the framework. The process of the implementation of the MAS to manage an experimental scenario is shown in Section 4. In Section 5 we discuss the findings, limitations, and future directions.

2. Related Work

There are multiple toolkits, platforms, and frameworks available for creating a MAS. In this section, we will describe a few of them. The works here by no means represent the only alternatives for creating MASs. For other options, refer to surveys on this topic, such as the one presented in [7].

The works reviewed here implement common aspects of a MAS to work in an actual environment, such as communication, interoperability, storage, security, and resource discovery. Therefore, the users can focus on the definition of the agents and their behaviour, how they are organized, and how they interact to solve a problem.

Among the popular MAS frameworks, JADE (also known as Java Agent Development Framework) [8] was created to address the problem of interoperability and provide an environment for the development of agents. It has no domain dependent requirement as seen in the other solutions. According to the authors, such dependencies were obstacles for the adoption of MAS technologies at the time of its conception. JADE simplifies the implementation of multi-agent systems through a middleware compliant with the FIPA (Foundation for Intelligent Physical Agents) specifications [9], a standard proposed for interoperability of agents. The JADE authors argue that is industry-driven and currently the most known FIPA-compliant agent platform in the academic and industrial community.

The A-Globe platform [10] is designed for testing experimental scenarios featuring agents' position that requires a Geographical Information System (GIS, though agents may suffer from communication inaccessibility either because of the spatial distance of the agents or by broken links. Because it is a closed-environment, interoperability is not one of the concerns of this platform. Hence, it is not fully compliant with the FIPA-specifications on inter-platform communication, albeit it provides compliance with the Agent Communication Language (ACL), a structure for composing messages exchanged by agents.

Based on the fulfillment of requirements such as robustness, security, and the ability to ensure that a partial solution can be executed when an optimized one is not found due to constraints, DARPA funded a project called Cougaar [11] (Cognitive Agent Architecture). This agent platform was created to offer specialized support for logistics-related problems. This platform is also not FIPA-compliant. It aims to facilitate the development of agent-based applications that are complex, large scale and distributed.

Other solutions offer more components built on top of the existing agent-based platforms. The Jadex BDI Agent System [12] follows the Belief Desire Intention (BDI) model [13] and facilitates intelligent agent construction over other middleware such as JADE. It has been used to build applications in different domains, such as simulation, scheduling, and mobile computing. The programming model of Jadex allows for designing an application as hierarchical decomposition of components interacting via services and thus helps to make complexity manageable. These components can be used in concurrent and dynamic distributed systems. Another example is JaCaMo [14]. It is an interpreter for an extended version of AgentSpeak, a BDI agent-oriented logic programming language [15]. It is a platform that integrates three projects with different MAS-related paradigm models for development: Jason [16] for agent development; Moise [17] for agent organization; and CArtAgO [18] for environment-oriented programming.

Lastly, the MadKit [19] offers a modular and scalable multiagent platform. Its central aspect is the ability to organize agents in groups and roles aiming at the development of artificial societies. It is closely related to our approach in the sense that has a simulator component. Nonetheless, the simulation and evaluation are not seamless in the workflow, leaving the responsibility of this integration to the user.

Previous work has often presented MAS tools that provide a base infrastructure to implement agents. These approaches also looked to help address other specific problems, such as interoperability by defining standards, robustness and security. In some cases, the work sought to introduce approaches to enrich the design process, provide clear definition of how agents are organized and what their roles are. However, these solutions lack a component that would allow a developer to assess the behaviour of the agents in a simulated environment before actual deployment. Further, this step should be as seamless as possible, i.e., it should take no or few modifications to alter agents in the simulated environment in order to make them run in the actual environment. This is the gap that ARPS fills.

Agent-Based Modelling (ABM)

One of the main solutions available to simulate complex systems is known as Agent-based modelling (ABM). It is employed in domains such as social sciences, biology, ecology, engineering, and economics. There are many platforms to implement ABM [20,21]. Among them, we list a few widely adopted examples. The Swarm package [22] provides object-oriented libraries of reusable components for building models and analyzing, displaying, and controlling experiments on those models. NetLogo [23] is a software that provides packages to simulate multi-agent in environments. It has a significant user community, it is highly documented, and they provide many demos. MASON [24] is a discrete event multi-agent simulation toolkit. It was designed to serve as the basis for a wide range of multi-agent simulation tasks ranging from swarm robotics to machine learning to social complexity environments. The Repast platform, initially developed as a tool to be used in social sciences, is a family of agent-based modelling and simulation platforms. Currently, not only does it provide a package to be used in regular environments, such as desktops, and laptops [25], but also has an advanced version to be used in HPC environments [26] to simulate more demanding scenarios.

The ABM and MAS concepts are closely related to each other since both are agent-based. The difference is that the modelling in ABM aims to gain insight about emergent properties in complex adaptive systems while MAS focus on actual agents [27,28].

Our framework aims to provide an environment to combine ABM properties, such as the ability to model agents in a simulated environment and observe their behaviour, with MAS's ability to implement and deploy the agents evaluated during the simulation in the action environment. The integration of both approaches can yield positive results. The framework in [29], features this combination. It enables mobile agents to work simultaneously both in the actual and virtualized agent platform to enable large-scale simulation to observe unknown emergence behaviour. This is accomplished by having the agents in the actual environment collecting data in real-time, and improving the simulation, which in turn can have its outcome used in decision making. In our case, we need a separate simulator component to enable the study of the agents' behaviour without disrupting the environment before their deployment. This then enables a developer to leverage the construction of reliable physical agents by evaluating their interactions and the effects of their actions. Because simulation can involve the compression of time, it is possible to simulate many different scenarios efficiently. Also, during the simulation, it is possible to create unexpected scenarios to see how the agents perform. This can be desirable in areas that already employ ABM for resource management during a disaster, such as [30], that could be extended to have software agents that could direct resources where it is necessary. We are aiming to combine the characteristics of both simulation and development to offer an alternative for creating solutions in resource management using MAS.

Figure 1 illustrates our workflow to achieve this integration. A user/admin can create the environment, its resources, and the agents' models, and define how they are organized. The MAS environment is generated, and the user can simulate a scenario. After analyzing the simulation results, the user can decide to refine the models or apply the agents' models to the actual environment. The output of this environment can also be used as feedback to improve the existing models and create a more reliable simulation.

As we have discussed, none of the previoius MAS frameworks have a simulator component that allows this workflow without having to reimplement the concepts defined during the simulation into the actual environment.

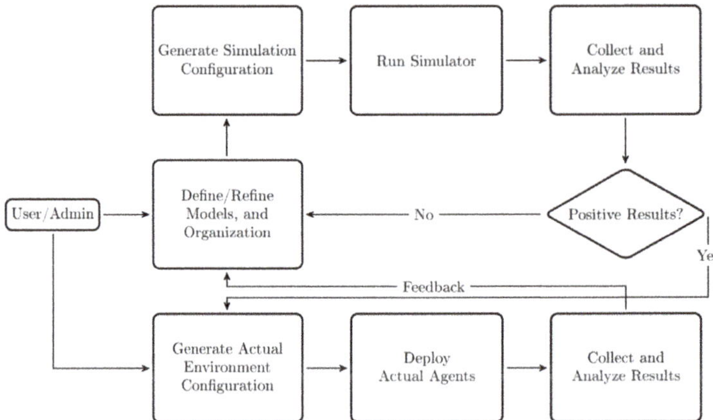

Figure 1. Workflow for MAS conception, evaluation, and deployment.

3. Our Approach

Below we describe the background of the framework, our design choices, and present the architecture.

3.1. Background

Data centres present complex dynamic environments where many resources are allocated to provide guaranteed, reliable services. These resources are not only related directly to the computer systems, such as processing power, storage capacity, and network, but also to the supporting equipment and environmental control. Data centre administrators face a daunting task in trying to manage them optimally. There are impacts when these resources are misconfigured or overallocated on the total cost of ownership because of excess power consumption or idle resources. Data centre operators can also incur financial penalties due to the broken guarantees related to service delivery.

One proposed way to tackle this problem is the adoption of autonomic computing architectures and strategies. An autonomic approach aims to embody the idea of self-management, which in turn can be realized by decomposing it into other sub-properties, such as self-configuration, self-healing, self-optimization, and self-protection [31]. This separation of concerns can be managed by decentralized autonomous agents that may interact with each other, optimizing local resources to achieve global optimization, as exemplified in [32], where a MAS manages the number of hosts available to process workloads depending on the demand.

One problem faced in the employment of MASs in data centres is the impracticality of having an actual data centre available to evaluate the effectiveness of the policies governing agents due to costs and security concerns. In some cases, data centre simulators, such as that described in [33], have been developed to evaluate the possible impacts of the autonomic agents policies in the data centre before implementing and deploying them in the real environment. Even so, there is a gap, however, in the development between the simulation step and the implementation and deployment of the actual agents. To address this problem, a framework was developed using initially the concepts of the aforementioned simulator with the additional feature of employing the assessed policies during the simulation to drive the software agents that will manage the actual resources in the environment.

Contrary to most of the works related to MAS platforms presented in the previous sections, our approach was first developed to solve a domain-specific problem focused on self-management of resources. During the development, the generic components were extracted to allow more flexibility during the implementation and the test of the proposed solutions. This resulted in the general purpose ARPS framework to enable MAS.

The generalization of the scenarios where our MAS framework can apply, although it is not limited by, is illustrated in Figure 2. As can be seen, there is a complex system with resources (R_n), grouped by environments ($Environment_m$), that are affected by external events (E). The events occur at a variable or fixed interval in this system. Without any management, these resources can be in a suboptimal state. To overcome this problem, agents (A_i), driven by policies, are employed to monitor resources or modify them using available touchpoints. The agents can be reactive, proactive. They can act in isolation, or they can communicate with each other for cooperation, coordination, or negotiation.

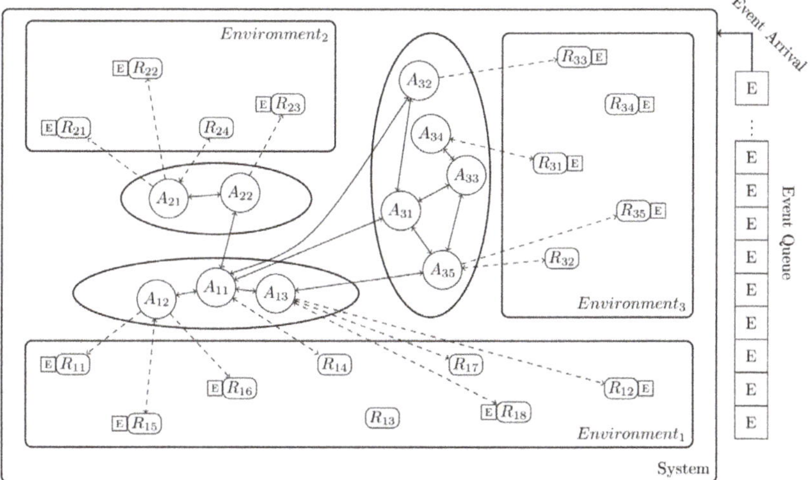

Figure 2. Scenario for resources being modified by external events while agents manage them.

The framework is being currently developed in Python 3, a high-level multiplatform language that can be deployed in myriad host platforms, including Internet of Things (IoT) devices [34,35]. The users can install the framework from source code or as a Python package, available at https://pypi.org/project/arps/. This means that, when implementing the agents for a specific domain, the user needs to implement all the components using Python. Albeit there is no single programming language that can be applied in every domain effectively, Python has been suggested as an alternative for a general programming language to be adopted by the scientific community and it has been used by researchers in areas not related to technology or engineering, such as psychology and astronomy [36,37].

3.2. Architecture

The ARPS framework is composed of four main components: agent manager, agents, discovery service/yellow pages service and Discrete Event Simulator (DES). The architecture and relationship of the first three are illustrated in Figure 3. The agent manager has three main aspects: the management of the availability of resources and policies related to an environment, agent life-cycle, and simulation when it is running for this purpose. It acts as a container for the agents. This container groups resources logically by some criteria, like resource similarity, or accessibility. Agent managers can be distributed across systems. The agent is an entity driven by policies that will manage one or more resources. It can interact with all other agents available by the discovery service. Since each agent manager can be deployed in distributed manner, the discovery service is a directory where its agents are registered when created and their location is made available to be discovered by other requesters, so the agent from one agent manager can exchange messages with agents from others agent managers. Lastly, the DES is a component available for the evaluation of the agents, and its integration with the others will be explained in the following sections.

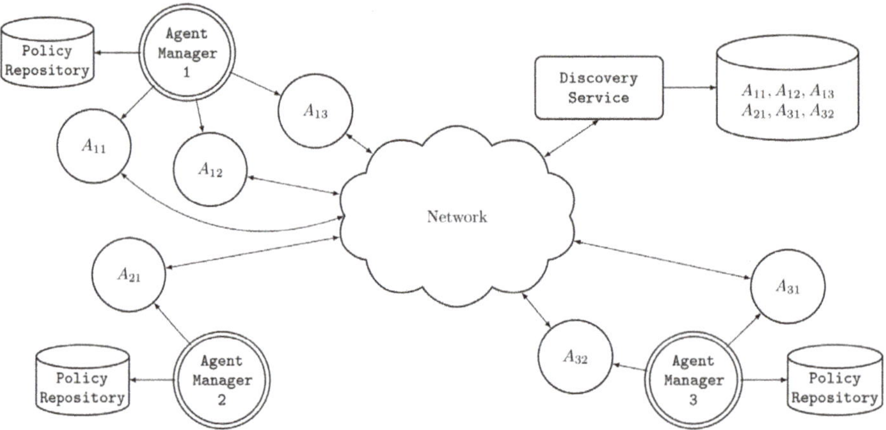

Figure 3. Architecture of the ARPS framework.

3.2.1. Interoperability

The agent manager, agents, and discovery service implement the RESTful architecture style for interoperability. This is done using HTTP methods, and the payload is in JSON format. This RESTful API employs uniform resource access using the format http://hostname:port/resource/, where the *hostname:port* are the host name and listening port to access the API respectively, and the *resource* is a web resource. The HTTP methods (GET, POST, PUT, and DELETE) are available for the clients. The payload of the POST and PUT methods, as well as the possible HTTP response codes, were omitted for simplification purposes. Their descriptions are available at https://gitlab.com/arps/arps/wikis/home. Below we present the API to perform the requisitions.

The agent manager API is intended to fulfill requests by the user. The summary in Table 1 shows how the policies that can be used to drive agents' behaviour can be listed, the agents that are currently running in its environment, the touchpoints that agents can monitor/control, and, when there are agents created just for the purpose of monitoring resources, a timestamped log containing the state of the resources can be retrieved.

The life-cycle of the agents is controlled by creation, modification, inspection, and termination methods, as seen in Table 2.

Table 1. REST API for environment.

HTTP Method	URI	Description
GET	/list_agents	show agents in the environment
GET	/list_policies	show policies available for agent creation/modification
GET	/list_touchpoints	show sensors and actuators available in the system for monitoring
GET	/monitor_logs	retrieve the state of the sensor and actuators monitored by the agents created to this end

Table 2. Agent Manager API for agent life-cycle.

HTTP Method	URI	Description
POST	/agents	Creates an agent; policies and the fixed interval is specified in the JSON body of the message
PUT	/agents/[agent_id]	Modify the agent's policies or relationship with other agents
GET	/agents/[agent_id]?[params=values]	Retrieve the internal state of the agent; parameters are used to specify the type of the content returned
DELETE	/agents/[agent_id]	Terminate the agent

Lastly, the REST API is available only when the agent manager is created to run the simulator is seen in Table 3.

Table 3. Agent Manager API in simulator mode.

HTTP Method	URI	Description
GET	/sim/run	Run simulation
GET	/sim/stop	Stop simulation
GET	/sim/status	Retrieve current status
GET	/sim/result	Retrieve result of the simulation in CSV format
GET	/sim/save	Retrieve the organization of the system

The agents have their API described in Table 4. It is not only accessed with the intent of message exchange among the agents, but it can also be called by users or other applications.

Table 4. Agent API.

HTTP Method	URI	Description
PUT	/policy	Add or remove a policy accordingly to the content of the message body
PUT	/meta_agent	Establishes or removes the relationship between two agents
PUT	/action	Requests a user defined action to be executed by the agent accordingly to the content of the message body
GET	/info	Request information regarding the agent, such as current policies, touchpoints, and relationships
GET	/sensors	Request current state of the sensors available to the agent
GET	/actuators	Request current state of the actuators available to the agent

The API provided by the discovery service, shown in Table 5, is used to register, unregister, and list the running agents in the system.

Table 5. Discovery Service.

HTTP Method	URI	Description
GET	/agents	List all agents
PUT	/agents/[agent_id]	Register the agent
GET	/agents/[agent_id]	Retrieve the location of the agent: hostname and port
DELETE	/agents/[agent_id]	Unregister the agent

We are aware of the existence of FIPA-HTTP to make compliant FIPA agents using HTTP [38]. This standard, however, only supports the POST method with the semantics of the message embedded in a multi-part content sent in the body of the message. This defeats the purpose of the semantics presented by the HTTP methods, and the already available HTTP status codes related to each HTTP method. Some studies that support this view of interoperability in MAS, using Resource Oriented Architecture (ROA), can be found in [39–41],

There are many advantages of using RESTful architecture style for interoperability. The system is open, so applications that the system designers did not take into account a priori can be integrated into the existing system since the web resources are accessed uniformly. Secure communication can be implemented over HTTPS. A cache can be used to save bandwidth and to optimize system communication. Visualization through different devices can be easily implemented by third-party entities across different devices. The support of the MAS for the REST API can also be used to extend the API and make it compliant with FIPA-HTTP if required.

3.2.2. Agent Model

The intelligent agent model, as described by Russel & Norvig [42], is used to define the agent. Thus, besides the conventional definition of the agent, as an entity perceiving the environment through sensors and modifying it through actuators, it has also the components to achieve reasoning. To this end, the agent's behaviour is driven by a set of policies that are executed by its control loop. A policy can be activated either by interaction with other agents, or internally by a fixed interval. Each policy has access to the touchpoints available in the environment provided by the agent during its creation. Another characteristic is that policies can be dynamically added or removed.

The policies can be reactive or proactive. The reactive policy follows the Event Condition Action (ECA) rule. This model is used to create simple reflex agents. Thus, the agent monitors the environment, and, when the current state matches a condition, a predefined action is performed. Alternatively, the agents can optimize the environment continuously. In this case, their behaviour can be defined in terms of proactive policies. To this end, the user can extend the interface to implement goal-based or utility-based policies.

Another characteristic is related to how agents are organized. The framework defines that an agent has a unidirectional relationship with other agents. This relationship is also dynamic and can be created or removed during the agent lifetime. Since this model does not impose any form of organization and agents communicate using a peer-to-peer architecture, they can be organized hierarchically, or horizontally. Agents can have a relationship established with any other agent available through the yellow pages service. Currently, there is no support to enforce groups or roles.

3.2.3. Simulation

During the simulation, only a single instance of an agent manager is necessary since it will act as the gateway to all other virtual agent managers that would be available in the system, as seen in Figure 4a. Similarly to the actual environment, the format used by the API is http://hostname:port/agent_manager/resource/, where *agent_manager* antecedes *resource* to identify the virtual agent manager that will perform the requisition.

In the actual environment, the transport system used to exchange messages between agents is implemented using the HTTP protocol, as seen previously. It relies on the physical network to function. During the simulation, however, this protocol is substituted by a global bus that will serve as a discovery service and communication layer at the same time. The content sent from one agent is put directly into the queue of messages of the receiver agent. The difference between real and simulated communication is seen in Figure 4b. This has the advantage of removing the overhead of communication. Additionally, it gives the freedom to the user to apply models into this layer to improve the reliability of the agents to evaluate them under intermittent communication, increased latency, or corrupted messages.

The resource is another component that has to be overridden by the models that describe how they behave in the real environment. For example, in our case, we can model the behaviour of the computational resources to have a certain load accordingly to the task that arrived in the system. As well, our model supports resources that affect other resources indirectly. Revisiting the previous example, we can create a model that increases the energy consumption based on the load in the system when the computational resources are utilized.

The main component of the DES is the events generator. It supports both deterministic and stochastic models. The former uses a log file containing the events—when the last event is completed the simulation is terminated—while the latter uses a stochastic generator implemented by the user—the termination needs to be invoked explicitly. Each event has two actions. The main action that is executed every step while the event is still unfinished, and a post-action, when the event exits the system. Figure 4c showing the resource being modified by an event during the simulation and in Figure 4d showing the queue of deterministic events illustrates the DES component working along with the agents

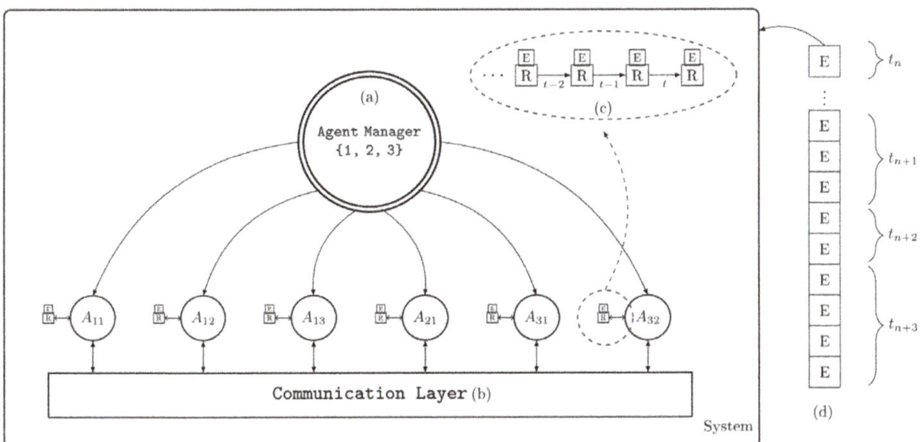

Figure 4. Agent Manager in Simulator Mode.

When finished, the results of the simulation are made available in the CSV format. We have chosen this format since it is supported by a myriad of statistical tools, such as Pandas (https://pandas.pydata.org/), R (https://www.r-project.org/), or spreadsheet-like apps such as LibreOffice (https://www.libreoffice.org/). We believe these third-party tools are better equipped to fulfill the needs of the user.

4. Demonstration of the Framework

In this section, we will illustrate how the ARPS framework can be used to solve the problem of resource management. In the next section, we describe the components needed to be implement to enable the MAS. Following this, we present an example in a specific domain: management of resources in data centres.

4.1. ARPS Framework Usage

The modelling of the MAS for resource management can be accomplished by following the steps:

1. Collect data about the resources in the environment using the available sensors.
2. Create a model of the environment, construct the resources behaviour, their relationship with each other.
3. Create the actuators and the policies to drive the agents based on the available resources.
4. Create a model of the events that will modify the environment.
5. Run the simulation, take a snapshot of the environment containing the agents deployed along with their policies, and collect the results in the CSV format.
6. Use more suitable external tools to evaluate the results.
7. Use a snapshot to deploy the agent managers and their agents in the real environment.
8. Modify the environment model, the policies, or the events and re-run step 4 to improve the agents in the actual environment.

The usage of the framework can be summarized as the implementation of some interfaces, and the creation of the configuration files needed by the agent managers and agents.

We have created the abstraction of the resources that provides the touchpoints to be accessed by sensors and actuators, seen in Figure 5.

Starting with the sensors and a minimal set of the configuration files (files that will be described in detail later in this section), it is possible to execute the agents in the actual environment to collect data since a set of monitoring policies related to the implemented resources are made available automatically. Therefore, it is possible to initially create agents with the only the purpose of gathering data periodically. This data provides insight about the actual resources . This will be essential during the modelling of the environment used by the simulator. This model can be later updated when comparing the simulated environment with the actual environment.

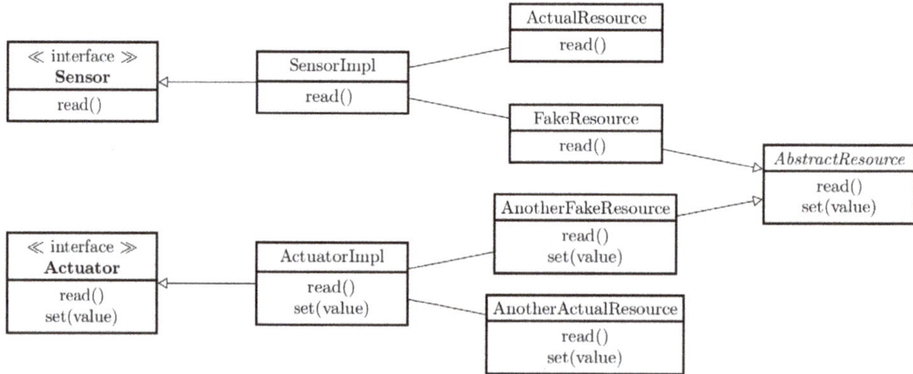

Figure 5. Resources, Sensors, and Actuators structure diagrams.

Once the resources and sensors are implemented, the next step is the implementation of the actuators. The actuators are used by the policies that drive the behaviour of the agents. These policies can be created by implementing the Policy interface, illustrated by a reflex policy in Figure 6. Since our approach is based on autonomic computing concepts [43], other types of policies, employing utility functions, or goal-based approaches can be implemented. The method *condition(event)* can be implemented to use optimization algorithms, defined by the user, that will compute the ideal parameters and then use the results to modify the resources through their actuators. Goal-based approaches would require the user to model the states and use the method *condition(event)* to search for the best action to perform to achieve a better state. Both policies would be executed as periodic policies, so the event is just a time event indicating that it is time for the policy to evaluate the current state and executes its actions based on the perceived environment.

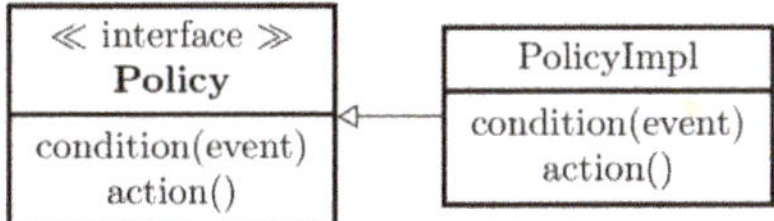

Figure 6. Policy structure diagram.

Since the policies encapsulate the perception and modification of the resources, and these resources can be modified by external events, the next step is the modelling of the these events. The events arrival in the system can be stochastic or deterministic and it can be created by implementing the interface EventQueueLoader. The events are encapsulated in the SimEvent class, which in turn contains the SimTask that will be executed during the simulation. The SimTask *main* method should provide the behaviour that will modify a resource, while the *pos* method will contain all the finalization process to be executed, like resource release, after the task is finished. The relationship between the DES components is illustrated in Figure 7.

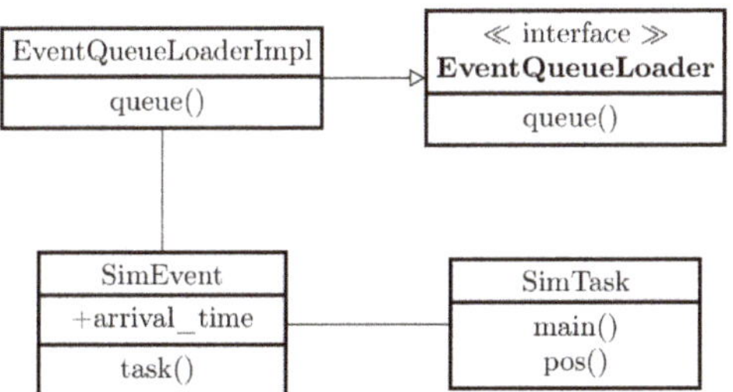

Figure 7. Simulator structure diagram.

Given the structure present in Figure 8, the final step before running the MAS in the actual or simulated environment is the creation of the configuration files, in JSON format, containing the implemented files previously described in this section. Both configuration files are similar in their structure. The file *simulation.conf* contains the paths of the files related to the fake resources, and the DES component classes, while the file *real.conf* contains only the actual resources. The remaining classes, like policies, sensors, and actuators, remain the same for both configuration files. Therefore, the transition of the agents from the simulated environment to the real environment can be done seamlessly.

After the instantiation of the agent manager in simulation mode, using the *simulation.conf* configuration file, the REST API presented in the previous section can be used to create the agents, organize them, and run, stop, and collect the result of the simulation.

Based on the result, only the policies classes need to be modified. Then, using the *real.conf* configuration file, the agents can be deployed in the actual environment. According to the behaviour observed in the actual environment, improvements can be made in models used by the simulated environment. Thus, it is only necessary to modify existing fake resources, simulation tasks and policies. Incrementally, new resources, touchpoints, and policies can be added into the MAS to improve the system.

```
src
├─ resources
│   ├─ ActualResource
│   ├─ FakeResource
│   ├─ AnotherActualResource
│   └─ AnotherFakeResource
├─ touchpoints
│   ├─ SensorImpl
│   └─ ActuatorImpl
├─ policies
│   └─ PolicyImpl
├─ des
│   ├─ EventQueueLoaderImpl
│   ├─ SimEvent
│   └─ SimTask
└─ conf
    ├─ simulation.conf
    └─ real.conf
```

```
simulator.conf:
{
    ...
    'resources': ['resources/FakeResource',
                  'resources/AnotherFakeResource'],
    'touchpoints': ['touchpoints/SensorImpl',
                    'touchpoints/ActuatorImpl'],
    'policies': ['policies/PolicyImpl'],
    'des': ['des/EventQueueLoaderImpl',
            'des/SimTask', 'des/SimEvent]
    ...
}
-----------------------------
real.conf:
{
    ...
    'resources': ['resources/ActualResource',
                  'resources/AnotherActualResource'],
    'touchpoints': ['touchpoints/SensorImpl',
                    'touchpoints/ActuatorImpl'],
    'policies': ['policies/PolicyImpl'],
    ...
}
```

Figure 8. The minimal structure required.

4.2. Example: Data Centre Management

As previously discussed in Section 3, we are researching ways to address the problems in trying to optimize energy efficiency in the data centre while ensuring that other performance metrics are met using MAS. To illustrate this, we provide a minimal example to cover the basic setup of the ARPS framework. The source code is available at https://gitlab.com/arps/arps/tree/main/arps/examples/computational_resources_management.

This example by no means provides an in-depth analysis of resource management in data centres and only serves to illustrate how the MAS can be realized. To keep this example simple, we will manage only two resources from a single host: the CPU and the Energy Monitor. Additional resources, such as temperature sensors of the environment, cooling system, and multiple hosts, can be added iteratively as the need for better understanding the system arises.

The CPU provides means to both read its current utilization or modify its frequency. It is driven by two governors: performance and powersave. This enables dynamic frequency scaling. The maximum frequency is 2.9 GHz in each one of its four cores and when in performance mode, the frequency is adjusted dynamically. Conversely, in powersave mode, the frequency stays closer to the minimum frequency available. The Energy Monitor only provides an interface to read the estimated power consumption in watts using the tool *PowerTOP* [44]. This tool was developed by Intel, and provides estimates on power consumption.

To understand how energy consumption is affected when the CPU has a certain workload, we collected the data using the sensors. As described in the previous section, when a resource is implemented, a monitor policy related to it is made available by the Agent Manager. In this case, it is possible to create agents executing the CPUMonitorPolicy and EnergyMonitorPolicy policies periodically. The collected data were summarized into two charts, where the Y-axis represents the normalized value of the CPU workload and the estimated power. The X-axis is the time in seconds. In Figure 9, the CPU is using a performance governor, while in Figure 10 the CPU is using powersave governor. As expected, there is a correlation between workload and energy consumption in performance mode since the CPU adjusts its frequency during the execution. The same does not apply in the powersave mode, where the energy consumption almost never goes over a certain value.

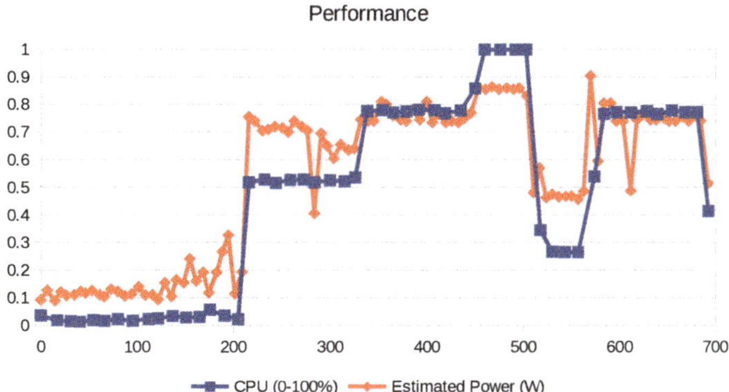

Figure 9. Correlation of energy consumption and CPU workload using performance governor.

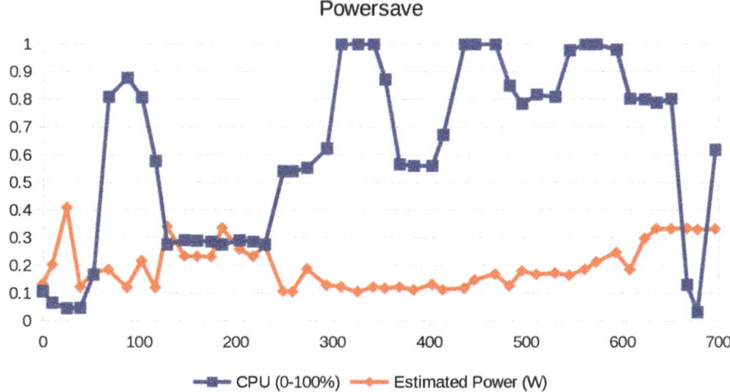

Figure 10. Correlation of energy consumption and CPU workload using powersave governor.

To reduce energy consumption, for example, we can use reactive policies implementing the method *condition(event)* of the Policy interface to evaluate the workload in the CPU. The action performed by the policy can be the adjustment of the CPU's frequency to a lower level. In this example, we considered two different strategies for saving energy. The strategies involve three policies; these policies are presented in Figure 11. The first adjusts the CPU to powersave mode when the utilization is over 80% while in the second sets the CPU to powersave mode when the utilization is over 50%. The third returns the CPU to performance mode when the utilization is under 50. The first strategy is composed of policies PowersaveDynamicScalingPolicy1 and PerformanceDynamicScalingPolicy; strategy two uses PowersaveDynamicScalingPolicy2.

We have chosen batch jobs to represent the external tasks that modify the environment to evaluate how the agents will manage energy consumption. In this case, we opted to use the deterministic model of the job arrival using an external file of events, where each row represents one event arriving in the system containing the arrival time, duration of the task, and workload. The SimTask *main* method, seen in Figure 12, allocates the CPU resource, and consequently causes additional power consumption proportionately to the workload when the CPU is performance mode and more steady when the CPU is in powersave mode, as mentioned previously. The SimTask *pos* method would release the resource as soon as the task is completed.

```
class PowersaveDynamicScalingPolicy1(Policy):
    def condition(self, event):
        return self.cpu.read() > 80

    def action(self):
        self.cpu_freq.set(governor='powersave', max_frequency=0.9)

class PowersaveDynamicScalingPolicy2(Policy):
    def condition(self, event):
        return self.cpu.read() > 50

    def action(self):
        self.cpu_freq.set(governor='powersave', max_frequency=0.9)

class PerformanceDynamicScalingPolicy(Policy):
    def condition(self, event):
        return self.cpu.read() <= 50

    def action():
        self.cpu_freq.set(governor='performance', max_frequency=2.9)
```

Figure 11. Example of implementation of a reactive policy.

```
class CPUTask(SimTask):
    def main(self):
        if not self.acquired:
            self.acquired = True
            self.cpu.value += self.wordload

        if self.cpu.frequency == 'powersave':
            estimate = self.powersave_estimate(self.cpu.value)
        elif self.cpu.frequency == 'performance':
            estimate = self.performance_estimate(self.cpu.value)
        self.energy_monitor.value = estimate

    def pos(self):
        self.cpu.value -= self.wordload
```

Figure 12. Example implementation of a task to update the energy monitor based on CPU workload.

The first analysis suggests that the average energy consumption with the first strategy is 9 Watts while for the second one it is 7.57 Watts. Without taking any other resources correlated to energy consumption and CPU, we can infer that the second one is better. However, other resources or metrics, such as a metric on Quality of Service (QoS), can be affected by the CPU dynamic frequency scaling of strategy two; these could be included in the model for further investigation.

When comparing the policies in the real environment, a similar behaviour is observed. The average energy consumption in the first policy is 13.75 Watts while in the second one is 12.34 Watts. These results come from a very limited set of experiments and comparisons and so no significance can be determined. Again, this was meant as example to illustrate how the ARPS framework can be used.

5. Discussion

We presented ARPS, a framework to enable evaluation of a MAS before deployment in an actual environment. This framework is the result of a general purpose MAS solution developed to be as flexible as possible to provide management of a data centre. It has proven useful in a specific domain, and we believe that others in the MAS community can benefit from this additional option when developing solutions to their problems. To this end, we illustrated how it could be applied using our specific case.

Due to the decision of using the RESTful architecture style, agents can be deployed manually by a user in an environment that would contain a single agent, without the need of an agent manager.

This unintended feature is possible because agents are self-contained and all the operations are made available through the API. The interaction can be done using web clients widely available on the Internet. The only requirement is that the platform can run Python applications. Thus, a user can instantiate an agent to control an autonomous robot developed in a platform such as a RaspberryPI, or it can deploy an agent in devices of an IoT environment.

We have identified other features that could enrich this framework. The inclusion of a Policy Management Tool can ease the process of creating new policies dynamically using a high-level language. Also, new resources and their touchpoints could be made available on the fly or dynamically removing unavailable resources. This could improve the integration of simulation and deployment even more. We plan to add a learning component to make agents more robust to complex adaptive environments. Further, reliability characteristics, such as fault tolerance, and security concerns, are being addressed and will be included in future versions.

Currently, the framework has some limitations related to features that were set aside due to low priorities. The framework does not come with a visualization component. Thus, it is not possible to visualize the running agents both during the simulation or within the real environment. Also, because we were aiming to have actual agents, we did not evaluate the usage of this framework applied to problems entailing only agent-based modelling simulation (ABMS). Since ABMS and MAS concepts overlaps, this MAS framework could be adapted to be used as ABMS tool. We do not know, however, how it performs when millions of agents are deployed in the environment. The framework does include statistical tools to analyze the results of the simulation or to assess the system in the actual environment since we believe that each user has their own needs and preference for tools. As mentioned, all the data gathered via the framework is made available in CSV format and so other applications that can provide more suitable tools can be used to extract insightful information from the data. Lastly, the framework is not FIPA compliant yet.

Author Contributions: Conceptualization, T.C.P. and M.B.; Methodology, T.C.P. and M.B.; Software, T.C.P.; Supervision, M.B.; Visualization, T.C.P.; Writing—original draft, T.C.P. and M.B.; Writing—review & editing, T.C.P. and M.B.

Funding: This research is partially supported by CNPq, Brazil grant number 249237/2013-0.

Conflicts of Interest: The authors declare no conflict of interest.

References

1. Wangermann, J.P.; Stengel, R.F. Optimization and coordination of multiagent systems using principled negotiation. *J. Guid. Control Dyn.* **1999**, *22*, 43–50. [CrossRef]
2. Roche, R.; Blunier, B.; Miraoui, A.; Hilaire, V.; Koukam, A. Multi-agent systems for grid energy management: A short review. In Proceedings of the 36th Annual Conference on IEEE Industrial Electronics Society (IECON 2010), Glendale, AZ, USA, 7–10 November 2010. doi:10.1109/iecon.2010.5675295. [CrossRef]
3. Hadeli; Valckenaers, P.; Kollingbaum, M.; Brussel, H.V. Multi-Agent Coordination and Control Using Stigmergy. *Comput. Ind.* **2004**, *53*, 75–96. doi:10.1016/s0166-3615(03)00123-4. [CrossRef]
4. Sislak, D.; Rehak, M.; Pechoucek, M. A-globe: Multi-Agent Platform with Advanced Simulation and Visualization Support. In Proceedings of the 2005 IEEE/WIC/ACM International Conference on Web Intelligence (WI'05), Compiegne, France, 19–22 September 2005. doi:10.1109/wi.2005.18. [CrossRef]
5. Pěchouček, M.; Mařík, V. Industrial Deployment of Multi-Agent Technologies: Review and Selected Case Studies. *Auton. Agents Multi-Agent Syst.* **2008**, *17*, 397–431. doi:10.1007/s10458-008-9050-0. [CrossRef]
6. Open Source Initiative. The MIT License. 1988. Available online: https://opensource.org/licenses/MIT (accessed on 20 November 2016).
7. Kravari, K.; Bassiliades, N. A Survey of Agent Platforms. *J. Artif. Soc. Soc. Simul.* **2015**, *18*, 1–11. doi:10.18564/jasss.2661. [CrossRef]

8. Bellifemine, F.; Bergenti, F.; Caire, G.; Poggi, A. Jade—A Java Agent Development Framework. In *Multi-Agent Programming*; Springer: Boston, MA, USA, 2005; pp. 125–147. doi:10.1007/0-387-26350-0_5. [CrossRef]
9. Poslad, S. Specifying Protocols for Multi-Agent Systems Interaction. *ACM Trans. Auton. Adapt. Syst.* **2007**, *2*, 15. doi:10.1145/1293731.1293735. [CrossRef]
10. Šišlák, D.; Rehák, M.; Pěchouček, M.; Rollo, M.; Pavlíček, D. A-globe: Agent Development Platform with Inaccessibility and Mobility Support. In *Software Agent-Based Applications, Platforms and Development Kits*; Springer: Birkhäuser Basel, 2005; pp. 21–46. doi:10.1007/3-7643-7348-2_2. [CrossRef]
11. Helsinger, A.; Thome, M.; Wright, T. Cougaar: A scalable, distributed multi-agent architecture. In Proceedings of the 2004 IEEE International Conference on Systems, Man and Cybernetics (IEEE Cat. No.04CH37583), The Hague, The Netherlands, 10–13 October 2004. doi:10.1109/icsmc.2004.1399959. [CrossRef]
12. Pokahr, A.; Braubach, L.; Lamersdorf, W. Jadex: A BDI Reasoning Engine. In *Multi-Agent Programming*; Springer: Boston, MA, USA, 2005; pp. 149–174. doi:10.1007/0-387-26350-0_6. [CrossRef]
13. Rao, A.S.; Georgeff, M.P.; BDI Agents: From Theory to Practice. In Proceedings of the ICMAS: International Conference on Multi-Agent Systems, San Francisco, CA, USA, 1995, Volume 95, pp. 312–319.
14. Boissier, O.; Bordini, R.H.; Hübner, J.F.; Ricci, A.; Santi, A. Multi-Agent Oriented Programming with Jacamo. *Sci. Comput. Program.* **2013**, *78*, 747–761. doi:10.1016/j.scico.2011.10.004. [CrossRef]
15. Rao, A.S. AgentSpeak(L): BDI agents speak out in a logical computable language. In *Lecture Notes in Computer Science*; Springer: Berlin/Heidelberg, Germany, 1996; pp. 42–55. doi:10.1007/bfb0031845. [CrossRef]
16. Bordini, R.H.; Hübner, J.F.; Wooldridge, M. *Programming multi-agent systems in AgentSpeak using Jason*; John Wiley & Sons: Chichester, WS, UK, 2007; Volume 8.
17. Hübner, J.F.; Boissier, O.; Kitio, R.; Ricci, A. Instrumenting Multi-Agent Organisations with Organisational Artifacts and Agents. *Auton. Agents Multi-Agent Syst.* **2009**, *20*, 369–400. doi:10.1007/s10458-009-9084-y. [CrossRef]
18. Ricci, A.; Piunti, M.; Viroli, M.; Omicini, A. Environment Programming in CArtAgO. In *Multi-Agent Programming*; Springer: Boston, MA, USA, 2009; pp. 259–288. doi:10.1007/978-0-387-89299-3_8. [CrossRef]
19. Gutknecht, O.; Ferber, J. MadKit. In Proceedings of the Fourth International Conference on Autonomous Agents—AGENTS '00, Barcelona, Spain, 3–7 June 2000; pp. 78–79. doi:10.1145/336595.337048. [CrossRef]
20. Allan, R.J. *Survey of Agent Based Modelling and Simulation Tools*; Technical Report, DL-TR-2010-007; Computational Science and Engineering Department, STFC Daresbury Laboratory: Daresbury, Warrington, England, 2010.
21. Railsback, S.F.; Lytinen, S.L.; Jackson, S.K. Agent-Based Simulation Platforms: Review and Development Recommendations. *Simulation* **2006**, *82*, 609–623. doi:10.1177/0037549706073695. [CrossRef]
22. Minar, N.; Burkhart, R.; Langton, C.; Askenazi, M. *The Swarm Simulation System: A Toolkit for Building Multi-Agent Simulations*; Working Papers 96-06-042; Santa Fe Institute: Santa Fe, NM, USA, 1996.
23. Sklar, E. Netlogo, a Multi-Agent Simulation Environment. *Artif. Life* **2007**, *13*, 303–311. doi:10.1162/artl.2007.13.3.303. [CrossRef] [PubMed]
24. Luke, S.; Cioffi-Revilla, C.; Panait, L.; Sullivan, K.; Balan, G. Mason: A Multiagent Simulation Environment. *Simulation* **2005**, *81*, 517–527. doi:10.1177/0037549705058073. [CrossRef]
25. North, M.J.; Collier, N.T.; Ozik, J.; Tatara, E.R.; Macal, C.M.; Bragen, M.; Sydelko, P. Complex Adaptive Systems Modeling With Repast Simphony. *Complex Adapt. Syst. Modeling* **2013**, *1*, 3. doi:10.1186/2194-3206-1-3. [CrossRef]
26. Collier, N.; North, M. Parallel Agent-Based Simulation with Repast for High Performance Computing. *Simulation* **2012**, *89*, 1215–1235. doi:10.1177/0037549712462620. [CrossRef]
27. Niazi, M.; Hussain, A. Agent-Based Computing from Multi-Agent Systems to Agent-Based Models: A Visual Survey. *Scientometrics* **2011**, *89*, 479–499. doi:10.1007/s11192-011-0468-9. [CrossRef]
28. Drogoul, A.; Vanbergue, D.; Meurisse, T. Multi-agent Based Simulation: Where Are the Agents? In *Multi-Agent-Based Simulation II*; Springer: Berlin/Heidelberg, Germany, 2003; pp. 1–15. doi:10.1007/3-540-36483-8_1. [CrossRef]
29. Bosse, S.; Engel, U. Augmented Virtual Reality: Combining Crowd Sensing and Social Data Mining with Large-Scale Simulation Using Mobile Agents for Future Smart Cities. In Proceedings of the 5th International Electronic Conference on Sensors and Applications, Canary Islands, Tenerife, 25–27 September 2019; Volume 4, p. 49.

30. Azimi, S.; Delavar, M.; Rajabifard, A. Multi-agent simulation of allocating and routing ambulances under condition of street blockage after natural disaster. *Int. Arch. Photogramm. Remote Sens. Spat. Inf. Sci.* **2017**, *42*, 325–332. [CrossRef]
31. Kephart, J.; Chess, D. The Vision of Autonomic Computing. *Computer* **2003**, *36*, 41–50. doi:10.1109/mc.2003.1160055. [CrossRef]
32. Tesauro, G.; Chess, D.M.; Walsh, W.E.; Das, R.; Segal, A.; Whalley, I.; Kephart, J.O.; White, S.R. A multi-agent systems approach to autonomic computing. In Proceedings of the Third International Joint Conference on Autonomous Agents and Multiagent Systems, New York, NY, USA, 19–23 July 2004; Volume 1, pp. 464–471.
33. Norouzi, F.; Bauer, M. Autonomic Management for Energy Efficient Data Centers. In Proceedings of the Sixth International Conference on Cloud Computing, GRIDs, and Virtualization (Cloud Computing 2015), Nice, France, 22 March 2015; pp. 138–146.
34. Schraven, M.; Guarnieri, C.; Baranski, M.; Müller, D.; Monti, A. Designing a Development Board for Research on IoT Applications in Building Automation Systems. In Proceedings of the International Symposium on Automation and Robotics in Construction (ISARC), Banff, AB, Canada, 2019; Volume 36, pp. 82–90. doi:10.22260/ISARC2019/0012. [CrossRef]
35. Andročec, D.; Tomaš, B.; Kišasondi, T. Interoperability and lightweight security for simple IoT devices. In Proceedings of the 2017 40th International Convention on Information and Communication Technology, Electronics and Microelectronics (MIPRO), Opatija, Croatia, 22–26 May 2017; pp. 1285–1291. doi:10.23919/MIPRO.2017.7973621. [CrossRef]
36. Perkel, J.M. Programming: Pick Up Python. *Nature* **2015**, *518*, 125–126. doi:10.1038/518125a. [CrossRef] [PubMed]
37. Ayer, V.M.; Miguez, S.; Toby, B.H. Why Scientists Should Learn To Program in Python. *Powder Diffr.* **2014**, *29*, S48–S64. doi:10.1017/s0885715614000931. [CrossRef]
38. Foundation for Intelligent Physical Agents (FIPA). FIPA Agent Message Transport Protocol for HTTP Specification. 2002. Available online: http://www.fipa.org/specs/fipa00084/SC00084F.html (accessed on 10 June 2019).
39. Ciortea, A.; Boissier, O.; Zimmermann, A.; Florea, A.M. Give Agents Some REST: A Resource-oriented Abstraction Layer for Internet-scale Agent Environments. In Proceedings of the 16th Conference on Autonomous Agents and MultiAgent Systems (AAMAS '17), São Paulo, Brazil, 8–12 May 2017; International Foundation for Autonomous Agents and Multiagent Systems: Richland, SC, USA, 2017; pp. 1502–1504.
40. Braubach, L.; Pokahr, A. Conceptual Integration of Agents with WSDL and RESTful Web Services. In *Lecture Notes in Computer Science*; Springer: Berlin/Heidelberg, Germany, 2013; pp. 17–34. doi:10.1007/978-3-642-38700-5_2. [CrossRef]
41. Pablo, P.V.; Holgado-Terriza, J.A. *Integration of MultiAgent Systems with Resource-Oriented Architecture for Management of IoT-Objects*; IOS Press: Amsterdam, The Netherlands; Volume 23: Intelligent Environments 2018; pp. 567–576. [CrossRef]
42. Russell, S.J.; Norvig, P. *Artificial Intelligence: A Modern Approach*; Pearson Education: Upper Saddle River, NJ, USA, 2009; Chapter 2.
43. Kephart, J.O.; Walsh, W.E. An artificial intelligence perspective on autonomic computing policies. In Proceedings of the Fifth IEEE International Workshop on Policies for Distributed Systems and Networks; Yorktown Heights, NY, USA, 9 June 2004. doi:10.1109/POLICY.2004.1309145. [CrossRef]
44. Intel Open Source. PowerTOP: Linux Tool to Diagnose Issues with Power Consumption and Power Management. 2007. Available online: https://01.org/powertop/ (accessed on 12 July 2019).

Article

Consensus Algorithms Based Multi-Robot Formation Control under Noise and Time Delay Conditions

Heng Wei [1,*], Qiang Lv [1], Nanxun Duo [1], GuoSheng Wang [1] and Bing Liang [2]

[1] Weapons and Control Department, Academy of Army Armored Forces, Beijing 100072, China; rokyou@live.cn (Q.L.); lizdnx@mail.nwpu.edu.cn (N.D.); gswang@126.com (G.W.)

[2] School of Information Engineering, Jiangxi University of Science and Technology, Ganzhou 341000, China; lbwgs@126.com

* Correspondence: wh_killer@foxmail.com

Received: 2 January 2019; Accepted: 6 March 2019; Published: 11 March 2019

Abstract: In recent years, the formation control of multi-mobile robots has been widely investigated by researchers. With increasing numbers of robots in the formation, distributed formation control has become the development trend of multi-mobile robot formation control, and the consensus problem is the most basic problem in the distributed multi-mobile robot control algorithm. Therefore, it is very important to analyze the consensus of multi-mobile robot systems. There are already mature and sophisticated strategies solving the consensus problem in ideal environments. However, in practical applications, uncertain factors like communication noise, communication delay and measurement errors will still lead to many problems in multi-robot formation control. In this paper, the consensus problem of second-order multi-robot systems with multiple time delays and noises is analyzed. The characteristic equation of the system is transformed into a quadratic polynomial of pure imaginary eigenvalues using the frequency domain analysis method, and then the critical stability state of the maximum time delay under noisy conditions is obtained. When all robot delays are less than the maximum time delay, the system can be stabilized and achieve consensus. Compared with the traditional Lyapunov method, this algorithm has lower conservativeness, and it is easier to extend the results to higher-order multi-robot systems. Finally, the results are verified by numerical simulation using MATLAB/Simulink. At the same time, a multi-mobile robot platform is built, and the proposed algorithm is applied to an actual multi-robot system. The experimental results show that the proposed algorithm is finally able to achieve the consensus of the second-order multi-robot system under delay and noise interference.

Keywords: multi-robot; consensus problem; formation control; noise; time delay

1. Introduction

In recent years, with the continuous development of computer science, complex network theory and control theory, autonomous mobile robots have received more and more attention [1]. Compared to single mobile robots, multi-mobile robot systems have better stability, higher fault tolerance and higher work efficiency. As a result, they have better application prospects and higher research value in the fields of reconnaissance, patrol, rescue and environmental survey. Formation control of multi-mobile robots is the basis of multi-mobile robot systems, and has become a hotspot in the field of robotics [2].

As part of the design process of multi-robot formation control algorithm, many problems need to be considered, including robot model, external environmental interference, sensor measurement noise, algorithm control precision, and the controllability of different formations [3]. The existing formation control algorithms for multi-robots mainly include the leader-follower algorithm [4], the behavior-based algorithm [5], the graph theory-based method [6], the virtual structure method [7], and the artificial potential field method. The leader-follower algorithm has flexible motion strategy and scalability, but the algorithm cannot form stable and reliable feedback between the follower and the leader. Therefore, the control error of the follower will increase with interference from the environment. In particular, when the leader fails, it can cause the entire multi-robot system to crash. The behavior-based algorithm can effectively reduce the complexity of the entire formation control algorithm, but it has higher requirements in terms of sensor sensing ability and communication ability between robots, and cannot accurately quantify the behavior of robots during operation. Thus, it is difficult to guarantee the system's robustness using the behavior-based algorithm. The virtual structure method is convenient for designing the formation behavior of multi-robot systems, while due to the constraints of rigid structures, it lacks flexibility with respect to obstacle avoidance and formation transformation. The artificial potential field algorithm has a simple structure and can effectively avoid collisions and obstacles, but it is susceptible to interference when maintaining the formation, and it is difficult to perform precise formation control. Moreover, the potential energy function needs to be reset if the formation transformation is performed, leading to a lack of flexibility.

In view of the shortcomings of the traditional formation control algorithm, considering the increase in the number of robots in the multi-robot system and the continuous improvement of the data processing capability of a single robot, the distributed multi-robot control algorithm has attracted the attention of researchers. The distributed multi-robot system can make full use of the data processing resources of the robot and share the pressure of the central processing machine, which has great advantages in terms of flexibility and fault tolerance [8,9]. In addition, solving the consensus problem is the core of the distributed multi-robot control algorithm [10]. There are already mature and sophisticated strategies for solving the consensus problem in ideal environments [11,12]. However, in practical applications, uncertain factors like communication noise, communication delay and measurement error will still lead to many problems in multi-robot formation control. Some algorithms have considered some practical problems. Reference [13] studied the conditions of the system reaching consensus under uniform delay, when the communication structures of second-order multi-robot systems were a directed graph with spanning tree or a strongly connected graph, respectively. However, that paper does not consider the noise condition or consensus under different delay conditions. Reference [14] studied the consensus problem of second-order multi-robot systems under noisy conditions. A control protocol based on distributed sampling data was proposed to achieve system consensus, but the delay condition was not taken into account in the algorithm. Reference [15] studied the consensus of second-order multi-robot systems under non-uniform and multi-time delays using the frequency domain analysis method. Compared with the Lyapunov method, it has lower conservativeness, and the results were extended to higher-order multi-robot systems. However, it did not take noise into consideration, which is unavoidable in practical environments. Reference [16] studied the consensus of second-order multi-robot systems under uniform time delay and noise environments, and designed different control protocols for different types of noise, thus achieving the consensus of the system. These algorithms provide some basic solutions to the second-order system consensus problem, but the problems encountered by multi-robots in practical applications are far more varied than these. On the basis of these algorithms, this paper performs a more in-depth analysis, especially considering the consensus of the second-order system in which there are many different time delays and multiplicative noises in the system, laying the foundations for a formation control algorithm for second-order multi-robot systems that can be truly implemented in real robot systems.

In summary, this paper analyzes the consensus problem of second-order multi-robot systems under various delay and noise conditions. The system character equations are transformed into quadratic polynomials of pure imaginary eigenvalues based on frequency domain analysis, and then solved. Finally, its critical steady state is obtained and verified using Matlab numerical simulation. Compared with existing algorithms, this algorithm has lower conservativeness, and it is easier to extend the results to higher-order multi-robot systems. Since the omnidirectional mobile robot is a fully driven robot, and the horizontal and vertical directions can be separately controlled, it can be constructed as two one-dimensional second-order multi-robot systems. Therefore, experiments were carried out on a multi-omnidirectional mobile robot platform built in the laboratory using the proposed algorithm [17,18], which verifies the effectiveness of the proposed algorithm.

2. Pre-Preparation and Problem Description

2.1. Graph Theory

$G = \{V, E\}$ represents the communication topology between robots, in which each robot represents a node. V is a set of nodes. E is a set of edges, representing the connection state between robots. The topology map is represented by a Laplacian matrix, which is $L = D - A$. D is the degree matrix, which represents how many nodes are adjacent to each node. $A = [a_{ij}]$ is the adjacent matrix and $i, j \in V$. N_i represents all sets of nodes adjacent to the i node. If node j is adjacent to node i, then $a_{ij} > 0$. If $a_{ij} = a_{ji}$ for any $i, j \in V$, the graph is an undirected graph; otherwise, it is a directed graph. If there is a directed path on any two nodes in the graph, the directed graph G is strongly connected. If there is a directed path to a node in the graph to any other node, then the directed graph G contains a spanning tree. If the undirected graph G is strongly connected, it is called a connected graph. When the undirected graph G is a connected graph, its Laplacian L matrix contains a zero root, and the other eigenvalues are positive real numbers. When a directed graph G contains a spanning tree, its Laplacian L matrix contains a zero root, and the rest eigenvalue's real part are positive.

2.2. Problem Description

Suppose the system consists of n omnidirectional robots. The dynamic characteristics of the omnidirectional robot in the x direction are:

$$\begin{cases} \dot{x}_i(t) = v_i(t) \\ \dot{v}_i(t) = u_i(t) \end{cases} \tag{1}$$

where $x_i(t)$ is position, $v_i(t)$ is velocity and $u_i(t)$ is input control. If any i robot and j robot in the multi-robot system satisfy the identities as follows:

$$\lim_{t \to +\infty} [x_i(t) - x_j(t)] = 0 \tag{2}$$

$$\lim_{t \to +\infty} [v_i(t) - v_j(t)] = 0 \tag{3}$$

then the multi-robot system (1) has achieved consensus under the control protocol $u_i(t)$ Let the state vector of the i robot be $\delta_i(t) = [x_i(t), v_i(t)]^T$, then the multi-robot system state vector is $S(t) = [\delta_1(t), \delta_2(t), \delta_3(t), \ldots, \delta_n(t)]$. Rewrite system (1) as:

$$\dot{S}(t) = \Psi S(t) \tag{4}$$

where $\Psi = I \otimes A - L \otimes B$, $A = \begin{bmatrix} 0 & 1 \\ 0 & 0 \end{bmatrix}$, $B = \begin{bmatrix} 0 & 0 \\ k_1 & k_2 \end{bmatrix}$, $\otimes$ is Kronecker. When ideally without noise and delay, the control protocol designed in [13] is as follows:

$$u_i(t) = \sum_{j \in N_i} a_{ij} \{k_1[x_i(t) - x_j(t)] + k_2[v_i(t) - v_j(t)]\} \tag{5}$$

where $a_{ij} > 0$ is the topology weight of the communication between robot i and robot j, k_1 is the position scale factor that needs to be designed, k_2 is the velocity scale factor that needs to be designed. Lemmas 1 and 2 give the conditions that the coefficient matrix Ψ of control protocol (5) must satisfy when the communication topology of system (4) is undirected graph and directed graph, respectively.

Lemma 1. *When the communication topology of multi-robot system (4) is connected graph, the coefficient matrix $\boldsymbol{\Psi}$ has a double zero root, and the real part of other eigenvalues is negative.*

Proof. Let there be an orthogonal matrix $\boldsymbol{Q}$, such that:

$$\boldsymbol{Q}^T\boldsymbol{L}\boldsymbol{Q} = diag\{0, \lambda_2, \lambda_3, \ldots, \lambda_n\} \tag{6}$$

where $0, \lambda_2, \lambda_3, \ldots, \lambda_n$ is the eigenvalue of the Laplacian matrix $\boldsymbol{L}$, and $\lambda_i > 0$ (i = 2, 3, ..., n). Formula (7) is obtained from Formula (6):

$$(\boldsymbol{Q} \otimes \boldsymbol{I}_2)^T\boldsymbol{\Psi}(\boldsymbol{Q} \otimes \boldsymbol{I}_2) = diag\{\boldsymbol{A}, \boldsymbol{A} - \lambda_2\boldsymbol{B}, \boldsymbol{A} - \lambda_3\boldsymbol{B}, \ldots, \boldsymbol{A} - \lambda_n\boldsymbol{B}\} \tag{7}$$

The determinant of Formula (7) is obtained:

$$|diag\{\boldsymbol{A}, \boldsymbol{A} - \lambda_2\boldsymbol{B}, \boldsymbol{A} - \lambda_3\boldsymbol{B}, \ldots, \boldsymbol{A} - \lambda_n\boldsymbol{B}\}| = s^2\prod_2^n s^2 + \lambda_i k_2 s + \lambda_i k_1 = 0 \tag{8}$$

Because there is s^2 in Formula (8), there must be a double zero root in the eigenvalue. By solving polynomial equation $s^2 + \lambda_i k_2 s + \lambda_i k_1 = 0$, we can get:

$$s_1 = \frac{-\lambda_i k_2 + \sqrt{(\lambda_i k_2)^2 - 4\lambda_i k_1}}{2}$$

$$s_2 = \frac{-\lambda_i k_2 - \sqrt{(\lambda_i k_2)^2 - 4\lambda_i k_1}}{2}$$

Based on this analysis, when $(\lambda_i k_2)^2 > 4\lambda_i k_1$, obviously $-\lambda_i k_2 \pm \sqrt{(\lambda_i k_2)^2 - 4\lambda_i k_1} < 0$, so the eigenvalues s_1 and s_2 are negative. When $(\lambda_i k_2)^2 < 4\lambda_i k_1$, because $\lambda_i k_2 > 0$, so $-\lambda_i k_2 < 0$, the eigenvalues s_1 and s_2 have negative real parts. Lemma 1 is proved. □

Lemma 2. *When the communication topology of multi-robot system (4) is directed graph and contains spanning tree, $k_1 \in (0, k_0 k_2^2)$ the coefficient matrix $\boldsymbol{\Psi}$ has a double zero root, and the real part of other eigenvalues is negative. Where $k_0 = \min\limits_{\|\lambda_i\| \neq 0}\left\{\frac{\|\lambda_i\|^2 \mathrm{real}(\lambda_i)}{\mathrm{imag}(\lambda_i)}\right\}$.*

Proof. The characteristic determinant of system (4) is obtained by Formula (8), assuming that there are polynomial equations:

$$s^2 + s(a + bj) + k(a + bj) = 0 \tag{9}$$

where $a > 0$, $k, a, b \in R$. Let $s = jw$:

$$-w^2 - bw + ka + (aw + kb)j = 0 \tag{10}$$

Solving Formula (10), we can get:

$$\begin{cases} -w^2 - bw + ka = 0 \\ aw + kb = 0 \end{cases} \tag{11}$$

Thus, solved:

$$\begin{cases} k_1 = 0 \\ k_2 = \frac{a(a^2+b^2)}{b^2} \end{cases} \tag{12}$$

Document [13] proves that when $0 < k < \frac{a(a^2+b^2)}{b^2}$, the roots of Formula (9) are all on the left open half plane. Formula (8) is modified according to Formula (9):

$$s^2 \prod_2^n s^2 + \lambda_i k_2 s + \lambda_i \frac{k_1}{k_2} k_2 = 0 \tag{13}$$

The analysis shows that when $k_0 = \min_{\|\lambda_i\| \neq 0} \left\{ \frac{\|\lambda_i\|^2 \text{real}(\lambda_i)}{\text{imag}(\lambda_i)} \right\}$, $k_1 \in (0, k_0 k_2^2)$, the coefficient matrix $\boldsymbol{\Psi}$ has a double zero root, and the real part of other eigenvalues is negative. Lemma 2 is proved. □

3. Consensus Analysis of Multi-Robot with Various Delays and Noise Conditions

In the previous section, we analyzed the conditions under which second-order systems achieve consensus in an ideal environment. However, in real environments, due to noise interference and communication differences between different robotic hardware, the above control protocols need to be improved. Assuming that there are D kinds of different delays in the system, the multi-agent system (4) can be changed to:

$$\dot{\boldsymbol{S}}(t) = (\boldsymbol{I} \otimes \boldsymbol{A}) \cdot \boldsymbol{S}(t) - \sum_{d=1}^{D} (\boldsymbol{L}_d \otimes \boldsymbol{B}) \cdot \zeta(t) \cdot \boldsymbol{S}(t - \tau_d) \tag{14}$$

where $\zeta(t)$ is the communication noise or measurement noise between robots, τ_{ij} is the transmission delay, which represents the time taken by i robot to receive and process information transmitted by j robot, $\boldsymbol{L}_d$ is the Laplacian matrix corresponding to the sub-topological graph of the robot node when the delay is τ_d, and $\sum_{d=1}^{D} \boldsymbol{L}_d = \boldsymbol{L}$.

Theorem 1. *If system (14) is a connected graph, the system can achieve consensus when the system delay τ_d is less than $\tau_{\max}$ under the action of noise $\zeta(t)$. Among them:*

$$\begin{cases} \tau_{\max} = \left[\arctan\left(\frac{k_2}{k_1} w_{\max}\right)\right] / w_{\max} \\ w_{\max} = \sqrt{\frac{\lambda_{\max}^2 k_2^2 \zeta^2(t) + \zeta(t) \sqrt{\zeta^2(t) (\lambda_{\max}^2 k_2^2)^2 + 4 \lambda_{\max}^2 k_1^2}}{2}} \end{cases} \tag{15}$$

Proof. Using the frequency domain analysis method for analysis, the Laplace transform of Equation (14) can be obtained:

$$\boldsymbol{S}(s) = (s\boldsymbol{I}_{2n} - (\boldsymbol{I}_n \otimes \boldsymbol{A}) + \sum_{d=1}^{D} (\boldsymbol{L}_d \otimes \boldsymbol{B}) \zeta(t) e^{-\tau_d s})^{-1} \boldsymbol{S}(0) \tag{16}$$

Let $G_\tau(s) = s\boldsymbol{I}_{2n} - (\boldsymbol{I}_n \otimes \boldsymbol{A}) + \sum_{d=1}^{D} (\boldsymbol{L}_d \otimes \boldsymbol{B}) \zeta(t) e^{-\tau_d s}$, so the eigenvalues of the determinant $|G_\tau(s)|$ are the eigenvalues of the system. Lemma 1 proves that multi-robot system (4) achieves the conditions of consensus. According to Lemma 1, how can the eigenvalues of system (14) be kept in the negative half-plane under the interference of time delay τ and noise $\zeta(t)$ relative to the system (4)? Because the measurement noise and communication noise are uncertainties in real environments, it is impossible to carry out accurate quantitative analysis. Therefore, only when the system delay τ increases to a value under the action of noise $\zeta(t)$ does a non-zero eigenvalue of the system appear on the virtual axis for the first time, while the time delay τ is the critical value for the system to maintain stability.

Assuming that the eigenvalue of the system is on the imaginary axis, let $s = jw$ be the eigenvalue; then $\alpha = \alpha_1 \otimes [1,0]^T + \alpha_2 \otimes [0,1]^T$ is the eigenvector corresponding to the eigenvalue, and if $\|\alpha\| = 1$, $\alpha_1, \alpha_2 \in \mathbb{C}^n$, then:

$$\left[jw\boldsymbol{I}_{2n} - (\boldsymbol{I}_n \otimes \boldsymbol{A}) + \sum_{\text{d}=1}^{D} (\boldsymbol{L}_d \otimes \boldsymbol{B})\zeta(t)e^{-jw\tau_d} \right] \alpha = 0 \tag{17}$$

The imaginary eigenvalues of the system appear in pairs in conjugate form. This paper only analyzes the case where $w > 0$. Formula (17) left multiplied by α^H is:

$$\alpha^H \left[jw\boldsymbol{I}_{2n} - (\boldsymbol{I}_n \otimes \boldsymbol{A}) + \sum_{\text{d}=1}^{D} (\boldsymbol{L}_d \otimes \boldsymbol{B})\zeta(t)e^{-jw\tau_d} \right] \alpha = 0 \tag{18}$$

Because each line of the left matrix of Formula (17) is zero, so $jw\alpha_1 = \alpha_2$, and substituting it into Formula (18):

$$\sum_{\text{d}=1}^{D} \beta_d \zeta(t) e^{-jw\tau_d} = \frac{w^2}{k_1 + jwk_2} \tag{19}$$

where $\beta_d = \frac{\alpha^H (\boldsymbol{L}_d \otimes \boldsymbol{I}_2)\alpha}{\alpha^H \alpha}$. Replace A with B in Formula (19):

$$F(w) = \sum_{\text{d}=1}^{D} \beta_d \zeta(t) e^{jw\tau_d} = \frac{w^2}{k_1 - jwk_2} \tag{20}$$

Take module operation on both sides of the upper equal sign:

$$\|F(w)\| = \|\sum_{\text{d}=1}^{D} \beta_d \zeta(t) e^{-jw\tau_d}\| < \|\sum_{\text{d}=1}^{D} \beta_d \zeta(t)\| = \frac{\alpha^H (\boldsymbol{L} \otimes \boldsymbol{I}_2)\alpha}{\alpha^H \alpha} \zeta(t) \leq \lambda_{\max} \zeta(t) \tag{21}$$

Let $w_{\max} = \sqrt{\frac{\lambda_{\max}^2 k_2^2 \zeta^2(t) + \zeta(t)\sqrt{\zeta^2(t)(\lambda_{\max}^2 k_2^2)^2 + 4\lambda_{\max}^2 k_1^2}}{2}}$ get $w \leq w_{\max}$, upper formula establishment. From Formula (20):

$$\theta(w) = \text{argz}[F(w)] = \arctan(\frac{k_2}{k_1} w) \tag{22}$$

where $\theta(w) \in [0, 2\pi)$. Let $\tau(w) = \frac{\theta(w)}{w}, a = \frac{k_2}{k_1}$, deriving for $\tau(w)$, we can obtain:

$$M_1(w) = \frac{d\tau(w)}{dw} = \frac{1}{w^2} M_2(w) = \frac{1}{w^2} \left[\frac{aw}{a^2 w^2 + 1} - \arctan(aw) \right] \tag{23}$$

Deriving for $M_2(w)$ we can obtain:

$$\frac{dM_2(w)}{dw} = -\frac{2a^3 w^2}{(a^2 w^2 + 1)^2} < 0 \tag{24}$$

$M_2(w)$ is decreasing, so when $w > 0$, $M_2(w) < M_2(0) = 0$, so $M_1(w) < 0$; that is, $\tau(w)$ is also decreasing. So $\tau(w) \geq \tau(w_{\max}) = \tau_{\max}$. When $\tau_d < \tau_{\max}$, when $\tau_d < \tau_{\max}$, we can get:

$$\tau(w) = \frac{\theta(w)}{w} = \frac{\text{argz}\left(\sum_{\text{d}=1}^{D} \beta_d \zeta(t) e^{jw\tau_d} \right)}{w} \leq \frac{\max[w\tau_d]}{w} < \frac{w\tau_{\max}}{w} < \tau_{\max} \tag{25}$$

That is to say, it contradicts $\tau(w) \geq \tau(w_{\max})$. Therefore, when $\tau_d < \tau_{\max}$, the eigenvalues of the system can be maintained in the left half plane, and the consensus of system (14) can be achieved. Theorem 1 is proved. □

Theorem 2. *If system (14) is a directed graph and there is a spanning tree, the system can achieve consensus when the system delay τ_d is smaller than the $\tau_{\max}$ under the action of noise $\zeta(t)$, and $k_1 \in (0, k_0 k_2^2)$. Among them:*

$$\begin{cases} \tau_{\max} = \min\limits_{\|\lambda_i\| \neq 0} \left[\left[\arctan\left(\frac{k_2}{k_1} w_i\right) - \operatorname{argz}(\lambda_i) \right] / w_i \right] \\ w_i = \sqrt{\frac{\lambda_i^2 k_2^2 \zeta^2(t) + \zeta(t)\sqrt{\zeta^2(t)(\lambda_i^2 k_2^2)^2 + 4\lambda_i^2 k_1^2}}{2}} \end{cases} \tag{26}$$

where $\operatorname{argz}(\lambda_i) \in (-\frac{\pi}{2}, \frac{\pi}{2})$.

Proof. Lemma 2 proved that, when the communication topology of multi-robot system (4) is directed graph and contains spanning tree, $k_1 \in (0, k_0 k_2^2)$, the coefficient matrix $\boldsymbol{\Psi}$ has a double zero root, and the real part of the other eigenvalues is negative. The same analysis is performed using the frequency domain analysis method. Similar to the proof of Theorem 1, only when the system delay τ increases to the value under the action of noise $\zeta(t)$ does a non-zero eigenvalue of the system first appear on the imaginary axis, while the delay τ is the critical value for the system to maintain stability. Take modulo operation on Formula (20):

$$\|F(w)\| = \|\frac{w^2}{k_1 - jwk_2}\| \tag{27}$$

Let w be a function of $\|F(w)\|$; then the above formula can be written as follows:

$$w = \sqrt{\frac{\|F(w)\|^2 k_2^2 + \sqrt{(\|F(w)\|^2 k_2^2)^2 + 4\|F(w)\|^2 k_1^2}}{2}} \tag{28}$$

Then we can get that w is an incremental function about $\|F(w)\|$. From Formula (20):

$$\begin{cases} \operatorname{argz}[F(w)] = \arctan(\frac{k_2}{k_1} w) \\ \operatorname{argz}[F(w)] \leq \operatorname{argz}(\sum\limits_{d=1}^{D} \beta_d) + \max(w\tau_m) \end{cases} \tag{29}$$

So:

$$\arctan(\frac{k_2}{k_1} w) - \operatorname{argz}(\sum_{d=1}^{D} \beta_d) \leq \max(w\tau_m) \tag{30}$$

Because $\beta_d = \frac{\alpha^H (\boldsymbol{L}_d \otimes \boldsymbol{I}_2)\alpha}{\alpha^H \alpha}$, so $\sum\limits_{d=1}^{D} \beta_d = \lambda_i$,where λ_i is the non-zero eigenvalue of Laplace matrix $\boldsymbol{L}$. So $\|F(w)\| \leq \|\zeta(t)\lambda_i\|$. Because w is an incremental function about $\|F(w)\|$, so:

$$w(\|F(w)\|) \leq w(\|\zeta(t)\lambda_i\|) = w_i = \sqrt{\frac{\lambda_i^2 k_2^2 \zeta^2(t) + \zeta(t)\sqrt{\zeta^2(t)(\lambda_i^2 k_2^2)^2 + 4\lambda_i^2 k_1^2}}{2}} \tag{31}$$

When $\tau_d < \tau_{\max}$, we can get:

$$\begin{aligned} \max(w\tau_d) < w_i \tau_{\max} &= \min\left[\left[\arctan(\tfrac{k_2}{k_1} w_i) - \operatorname{argz}(\sum_{d=1}^{D} \beta_d)\right] / w_i\right] w_i \\ &\leq \arctan(\tfrac{k_2}{k_1} w) - \operatorname{argz}(\sum_{d=1}^{D} \beta_d) \end{aligned} \tag{32}$$

We can find that this contradicts Formula (30), so when $\tau_d < \tau_{\max}$, the eigenvalue of the system can be maintained in the left half plane, and the consensus of system (14) can be achieved. Theorem 2 is proved. □

4. Simulation Verification

In this section, two sets of Matlab/Simulink numerical simulation experiments are carried out to verify the consensus of the system described in Theorems 1 and 2 when the communication topology is undirected graph and directed graph under the conditions of noise and various delays.

Experiment 1. *Let system (14) consist of four robots whose communication topology is shown in Figure 1.*

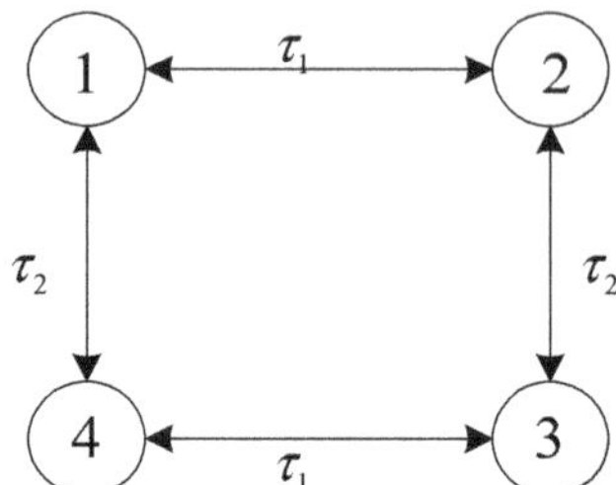

Figure 1. Experiment 1 system communication topology.

As can be seen from Figure 1, the time delay between robots 1 and 2 is τ_1, between robot 2 and robot 3 it is τ_2, between robot 3 and robot 4 it is τ_1, between robot 4 and robot 1 it is τ_2. If the adjacent communication weight a_{ij} is 1, then the Laplace matrix $\boldsymbol{L}$ is:

$$\boldsymbol{L} = \begin{pmatrix} 2 & -1 & & -1 \\ -1 & 2 & -1 & \\ & -1 & 2 & -1 \\ -1 & & -1 & 2 \end{pmatrix} \tag{33}$$

We can get $\lambda_{\max} = 4$. Assume that the communication noise or measurement noise is white noise with a maximum amplitude of two. According to Theorem 1, $\tau_{\max} = 0.226$.

In the first group of Experiment 1, set $\tau_1 = 0.21$, $\tau_2 = 0.22$, and the initial posture is assumed to be (1,0), (2,0), (3,0), (4,0). The simulation results are shown in Figure 2.

To verify Theorem 1 and compare with the first group of experiments, in the second group of experiments, set $\tau_1 = 0.23$, $\tau_2 = 0.24$ under the same conditions. The simulation results are shown in Figure 3.

According to Experiment 1, system (14) satisfying lemma 1 can achieve consensus when all τ_d are less than $\tau_{\max}$, and the system will diverge when all τ_d are greater than $\tau_{\max}$, which cannot achieve consensus; thus Theorem 1 is verified.

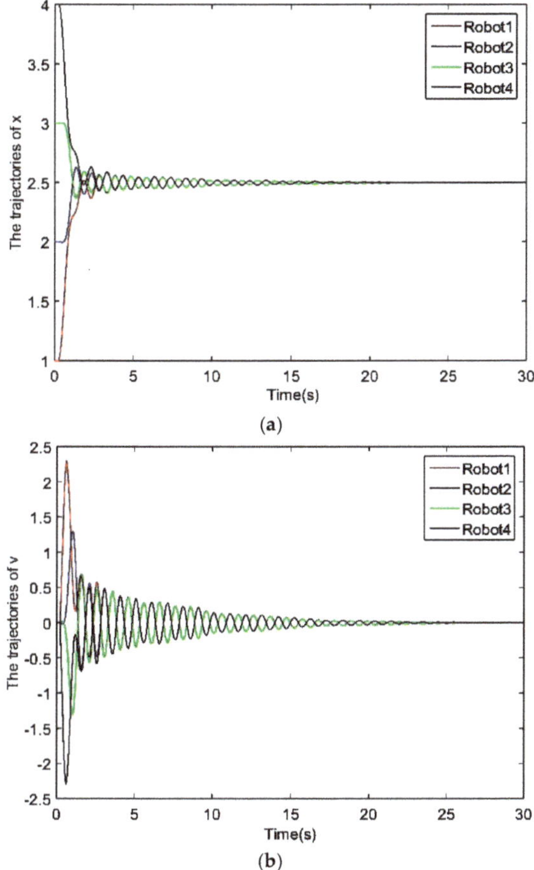

(a)

(b)

Figure 2. Experiment 1 Group 1 simulation results; x is position, v is velocity. (**a**) Trajectory of x changing with time; (**b**) Trajectory of v changing with time.

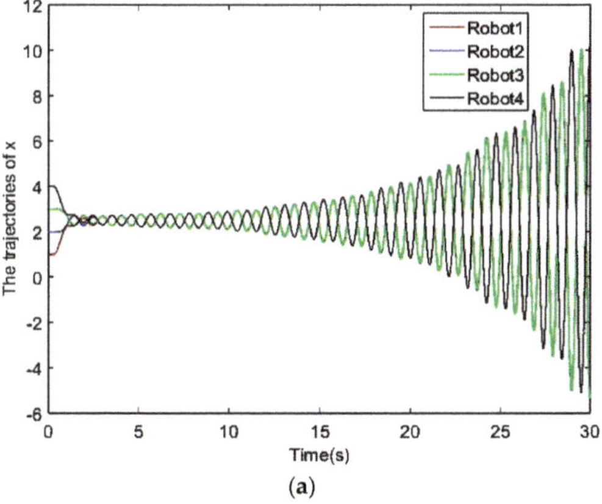

(a)

Figure 3. *Cont.*

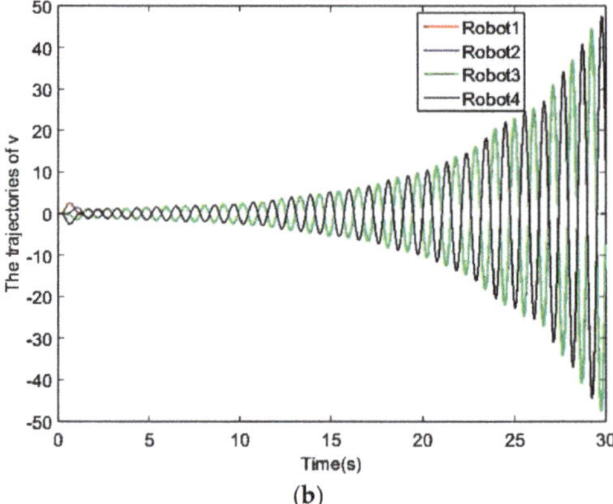

(b)

Figure 3. Experiment 1 Group 2 simulation results; x is position, v is velocity. (**a**) Trajectory of x changing with time; (**b**) Trajectory of v changing with time.

Experiment 2. *Let system (14) consist of four robots whose communication topology is shown in Figure 4.*

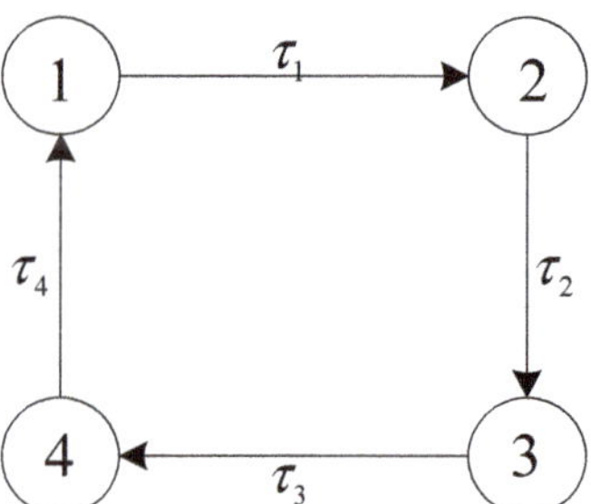

Figure 4. Experiment 2 system communication topology.

The time delay from Robot 1 to Robot 2 is τ_1, from Robot 2 to Robot 3 it is τ_2, from Robot 3 to Robot 4 it is τ_3, from Robot 4 to Robot 1 it is τ_4. If the adjacent communication weight a_{ij} is 1, then the Laplace matrix $\boldsymbol{L}$ is:

$$\boldsymbol{L} = \begin{pmatrix} 2 & & -1 & \\ -1 & 1 & & -1 \\ & -1 & 2 & \\ -1 & & -1 & 1 \end{pmatrix} \tag{34}$$

Then $k_0 = \min\limits_{\|\lambda_i\| \neq 0} \left\{ \frac{\|\lambda_i\|^2 \text{real}(\lambda_i)}{\text{imag}(\lambda_i)} \right\} = \frac{5\times 2}{1} = 10$, so $k_1 \in (0, 10k_2^2)$. Assume that the communication noise or measurement noise is white noise with a maximum amplitude of two. Set $k_1 = 1$, $k_2 = 1$, according to Theorem 2, $\tau_{\max} = 0.137$.

In the first group of experiment 2, set $\tau_1 = 0.13$, $\tau_2 = 0.12$, $\tau_3 = 0.11$, $\tau_4 = 0.1$, and the initial posture is assumed to be (1,0), (2,0), (3,0), (4,0). The simulation results are shown in Figure 5.

To verify Theorem 2 and compare with the first group of experiments, in the second group of experiments, set $\tau_1 = 0.14$, $\tau_2 = 0.141$, $\tau_3 = 0.142$, $\tau_4 = 0.143$ under the same conditions. The simulation results are shown in Figure 6.

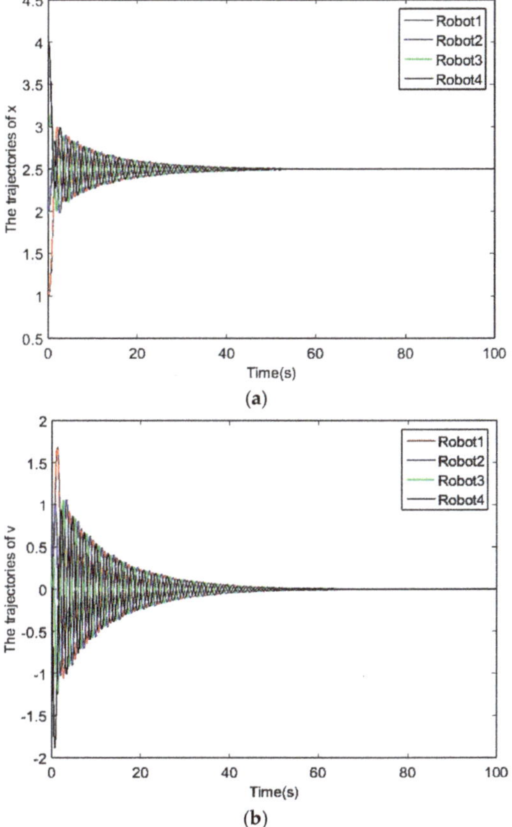

Figure 5. Experiment 2 Group 1 simulation results; x is position, v is velocity. (**a**) Trajectory of x changing with time; (**b**) Trajectory of v changing with time.

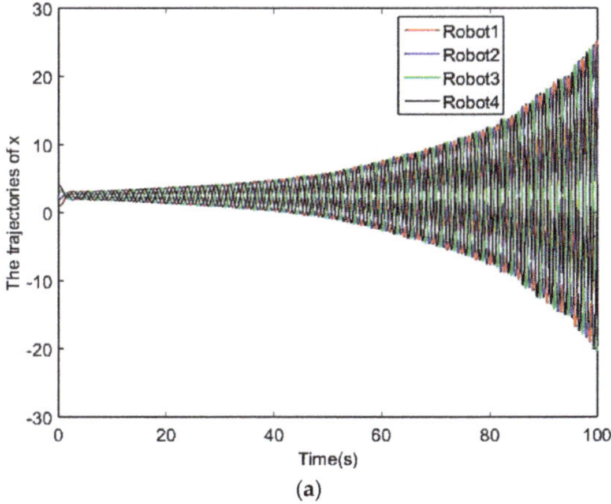

Figure 6. *Cont.*

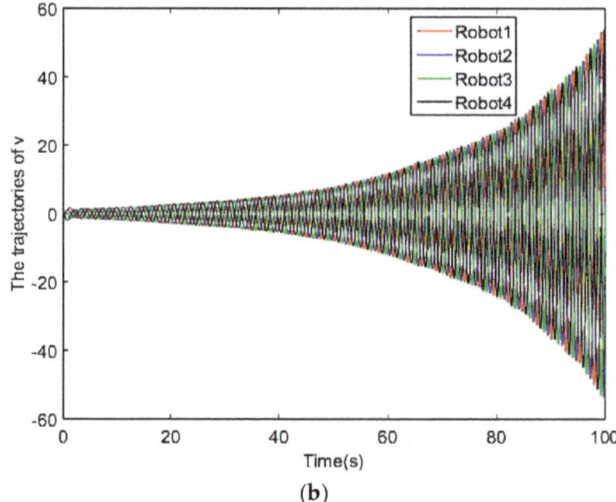

(b)

Figure 6. Experiment 2 Group 2 simulation results;x is position, v is velocity. (**a**) Trajectory of x changing with time; (**b**) Trajectory of v changing with time.

According to experiment 2, system (14) satisfying Lemma 2 can achieve consensus when all τ_d are less than $\tau_{\max}$, and the system will diverge when all τ_d are greater than $\tau_{\max}$, which cannot achieve consensus; thus Theorem 2 is verified.

5. Physical Experiment Verification

To verify the proposed formation control algorithm, we did the experiment based on a pre-constructed multi-mobile robot research platform built by our laboratory, which was constructed with a self-designed three-wheeled omnidirectional robot carrying an UWB (Ultra-Wide Band) ranging module. The system is shown in Figure 7, the omnidirectional robot is shown in Figure 8, and the performance parameters are shown in Table 1. Because the consensus of the second-order system is analyzed in the theoretical analysis part, the speed and position are consistent, and while the omnidirectional robot is a fully driven robot, the horizontal and vertical directions can be controlled separately. Because the velocity control in a given direction is a second-order system, therefore, multi-omni-directional mobile robots can be decomposed into two one-dimensional second-order multi-robot systems. Therefore, omni-directional robots are used to verify the proposed algorithm. In the experiment, it is possible to determine whether the algorithm is valid based on whether the speed and the position of the robot after final stabilization are consistent. In the data acquisition part, the external positioning data of the robot are collected by the UWB positioning system built by myself, and the speed of the robot itself is collected by the encoder on the wheel of the robot and transmitted to the central processing computer via Wi-Fi for processing. The ranging error between the UWB ranging modules used in the experiment is 7 cm. Experiments were carried out in an indoor environment with length × width of 4 m × 5 m in order to verify the effectiveness of the proposed algorithm.

Table 1. Omnidirectional robot parameters.

Parameter Name	Value
weight	1.5 Kg
diameter	235 mm
Maximum linear velocity	1.2 m/s
Maximum angular velocity	6.6 rad/s
Battery capacity	2800 mah
No-load Maximum Standby Time	2 h
Maximum Load Weight	3 Kg

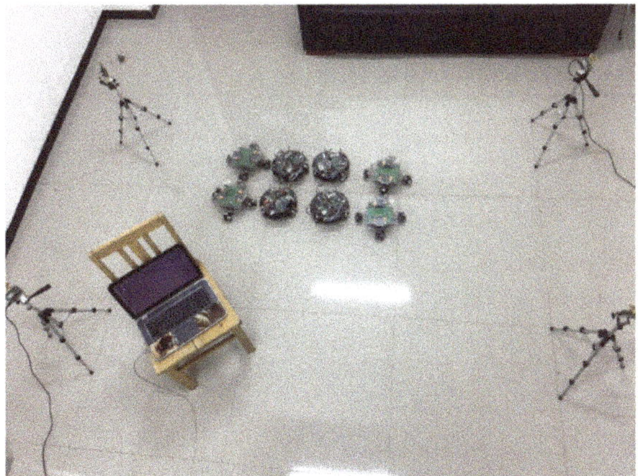

Figure 7. Multi-robot research platform.

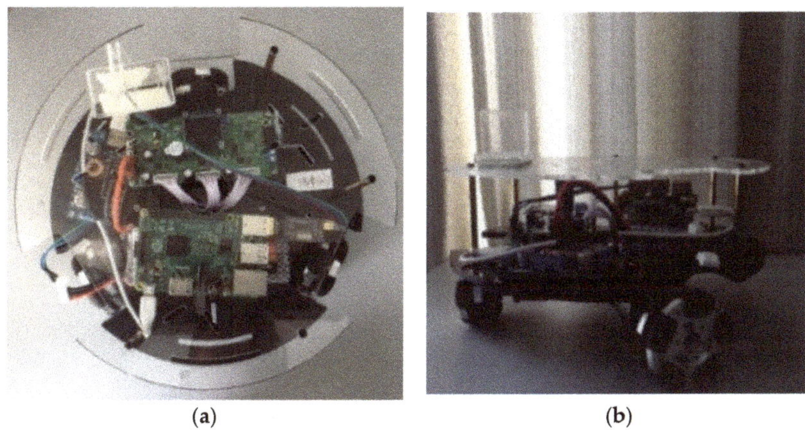

Figure 8. Omnidirectional robots. (**a**) Top view; (**b**) Side view.

It should be pointed out that when the proposed algorithm is applied to a practical multi-robot system, it is necessary to first determine the maximum communication delay between robots and the maximum amplitude of the noise environment, and then design k_1 and k_2 based on this. At the same time, it should be noted that this experiment mainly focuses on verifying whether the system can achieve consensus under the control law. The collision avoidance behavior of the multi-robot system is not the emphasis in this research. Therefore, the collision avoidance algorithm program is written in the bottom control program of the robot in this experiment. When the robot is about to collide, the formation algorithm program will be interrupted, and the collision avoidance behavior will be executed. When a safe distance between the robots has been reached, the formation algorithm program will continue to be executed [17,18].

The communication topology used in the experiment is shown in Figure 4. The adjacent communication weight a_{ij} is 1, so $k_1 \in (0, 10k_2^2)$. The central processor logs on each robot remotely through SSH, and obtains the communication delay between two robots whose communication weight A is not zero by PING command. The time delay between robots in the actual communication environment is time-varying, so take its maximum delay. We get $\tau_{a1} = 0.56s$, $\tau_{a2} = 0.043s$, $\tau_{a3} = 0.047s$, $\tau_{a4} = 0.061s$. The time taken for each robot to receive data and process them is $\tau_{b1} = 0.021s$, $\tau_{b2} = 0.02s$, $\tau_{b3} = 0.021s$, $\tau_{b4} = 0.021s$. Therefore, the time delay between robots is $\tau_1 = 0.077s$,

$\tau_2 = 0.063s$, $\tau_3 = 0.068s$, $\tau_4 = 0.082s$. Because this experiment is being performed in a laboratory environment, it is assumed that the communication noise is white noise with a maximum amplitude of 2. According to Formula (26) and the moving speed of omnidirectional robot, set $k_1 = 1$, $k_2 = 1.4$. Four omnidirectional robots were placed in arbitrary positions, (0.83,2.20,0), (0.35,1.74,0), (0.67,0.88,0), (0.56,0.48,0), respectively. In the practicality experiment, the robot cannot converge to one point, so Formula (2) is changed to:

$$\lim_{t \to +\infty} [x_i(t) - x_j(t) - F_p] = 0 \tag{35}$$

where F_p is the formation parameters and $p = 1, 2, \ldots, n$. Since the system only installs a UWB ranging sensor to provide positioning for the robot, the robot will perform pose determination before the experiment starts, and the robot's body coordinate system will be consistent with the global coordinate system. The experimental results are shown in Figure 9. The experimental video address can be found in reference [19].

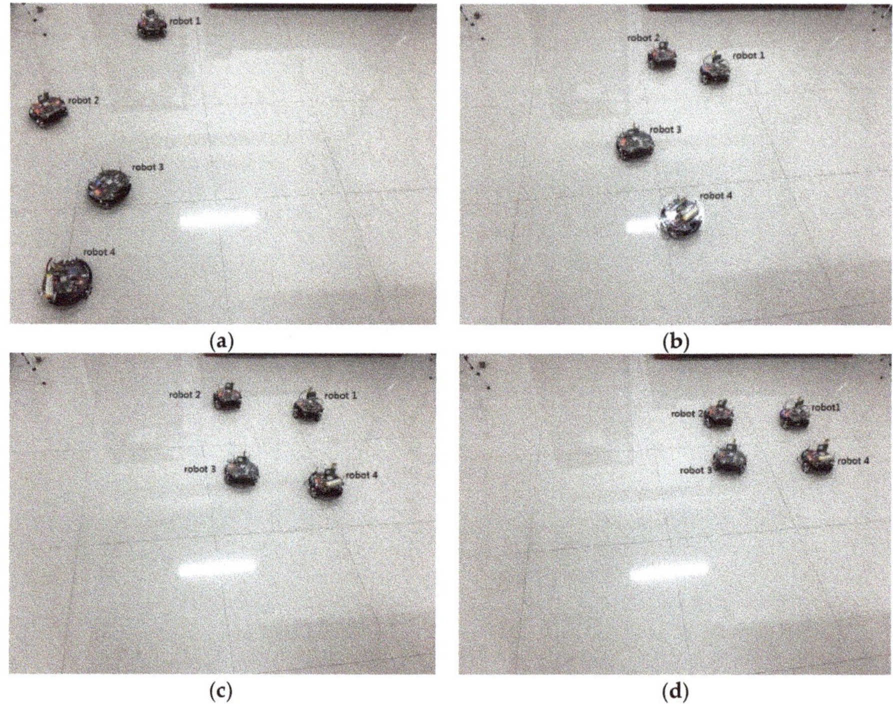

Figure 9. Experiment process screenshots. (**a**) T = 0 s; (**b**) T = 15 s; (**c**) T = 25 s; (**d**) T = 30 s.

The experimental data collected by the UWB positioning system are shown in Figure 10.

The experiments show that the multi-robot system can eventually achieve consensus and form a formation in a variety of time-delay and noise environments, which verifies the effectiveness of the proposed algorithm.

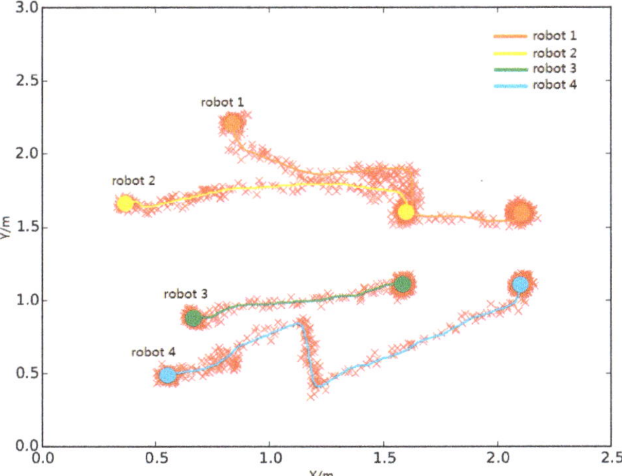

Figure 10. Experiment 1 data of UWB positioning system.

6. Conclusions

Aiming at the consensus of multi-mobile robots under uncertain conditions such as communication delay, communication noise and measurement noise, we used the frequency domain analysis method, transformed the characteristic equation into the quadratic polynomial of the pure imaginary eigenvalue, and then obtained the conditions for achieving consensus under various time delay and noise conditions for a second-order multi-robot system. That is, when the time delays of all robots are less than the maximum time delays, the system can achieve consensus. In this paper, based on two aspects of system communication topology—directed graph and undirected graph—the results were verified by numerical simulation using MATLAB/Simulink, verifying the correctness of the theoretical derivation of the proposed algorithm. Finally, a multi-robot research platform was built, and formation control experiments were carried out in a real laboratory environment. The experimental results showed that the proposed algorithm could effectively make the second-order multi-mobile robot systems consistent. This paper only analyzes the consensus problem of second-order systems, while most existing multi-mobile robot systems are higher-order systems. Therefore, the consensus analysis of higher-order systems under noise and time delay conditions will be the focus of our next research.

Author Contributions: H.W. wrote this article and accomplished the algorithm analysis; Q.L. was the principle researcher on this project and provided many valuable suggestions; N.D. performed the simulation; G.W. established the physical experimental system; and B.L. performed the experiment and analyzed the experimental data.

Funding: This research received no external funding.

Acknowledgments: The work in this paper is supported by the National Natural Science Foundation of China (Grant No. 61663014).

Conflicts of Interest: The authors declare no conflict of interest.

References

1. Nelson, E.; Corah, M.; Michael, N. Environment model adaptation for mobile robot exploration. *Auton. Robots* **2018**, *42*, 257–272. [CrossRef]
2. Alonso-Mora, J.; Montijano, E.; Nägeli, T.; Hilliges, O.; Schwager, M.; Rus, D. Distributed multi-robot formation control in dynamic environments. *Auton. Robots* **2018**, 1–22. [CrossRef]

3. Yang, X.; Hua, C.C.; Yan, J.; Guan, X.P. Adaptive Formation Control of Cooperative Teleoperators With Intermittent Communications. *IEEE Trans. Cybern.* **2018**, *3*. [CrossRef] [PubMed]
4. Trinh, M.H.; Zhao, S.; Sun, Z.; Zelazo, D.; Anderson, B.D.O.; Ahn, H.-S. Bearing-Based Formation Control of A Group of Agents with Leader-First Follower Structure. *IEEE Trans. Autom. Control* **2018**. [CrossRef]
5. Galceran, E.; Cunningham, A.G.; Eustice, R.M.; Olson, E. Multipolicy decision-making for autonomous driving via changepoint-based behavior prediction: Theory and experiment. *Auton. Robots* **2017**, *41*, 1367–1382. [CrossRef]
6. Li, X.; Xie, L. Dynamic Formation Control Over Directed Networks Using Graphical Laplacian Approach. *IEEE Trans. Autom. Control* **2018**, *56*, 21–35. [CrossRef]
7. Pan, W.W.; Jiang, D.P.; Pang, Y.J.; Li, Y.M.; Zhang, Q. A multi-AUV formation algorithm combining artificial potential field and virtual structure. *Acta Armamentarii* **2017**, *38*, 326–334. [CrossRef]
8. Xin, L.; Zhu, D. An Adaptive SOM Neural Network Method to Distributed Formation Control of a Group of AUVs. *IEEE Trans. Ind. Electron.* **2018**. [CrossRef]
9. Liu, Z.; Wang, L.; Wang, J.; Dong, D.; Hu, X. Distributed sampled-data control of nonholonomic multi-robot systems with proximity networks. *Automatica* **2017**, *77*, 170–179. [CrossRef]
10. Porfiri, M.; Roberson, D.G.; Stilwell, D.J. Tracking and formation control of multiple autonomous agents: A two-level consensus approach. *Automatica* **2007**, *43*, 1318–1328. [CrossRef]
11. Du, H.; Zhu, W.; Wen, G.; Duan, Z.; Lü, J. Distributed Formation Control of Multiple Quadrotor Aircraft Based on Nonsmooth Consensus Algorithms. *IEEE Trans. Cybern.* **2019**, *49*, 342–353. [CrossRef] [PubMed]
12. Li, S.; Er, M.J.; Zhang, J. Distributed Adaptive Fuzzy Control for Output Consensus of Heterogeneous Stochastic Nonlinear Multiagent Systems. *IEEE Trans. Fuzzy Syst.* **2018**, *26*, 1138–1152. [CrossRef]
13. Lin, P.; Jia, Y.; Du, J.; Yuan, S. Distributed Consensus Control for Second-Order Agents with Fixed Topology and Time-Delay. In Proceedings of the 26th Chinese Control Conference, Zhangjiajie, China, 26–31 July 2007; pp. 577–581. [CrossRef]
14. Cheng, L.; Wang, Y.; Hou, Z.G.; Tian, M.; Cao, Z. Sampled-data based average consensus of second-order integral multi-agent systems: Switching topologies and communication noises. *Automatica* **2013**, *49*, 1458–1464. [CrossRef]
15. Mengji, S.; Kaiyu, Q. Distributed Control for Multiagent Consensus Motions with Nonuniform Time Delays. *Math. Probl. Eng.* **2016**, 1–10. [CrossRef]
16. Wang, B.; Tian, Y.P. Consensus of second-order discrete-time multi-agent systems with relative-state-dependent noises. *Int. J. Robust Nonlinear Control* **2017**, *27*, 4591–4606. [CrossRef]
17. Lv, Q.; Wei, H.; Lin, H.; Zhang, Y. Design and implementation of multi robot research platform based on UWB. In Proceedings of the 2017 29th Chinese Control and Decision Conference (CCDC), Chongqing, China, 28–30 May 2017; pp. 7246–7251. [CrossRef]
18. Wei, H.; Lv, Q.; Wang, G.-S.; Lin, H.; Liang, B. Trajectory tracking control for heterogeneous mobile robots based on improved UWB ranging. *J. Beijing Univ. Aeronaut. Astronaut.* **2018**, *44*, 1461–1471. [CrossRef]
19. Available online: http://i.youku.com/i/UNDUxODcyNDI0NA==/videos?spm=a2hzp.8244740.0.0 (accessed on 3 January 2019).

Project Report

A Multi-Agent Based Intelligent Training System for Unmanned Surface Vehicles

Wei Han [1,2,*], Bing Zhang [2], Qianyi Wang [2], Jun Luo [1], Weizhi Ran [3] and Yang Xu [3]

1 School of Mechatronic Engineering and Automation, Shanghai University, Shanghai 200444, China; luojun@shu.edu.cn
2 System Engineering Research Institute, China State Shipbuilding Corporation Limited, Beijing 100036, China; zhangbing_857@163.com (B.Z.); wangqianyi815@sohu.com (Q.W.)
3 School of Computer Science and Engineering, University of Electronic Science and Technology of China, Chengdu 611731, China; 201722060227@std.uestc.edu.cn (W.R.); xuyang@uestc.edu.cn (Y.X.)
* Correspondence: hanwei_seri@163.com

Received: 10 February 2019; Accepted: 6 March 2019; Published: 15 March 2019

Abstract: The modeling and design of multi-agent systems is imperative for applications in the evolving intelligence of unmanned systems. In this paper, we propose a multi-agent system design that is used to build a system for training a team of unmanned surface vehicles (USVs) where no historical data concerning the behavior is available. In this approach, agents are built as the physical controller of each USV and their cooperative decisions used for the USVs' group coordination. To make our multi-agent system intelligently coordinate USVs, we built a multi-agent-based learning system. First, an agent-based data collection platform is deployed to gather competition data from agents' observation for on-line learning tasks. Second, we design a genetic-based fuzzy rule training algorithm that is capable of optimizing agents' coordination decisions in an accumulated manner. The simulation results of this study demonstrate that our proposed training approach is feasible and able to converge to a stable action selection policy towards efficient multi-USVs' cooperative decision making.

Keywords: unmanned surface vehicles; multi-agent system; training system; genetic-based fuzzy rule learning; intelligent autonomous control; modeling and simulation

---

1. Introduction

The modeling and design of multi-agent systems for applications in the evolving intelligence of unmanned systems is interesting and promising [1–4], especially in situations where traditional methods can be costly, dangerous, or even impossible to realize. Several applications can be found in a very broad range of domains such as energy [5–7], security [8], robotics [9], and resource management [10]. A multi-agent system consists of autonomous agents that interact in an environment to achieve specific goals [11]. An autonomous agent, in this sense, is able to perceive its environment and perform actions using actuators.

Over the years, there have been several studies that have proposed principles for designing multi-agent systems, as well as approaches to coordinate the individual behavior of agents [12–16]. In most multi-agent application domains, a priori specification of the optimal agents' behavior is difficult due to the complexity and/or dynamics of the environment [11]. In such environments, it is natural for agents to adapt or learn optimal actions that maximize performance on-line. However, one of the key challenges is the need to build a simulation platform that can be used for fast training so as to gather enough training data to promote the intelligent development process.

In the context of multi-surface vehicles' modeling and design [17–19], an unmanned surface vehicles (USVs) system is one of the development trends of modern weapon equipment [20].

The unmanned surface vehicle system has become an important means of information confrontation, precision strikes, and special combat tasks in future war [21]. Over the years, unmanned surface vehicles have long been applied to varying kinds of military applications in the real world. Some of the application domains include anti-mine warfare, submarine warfare, reconnaissance, and surveillance [22–24]. Due to the open and complex dynamic combat environment, the USV combat system must develop in the direction of autonomy and synergy, and the future surface unmanned combat should have a strong autonomous capacity, be able to be controlled autonomously, and complete complex and diverse combat tasks independently or collaboratively in a complex uncertain environment [25]. However, human knowledge is difficult to apply directly to the coordination of USVs.

We propose a multi-agent-based intelligent training system for unmanned surface vehicles (USVs). We focus on the problem of training agents of a multi-agent-based unmanned surface vehicles system where no historic data concerning the agent behavior is available. Using the team learning framework, we provide approaches for learning the rule base for multi-agent systems, designing the learning environment for simulating cooperative and competitive agents' behavior, and gathering competition data from agents' observation for off-line learning tasks. To this end, this paper first presents a decentralized coordination platform and simulator design based on a multi-agent architecture. Secondly, a USV-based agent model is presented. The paper also proposes a data collection platform for agent learning and USV training. Lastly, a fast learning and training algorithm design based on the agents' rule base is presented.

To evaluate our approach, we used an island-conquering scenario where two teams of unmanned surface vehicles compete to conquer islands in an environment. We model this case study as a partially-observable stochastic game where one team has to learn the behavior that maximizes their returns against a human-controlled team. Our empirical evaluations show that our approach to learning the knowledge base of a multi-USV system, when applied to the trained team, was able to find a knowledge base (KB) to achieve better performance.

We first present the problem this study seeks to address in Section 2. Section 3 discusses the multi-surface vehicle training system design. Finally, in Section 4, we discuss our experiments and analysis of the simulation results followed by the conclusion of our study in Section 5.

2. Multi-Agent-Based USVs' Training Problem

We first discuss the multi-agent-based intelligent training system learning problem of this study.

In general, a stochastic game is an extension of the Markov decision process (MDP) to the multi-agent context. In such a game, agents may have conflicting goals, and their joint actions determine rewards and transition between states. Stochastic games usually assume agents as having a complete view of their states, whereas the partial observation case is discussed under the more general partially-observable stochastic games (POSGs) domain. Our study is performed under the POSG case.

Theorem 1. *A POSG is a tuple* $< X, U_1...U_n, O_1...O_n, f, R_1...R_n >$ *where:*

- n *is the number of agents*
- X *is the finite set of states*
- $U_i, i = 1,...,n$ *is the finite set of actions available to the agents, which form the joint action set* $\mathbf{U} = U_1 \times ... \times U_n$
- $O_i, i = 1,...,n$ *is the finite set of observations of the agents. The joint observation is denoted as* $\boldsymbol{o} = \{o_1,...,o_n\}$
- $f : X \times \mathbf{U} \times X \rightarrow [0,1]$ *is the Markovian state transition function where* $f(x'|x,\boldsymbol{u})$ *denotes the probability of reaching state* x' *after the joint action* $\boldsymbol{u}$ *in state* x*. The joint action of all agents* $\boldsymbol{u}$ *at a stage* t *is denoted as* $\boldsymbol{u}_t = [u_{1,t},...,u_{n,t}]^\top$ $u_t \in \mathbf{U}, u_{i,t} \in U_i$
- $R_i : X \times \mathbf{U} \times X \rightarrow \mathbb{R}, i = 1,...,n$ *are the reward functions of the agents.*

In this study, we consider a team of multiple unmanned surface vehicles that must be trained to perform a given mission m in a mission space $\mathcal{D} \subset \mathbb{R}^n$ consisting of opposite forces. Let $H(m)$ be the objective function dependent on the mission. Given that a team T consists of N USVs, the dynamics of the ith USV is given by:

$$\begin{aligned} \dot{x}_i &= f_i(x_i(t), u_i(t)) \\ i &= 1, 2, 3..., N \end{aligned} \tag{1}$$

where $x_i(t) \in \mathbb{R}^n$ and $u_i(t) \in \mathbb{R}^m$ are the state and control inputs (available actions) of the USV, respectively, at time t. Depending on the USV model and mission type, a set of constraints (such as obstacle avoidance and the physical limitations of the USV) on the state $x_i(t)$ of a USV are imposed. The control inputs of a USV are bounded according to the limits of the USV's actuators $|u_i(t)| < u_{max}$.

The reward of an USV i after taking action $u_{i,t} \in U_i$ in stage t is denoted as $r_{i,t+1}$. The individual policies $h_i : X \times U_i \rightarrow [0, 1]$ of the USVs in a team form the joint policy $\mathbf{h}$. In the case of team learning, the joint policy is defined by the single learner. Since the reward of each USV in a team depends on the joint action, the respective USV rewards depend on the joint policy:

$$R_i^{\mathbf{h}}(x) = \mathbf{E}\left\{ \sum_{t=0}^{T} \gamma^t r_{i,t+1} | x_0 = x, \mathbf{h} \right\} \tag{2}$$

where γ is the discounting factor and T is the length of the horizon. The training objective for the intelligent training system is to find a policy that maximizes Equation (2) such that the trained team of USVs can outperform its rival USVs, while adapting to the changing dynamics of it's operating environment.

3. Multi-Agent-Based Training System Design

In this section, we discuss the multi-agent-based training system design of our study, as well as the ideas involved in training the agents. To start with, the section first presents the architecture of the multi-agent-based intelligent training system. Next, we present the USV agent model for cooperative training and competition. Finally, the section presents the agent data collection platform for gathering competition for the agent information sharing and learning algorithm.

3.1. System Architecture

The architecture of the system used in this study is presented in Figure 1. We decouple the agents' control from the main environment in order to allow different implementations of controllers. In this case, we consider the multi-agent simulation platform as a server, and any attached controller becomes a client. Each USV agent selects its action based on the observed information from the environment. Based on the data visualization model of the USV, the data collection platform forms and mounts the corresponding resource tree structure of the USV. Historical data are provided to the upper machine learning algorithm by the data collection platform after data fusion. The obtained algorithm controller is then loaded into the USV.

The multi-agent cooperative reasoner component serves as the policy used to select an action for a controlled agent using the agent's local observation. In this study, we model the reasoner as an FISdecomposed into a genetic fuzzy trees (GFT). Thus, the reasoner maintains a KB, which is used for the mapping fuzzy values of the observation of an agent to an action for execution in the POSG.

The learner component of the intelligent control model facilitates KB learning, tuning, or optimizing. It seeks to find the policy that exhibits the designed behavior of the controlled team as specified by the reward function. Thus, gradient-based approaches such as neural networks (NN) [26], as well as non-gradient-based approaches can be employed for this purpose. As already indicated in our usage of GFT for the reasoner component, we use the GA in this study to learn the fuzzy rules and tune the MFsof a GFT. Each candidate KB is applied to the control problem for the whole of a simulation episode (we use round and episode interchangeably) as found in the Pittsburgh approach. Hence, the rewards that are received each time step are aggregated and assigned as the fitness value of the candidate KB used at the end of the round/episode.

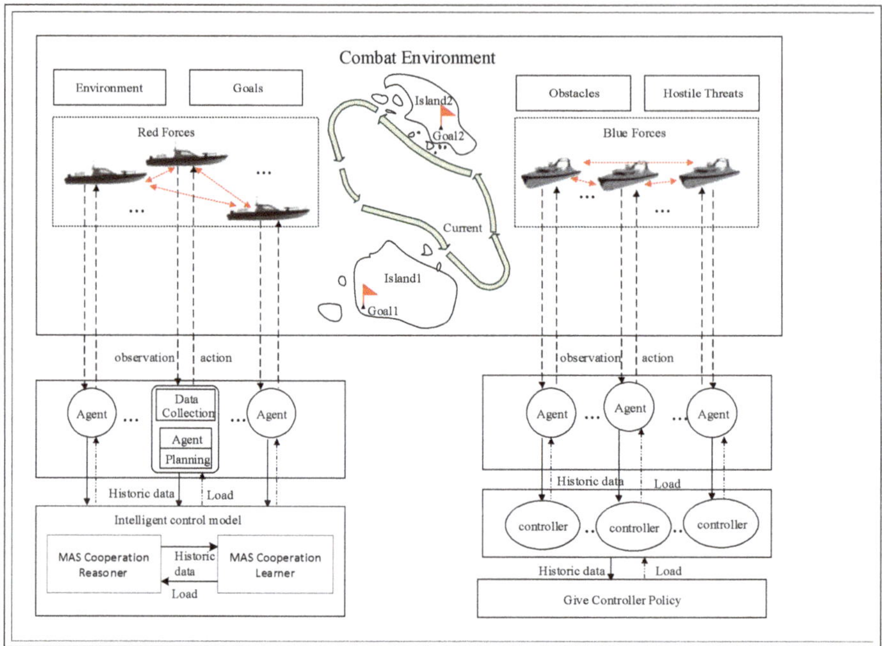

Figure 1. The architecture of the multi-agent based intelligent training system.

3.2. Agent Design

The USV agent model that is used by the intelligent training system for cooperative training and competition is equipped with a radar and weapon system. The radar of a USV can detect a hostile threat within it's detection range. In this work, the weapon system of a USV consists of a calibrated gun with limited forward turning angle and firing range. A USV can turn in both directions (left or right) and with limited speed and turning radius r. In this study, only surge, sway, and yaw are used to describe the USV's movement at sea, as shown Figure 2). Therefore, the kinematic relationship between the USV position in the global inertial frame XYZ and the boat body-fixed frame xyz can be defined as:

$$
\begin{aligned}
x &= u\cos(\psi) - v\sin(\psi) \\
y &= v\sin(\psi) + v\cos(\psi) \\
\psi &= \lambda
\end{aligned}
\tag{3}
$$

The location and orientation of USVs in the coordinates of the environment are represented by (x, y) and ψ in the Earth-fixed reference frame, while u, v, and λ represent the velocity of surge, sway, and yaw in the body-fixed reference frame, respectively.

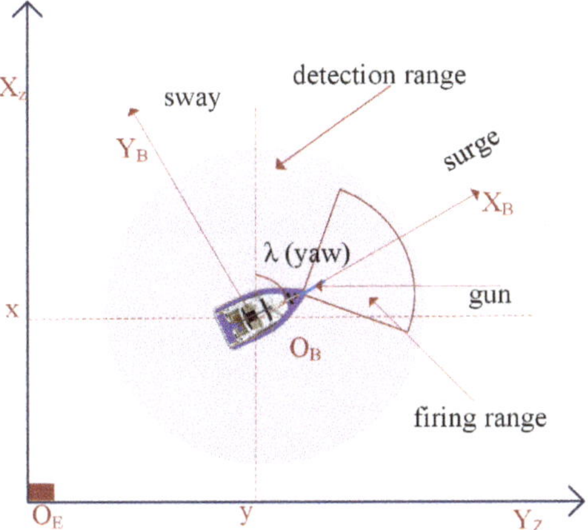

Figure 2. Schematic diagram of the USV model and planar motion where $X_E O_E Y_E$ is the Earth-fixed reference frame and $X_B O_B Y_B$ denotes the body-fixed reference frame.

3.3. Data Collection Model

Agents' observation can persist in the course of training to enable off-line operations. Data collected during training can be useful for agents' cooperative decision making. Furthermore, the gathering of training data means that different agents' learning algorithm can be employed both on-line and off-line for agents' performance improvement. Moreover, the data can be organized to enable communication between agents. The gathered domain data (in resource forms) can be used for purposes such as improving agents' performance using state-of-the-art deep reinforcement learning (DRL) algorithms. Hence, the data collector component is not only necessary, but also can extend the scope of learning algorithms that can be employed for intelligent training.

The data collection model is designed using a training and control platform consisting of loosely-coupled components and a common resource model. One such paradigm for implementing modular services is the Future Airborne Capability Environment's (FACE (http://www.opengroup.org/face)) common operating environment approach. In Figure 3, we show the components of FACE with regards to our study. Their descriptions, in the context of our study, are as follows:

- Historical training data level (HTDL): The HTDL layer consists of the entities involved in learning the KB or agents' policies for a POSG. In situations where the reasoner or policy output corresponds to composite tasks, planning and agent scheduling services may be implemented as modular components in this layer. This layer contains the historical resource data that are leveraged by the underlying learning algorithm to learn the KB.
- Cooperative decision data level: This level consists of entities that handle individual agent-level data for agent decision making. Data at this level are real-time data and observations from the agent.
- Individual agent decision model (IADM): A resource management service for hosting agents' observations composes this layer. Such a service exposes a uniform set of APIs that can be consumed by the training and control platform or the agents for reasoning or cooperation operations. Modeling the observations as resource data enables the data elements to be uniquely addressed through the APIs.
- Data collection services: Services in this layer receive data from the agents in their raw form and then perform adaptation to a common resource model for use in the training and control platform. Thus, they abstract the heterogeneous forms that agents may report from their local observation.

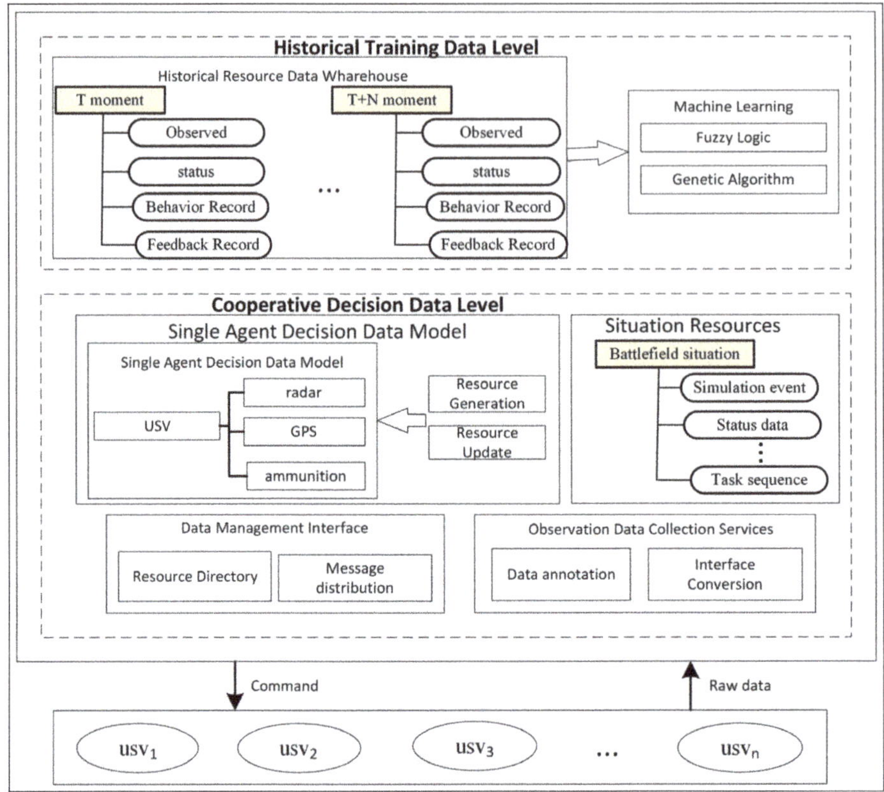

Figure 3. The architecture of the agent data collection platform based on Future Airborne Capability Environment (FACE) components.

3.4. Multi-Agent-Based USV Training Algorithm Design

The intelligent control model, as depicted in Figure 4, is responsible for performing learning and reasoning. It controls a team of agents in the simulation environment by sending agent actions to the environment every time step. Thus, at time step t, it sends the joint actions u_t to the environment for execution where the action for agent $i, i \leq n$, is $u_{i,t} \in u_t$. The intelligent control model consists of a learner and a reasoner that employs fuzzy logic and genetic algorithms (GA) known as genetic fuzzy systems (GFS) [27]. In this technique, GA is used to learn and tune the rule base and membership function of an FIS, respectively. To do this, an initial population of solutions, or strings, is created to encode the rule base and membership functions.

Classical approaches such as Michigan [28], Pittsburgh [29], and iterative rule learning [30] are mostly used to derive GFS fuzzy rules. We give an instance to illustrate how the rule base is encoded in this study using the Pittsburgh approach. Suppose X_1 and X_2 are input variables each with linguistic terms $\{term_1, term_2\}$ and output variable Y with terms $\{a_1, a_2\}$ with an arbitrary rule base of an FIS as follows:

1. IF X1 IS term1 AND X2 IS term1 THEN Y IS a1
2. IF X1 IS term1 AND X2 IS term2 THEN Y IS a2
3. IF X1 IS term2 AND X2 IS term1 THEN Y IS a2
4. IF X1 IS term2 AND X2 IS term2 THEN Y IS a1

Assume we assign codes zero and one to a_1 and a_2 respectively. The chromosome 0110 is obtained. This approach can represent an if-then rule in a single digit. This implies that each chromosome encodes possible outputs of a set of rules.

Similarly to tuning membership functions, each digit in the string corresponds to some endpoint of a membership function. GA can then be used to tune the parameters as part of the evolution process; such as found in [31], where initial parameters of a triangular MF $\mathbb{T}(\alpha, \beta, \gamma)$ are tuned using:

$$\alpha_{i+1} \leftarrow (\alpha_i + \delta_i) - \eta_i \tag{4}$$

$$\beta_{i+1} \leftarrow (\beta_i + \delta_i) \tag{5}$$

$$\gamma_{i+1} \leftarrow (\gamma_i + \delta_i) - \eta_i \tag{6}$$

where δ and η are the tuning coefficients, whereas α, β and γ parameterize $\mathbb{T}$. δ shifts the MF to the right or left, whereas η shrinks or expands the MF. Therefore, every MF has two tuning parameters that can be optimized using GA.

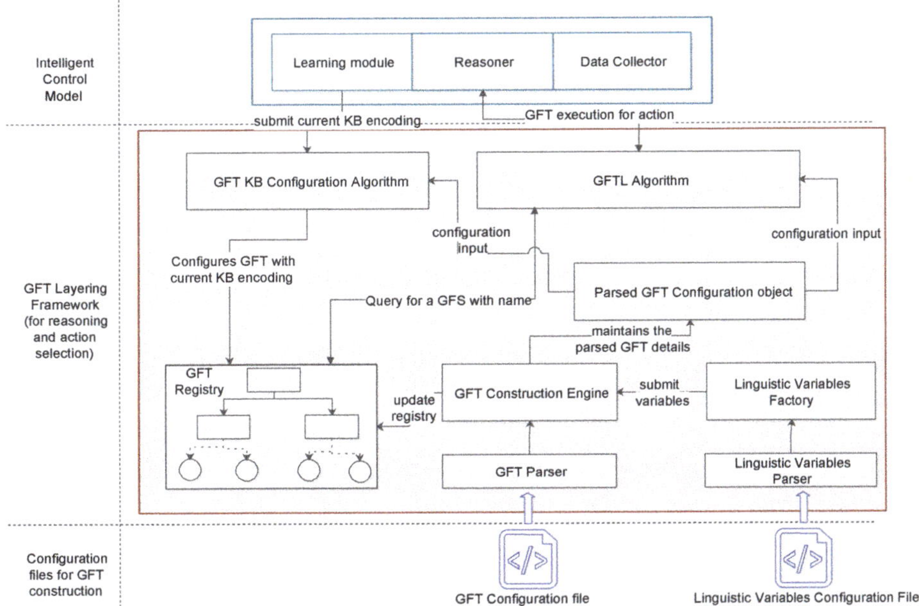

Figure 4. The architecture of the multi-agent-based intelligent training system. KB, knowledge base. GFT, genetic fuzzy trees.

Although a GFS can be used to find KBs that optimize the fitness function, it is inefficient to use a single GFS for a complex control problem. In such a case, there is an increase in GA search space complexity, the tendency to have redundant rules, and KB size. Therefore, using genetic fuzzy trees (GFT) in [2] helps to mitigate this problem. A GFT is, essentially, an ensemble of GFSs arranged in a tree structure according to a logical and conditional sequence of execution. A GFT enables the decomposition of a complex fuzzy control system (FCS) or GFS into logical sections with each node in the tree focusing on an aspect of the control problem. Each GFS in a GFT defines its own KB. Hence, the KB of a GFT consisting of $m \in \mathbb{R}$ GFSs $G = \{gfs_1, gfs_2, ..., gfs_m\}$ may have the genetic structure (represented as a concatenated string) $gfs_1^{rb} gfs_2^{rb} ... gfs_n^{rb} gfs_1^{mf} gfs_2^{mf} ... gfs_n^{mf}$ where gfs_i^{rb} and gfs_i^{mf}, $i \leq m$, denote the RB and MF string/genome, respectively, of a given GFS. This structure can be seen in Figure 5.

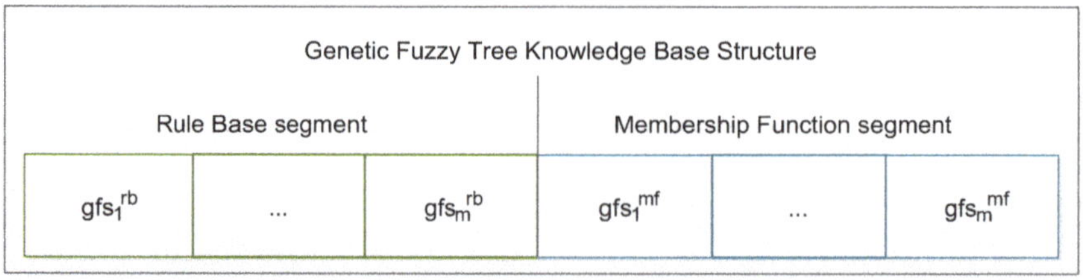

Genetic Fuzzy Tree Knowledge Base Structure

Rule Base segment			Membership Function segment		
gfs_1^{rb}	...	gfs_m^{rb}	gfs_1^{mf}	...	gfs_m^{mf}

Figure 5. The knowledge base genetic structure of a genetic fuzzy tree. Each chromosome or member of the GA population takes this form.

Algorithm 1 shows how beginning with an initial population, the KB of a GFT can be learned/optimized using the GA process. In Line 2, the initial population of the GFT KB is generated as the current population. This can be based on a predefined KB set or randomly generated. In Lines 3–4, the current population is passed on to the the FIS to be used for the control task. After all these KBs have taken turns performing the control task and have been evaluated, they are then subjected to the GA operators (Lines 6–17). This operation continues till the termination condition is reached. The resulting candidate KBs are returned as the best performing or most suitable of the KBs.

Algorithm 1 Procedure: GFS procedure.

Input: GA hyperparameters
Output: Best set of GFT KBs

```
1: Initialize t = 0
2: Generate initial population as P_0
3: Set current population C_t := P_0
4: Run the fuzzy control system using C_t
5: Evaluate the members of C_t
6: while Not termination condition do
7:     ℙ_t := selection(C_t)
8:     O_t := crossover(ℙ_t)
9:     C_temp := mutate(O_t)
10:    if elitism then
11:        C_temp := applyElitism(C_temp) //keeps a percentage of previous chromosomes
12:    end if
13:    t := t + 1
14:    C_t := C_temp
15:    Run the fuzzy control system using C_t
16:    Evaluate the members of C_t
17: end while
18: Return best-performing candidate solutions
```

Non-stationarity is an inherent problem in the multiple learning agents' environment. In this section, we present an approach for incorporating agent-induced non-stationarity awareness into a GFT based on the framework of [32].

Since the root of the GFT has the strongest impact on the learning process or action selection, we consider it to be the main component for addressing this problem. Unlike the other GFSs in the GFT, which can be modeled to address specific aspects of a control problem (domain dependent), we consider the root GFS to be a special case.

We regard the sub-trees that are formed under the root GFS as the elements of the BR function co-domain. Thus, a BR function, in the GFT case, selects a sub-policy to be used for selecting an action for the agent. The influence function is also represented as an FIS (or GFS in our case, since GA is applied to learn the KB), which takes PGFsas input variables and a BF as output the variable. This is motivated by the idea that an influence function maps beliefs to possible best responses and can be designed to use deductive reasoning. These are properties that an FIS exhibits due to fuzzy logic. Thus, the root of the GFT takes incomplete/partial observations and produces its belief of another agent's policy. This connotes that the terms of the output variables then serve as possible opponent strategies as perceived by the reasoning agent.

In the case of the PGFs that serve as input variables to the GFT root FIS, the history of observations concerning the opponent or another agent is maintained and used by each PGF variable to produce an input value. This PGF input value can then be fuzzified and used in the inference process. The problem that arises is then how to design a PGF to capture the adaptation dynamics of another agent. In the work of [33], an agent i estimated the model of agent j as:

$$\hat{\sigma}_j^i(u_j) = \frac{C_j^i(u_j)}{\sum_{\bar{u} \in U_j} C_j^i(\bar{u}_j)} \tag{7}$$

where $C_j^i(u_j)$ is the frequency that agent i observed agent j taking action u_j. Therefore, given the history of observations, Equation (7) can be extended to include the observation of agent i as:

$$\hat{\sigma}_j^i(o_i, u_j) = \frac{C_j^i(o_i, u_j)}{\sum_{\bar{u} \in U_j} C_j^i(o_i, \bar{u}_j)} \tag{8}$$

where $C_j^i(o_i, u_j)$ is the frequency that agent i observed agent j taking action u_j when its local observation was o_i. $\hat{\sigma}_j^i(o_i, u_j)$ becomes the input value of agent i's PGF for monitoring a specific action u_j of agent j. With the input variables (PGFs) and output variable (BF) determined for the root FIS, GA can be used to find the best response for agent-induced non-stationarity and thereby stabilize learning in the POSG.

4. Multi-Agent-Based Simulator Design

The simulator used for the experiment was developed using the Java programming language. The fuzzy control library [34] was extended for the design and encoding of fuzzy rules for learning purposes as implemented by the intelligent control model of our architecture. In this section, we present the experiment conducted for this study and follow it up with the analysis of the simulation results.

4.1. Environment Setup

To evaluate the effectiveness of the proposed learning system, a scenario of multiple boats competing for conquering more islands while engaging in combat was adopted. In this scenario, the environment set in a maritime setting consisted of N islands and two teams of unmanned surface vehicles (boats) referred to as blue and red forces. The boats were equipped with radar and guns for detecting and shooting enemies, respectively. The guns were set to have a fixed left-to-right traversal angle and shooting range, as shown in Figure 6. Since both teams had conflicting goals, a team achieved its goal by contending with opponent boats. Each team had information of the location and number of islands and their states, whether conquered or unconquered by the team. An island is said to be conquered by a team if a member of the team moves to the coordinate of the island and stays there for that time step or no opponent boats move to that particular island conquered by the team. If two opponent boats occupy an island at the same time, the island is not awarded to any team for the elapsed time steps. For our experiment, we implemented two controllers for both teams. The controller for the blue force was made to use fixed rules provided by humans, whereas the red force controller, which we sought to train, had to learn the best rules that maximized their performance. At the beginning of each time step, the simulation environment received a batch of commands from the (ally and enemy) controllers of both teams and updated the environment.

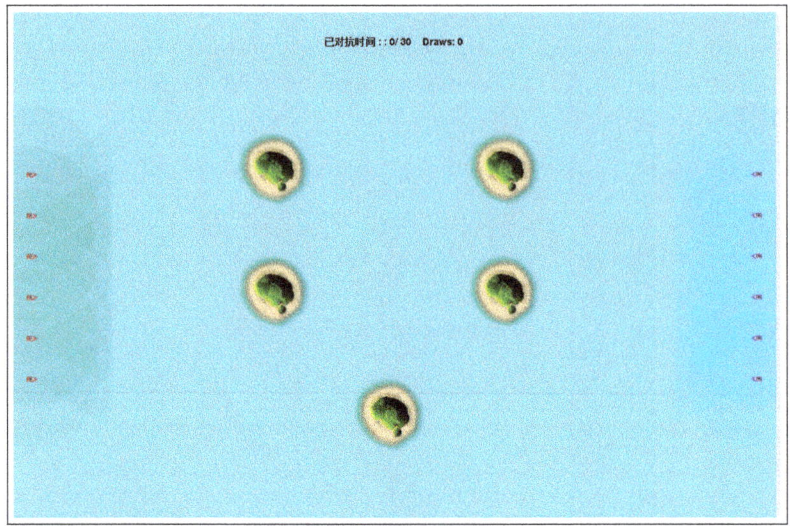

Figure 6. A snapshot of the island-conquering scenario. In the middle are the 5 islands to be conquered by both teams. The circles around the boats are their detection ranges, and the forward looking arc is the the firing regions of the boats.

The possible actions of each agent are also explained below:

- retreat: the agent moves towards its base.
- left: turn left and move.
- right: turn right and move.
- straight: move forward at the current heading.
- stop: stop moving.
- assist: move to help a teammate that is under enemy fire; a boat can assist a random or the closest teammate.
- units: causes a boat to team up with other ally boats to conquer an island.
- closest: the boat conquers an unconquered island that is closest to it.

4.2. Competition Objective

The goal of each team was to conquer all islands and destroy opponents, while staying alive. The performance of each team was evaluated after each time step using the function:

$$R_t^f = A_t^f \times p + I_t^f - D_t^{f'} \times p \tag{9}$$

where A_t^f is the destruction suffered by the opponent team f' as a result of attacks from team f in time step t, I_t^f the conquered island points received by team f, D_t^f the damaged caused by the opponent f' to the team being assessed f, and p the points awarded for boat attacks/destruction. The winner of an encounter is decided after the end of the episode. The team with the highest score is declared a winner of that encounter.

4.3. Data Collector for Training and Learning

As mentioned earlier, gathering of training data is useful in a number ways. The cooperation of agents can be enhanced when agents share their observation. Furthermore, data gathered during training can be used later by other learning algorithms such as deep learning methods for analysis and performance improvement. The resource model shown in Figure 7 is used for on-line data gathering during training. The data that are sent to a controller as feedback take this form. The main components are as follows:

- Simulation information: This node provides general information about the simulation for a given time step.
- Observation: This provides the observation made by all agents in the team. Each agent also reports opponents that have been detected, as well as the observable properties of each detected opponent.
- System data: All miscellaneous data concerning the simulation platform (e.g., time, CPU consumption, memory usage, etc.) may be provided under this resource element.

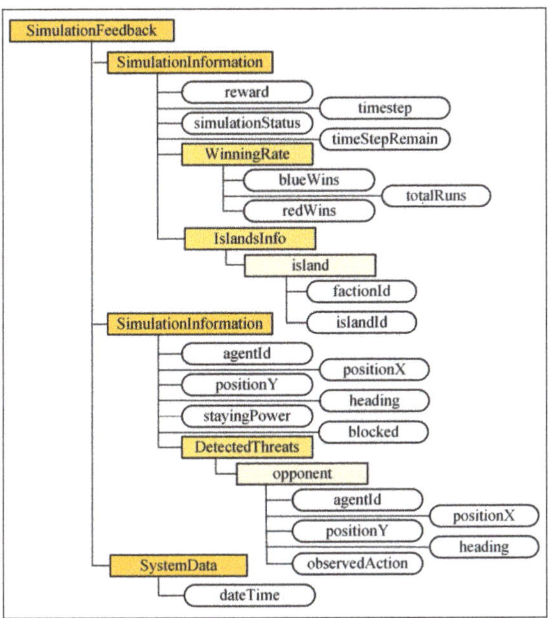

Figure 7. The simulation feedback resource structure of the island-conquering case study.

4.4. Training and Learning Algorithm

In order to provide a realistic virtual environment and opponent for the red force to train against, the blue force control algorithm uses human-defined rules. Table 1 shows an example encoding rules used for the blue force task selection during simulation. The GFT of the red force control algorithm is illustrated in Figure 8. The details of each of the GFSs of the GFT can also be seen in Table 2.

Table 1. Example encoded task selection rules used by the enemy team.

Unconquered	DetectedEnemies	TaskOutput
None	None	Retreat
None	Moderate	Intercept
None	Many	Intercept
Moderate	None	Conquer
Moderate	Moderate	Conquer
Moderate	None	Conquer
Moderate	Many	Intercept
Many	Many	Intercept
Many	Moderate	Conquer

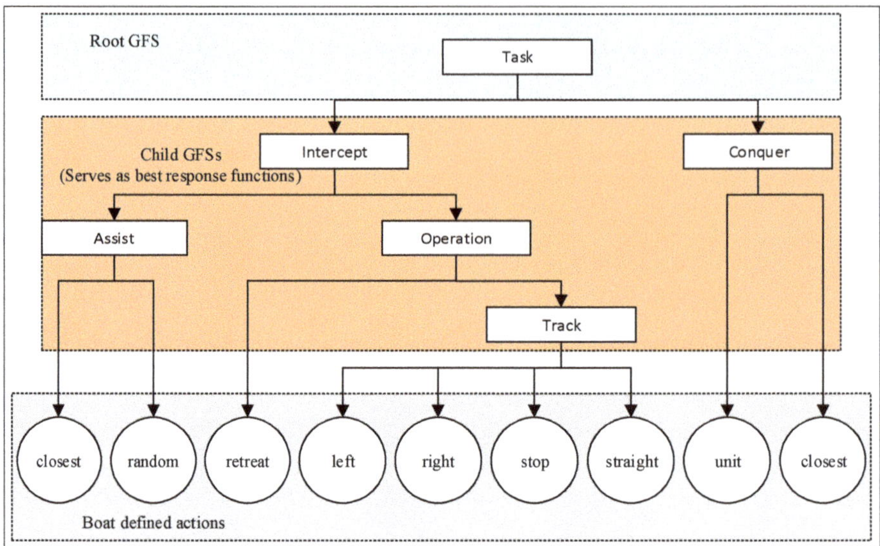

Figure 8. The genetic fuzzy tree structure used for training the ally team. The rectangles are FISs, whiles the circles represent predefined actions in the simulation.

Table 2. Details of the GFSsthat constitute the GFT used for training the ally team.

GFS	Input Variables	Output Variable	Output Variable Terms	RBSize	No. of MF Tuning Parameters
Task	UnconquredIslands PGF_attacked PGF_moved PGF_conquered PGF_retreated	BeliefFunction	conquering Intercepting	243	30
Intercept	DetectedEnemies InEnemyFiringRange	InterceptOutput	Assist Operation	6	6
Assist	DetectedEnemies FactionDetectedEnemies	InterceptOutput	assistClosest assistRandom	9	12
Operation	StayingPower DetectedEnemies	OperationOutput	track retreat	9	12
Track	DistanceToEnemy HeadingDifference	TrackOutput	left, right stop, straight	21	20
Conquer	UnconqueredIslands	UnconqueredIslands	conquer in units conquer closest	3	6

The Assignment GFS was designed to consider opponent-induced non-stationarity with the following PGFs:

- PGF_attacked: reports the confidence level that the detected enemy boats are attacking
- PGF_moved: reports the confidence level that the detected enemy boats are moving towards an island or the agent whose local observations are being used for reasoning
- PGF_conquered: computes the confidence level that the detected enemies are conquering an island
- PGF_retreated: evaluates the confidence level that the detected enemies are withdrawing to their initial positions or base.

The linguistic terms of each PGF are low, moderate, and high. Each PGF first computes the action probabilities for each detected agent. A softmax is then computed over the reported action probabilities, and the maximum is selected as the PGF input value. The belief function output variable selects the perceived opponent strategy in each IF-THEN rule. The other input variables are:

- UnconqueredIslands: The number of islands not conquered by the team to which a boat belongs.
- DetectedEnemies: The number of detected enemies by a boat.
- FactionDetectedEnemies: The number of detected enemies by a team.
- InEnemyFiringRange: Returns 1 if this boat is in the enemies' firing range, and 0 otherwise.
- HeadingDifference: The difference in heading between this boat and the target boat.
- DistanceToEnemy: The distance between this boat and the target boat.
- StayingPower: The current strength of a boat to sustain enemy attacks.
- TeammatesUnderFire: Reports the number of teammates a boat has observed to be under attack by opponent boats.

Triangular MFs are used to define the semantics of the fuzzy rules. The initial MF tuning parameters were sampled uniformly from $[-1.5, 1.5]$ for each input variable MF. The GA parameters we set for training the ally team controller are presented in Table 3.

Table 3. The GA parameters used for training the ally team.

Parameter	Value
Population size	20
Number of generations	101 (initial population included)
Crossover probability	0.7
Mutation probability	0.1
Mutation decay Schedule	Exponential decay (decay rate = 0.15)
Crossover operator	BLXcrossover (alpha = 0.15)
Mutation operator	Adaptive non-uniform mutation (b = 5)
Selection operator	Tournament selection

As indicated in Table 3, we combined a mutation probability schedule with an adaptive non-uniform mutation to control exploration and exploitation. The ally team learner was set to maintain the last 5 observations of each detected enemy boat.

4.5. Experiment Results and Validation Analysis

We have evaluated our intelligent training system using four distinct scenarios of the island conquering case study. The parameters used in Scenario 1 for both team are shown in Table 4. Furthermore, in Scenario 1, sinking an opponent and conquering islands were both worth two points each. In Scenario 2, the number of USVs in a team was reduced to three, and sinking an opponent point was reduced to one. The simulation for Scenarios 1 and 2 was run concurrently on two different machines for 2020 from 20 randomly-generated chromosomes as the initial rules for the blue force against the enemy team, which used a human-sampled rules sets. Figure 9 shows the snapshots of the competition as simulation progresses.

If the blue force began an episode knowing exactly what to do in the environment and the red force was yet to learn, then the blue force should perform better than the red force in the initial stages of the simulation. However, as the simulation progresses, the performance of the red force should be improving since the red force learns, while the blue force uses the static rule. Furthermore, if the blue blue force is using a different set of rules even though a static set of rules, then there is a tendency for a dynamic rise and fall in the performance measure by the strength of a simulating rule set.

Table 4. The fixed simulation parameters.

Parameter	Value
Number of boats per team	6
Number of islands	5
Simulation steps per episode	30
Time per step	3 s
Radar radius	140
Forward firing angle	20
Boat's velocity	4/system iteration
Firing range	100

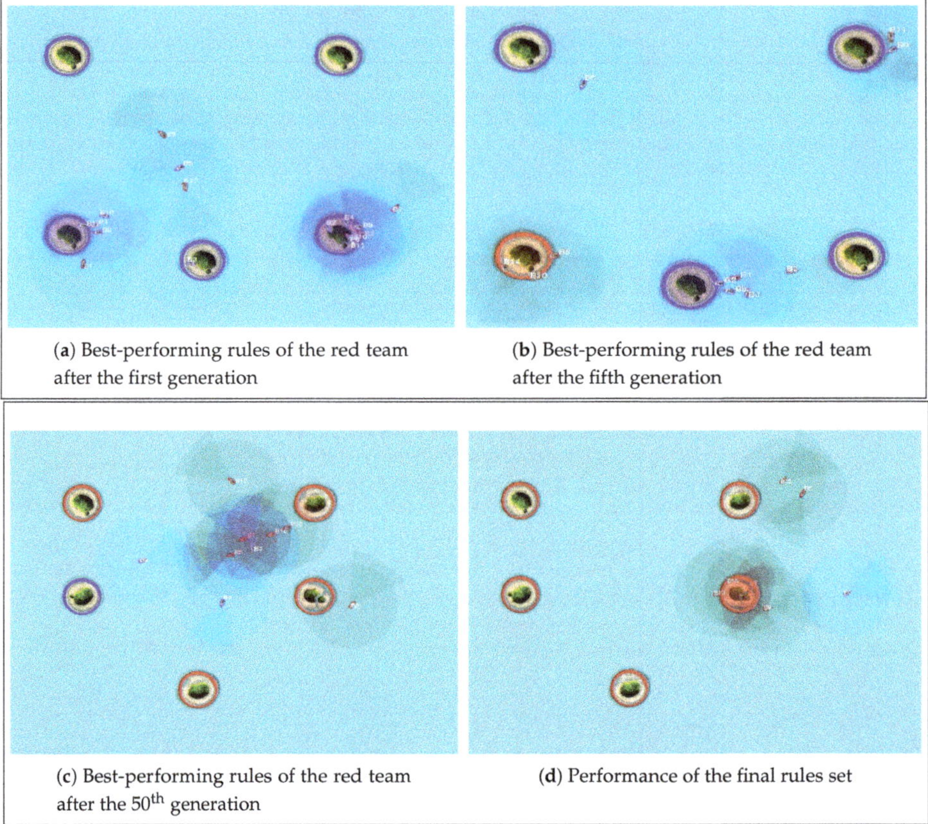

(**a**) Best-performing rules of the red team after the first generation

(**b**) Best-performing rules of the red team after the fifth generation

(**c**) Best-performing rules of the red team after the 50th generation

(**d**) Performance of the final rules set

Figure 9. Illustration of the best-performing rules as training progresses. At the initial stages, the blue team had total control over the red team. However, during the final stages of the training, the performance of the red surpasses that of the blue.

The observed simulation results of both teams during training are presented in Figure 10. As can be seen in Figure 10a,c, the blue force was performing better than the red force during the first generation of rules. However, after 10 and 13 generations, respectively, the red force began to outperform the blue force in the environment, and this is what caused the red force performance to rise in Figure 10a,c. In Figure 10b,d, we compare the highest score attained by both teams in each generation. Furthermore, the same analogy can be drawn from this graph, as in the the initial generations, the blue force seemed to be scoring high as compared to the counterpart red force.

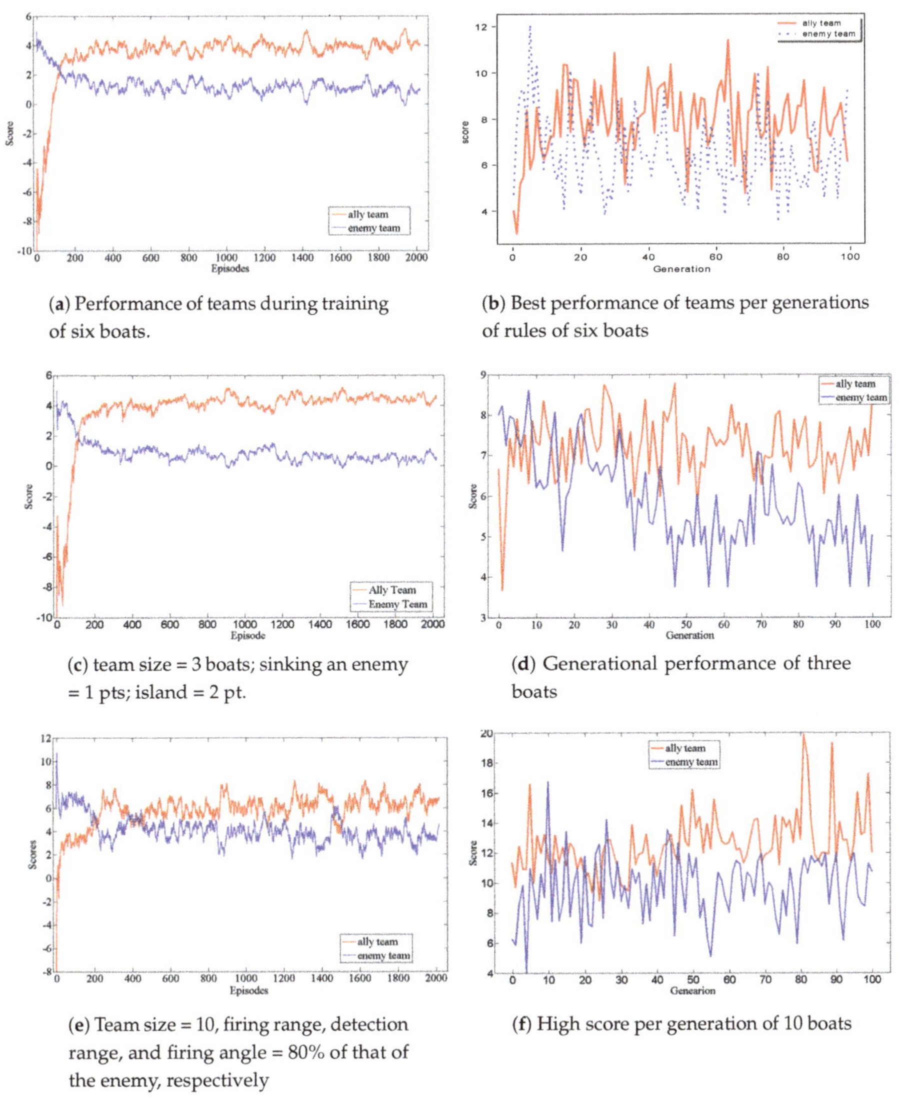

(**a**) Performance of teams during training of six boats.

(**b**) Best performance of teams per generations of rules of six boats

(**c**) team size = 3 boats; sinking an enemy = 1 pts; island = 2 pt.

(**d**) Generational performance of three boats

(**e**) Team size = 10, firing range, detection range, and firing angle = 80% of that of the enemy, respectively

(**f**) High score per generation of 10 boats

Figure 10. *Cont.*

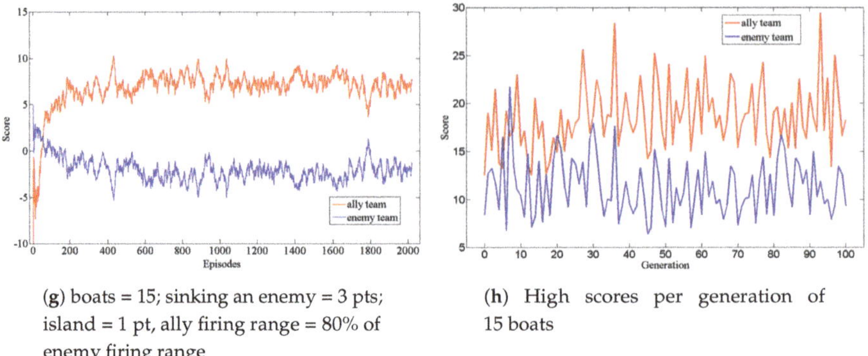

(**g**) boats = 15; sinking an enemy = 3 pts; island = 1 pt, ally firing range = 80% of enemy firing range

(**h**) High scores per generation of 15 boats

Figure 10. Illustration of the performance of the red and blue teams over 2020 (100 generations) episodes during training. Both series of the episodic performances are averages over 20 different sequences of episodes.

In the third simulation, the capabilities of the red force was constrained below that of its counterpart blue force. With each team consisting of 10 boats, the firing range, detection range, and firing radius of the ally team was set to 80% of the opponent team, respectively. Furthermore, we set the initial population of rules to that of the rule base of Simulation 1. Figure 10e,f shows the performance graphs of both teams.

Finally, a simulation of 15 boats with the objective of destroying more opponent boats while still countering islands was conducted. To this end, we assigned three points for destroying opponent boats and a point for conquering an island. All other parameters were reset to the initial fixed parameters with the exception of the firing range, which was maintained as 80% of the opponents' firing range. The initial population of rules in this scenario was the final population rules' set from the scenario. The performance graph is shown in Figure 10g,h.

The observed performance of these KBs run over one hundred episodes against the enemy team is presented in Figure 11. The blue line in Figure 11a represents the best performance of the enemy team in the three cases against the ally team's performances in all cases. In Figure 11b, the average scores over 100 episodes with increasing number of boats in a team is illustrated. The blue bar is the average scores of the the enemy teams, while the rest of the bars represent the performance of the ally team with varied team size and point allocation.

Whilst the blue force's performance degraded, the red force's reasoner was able to maintain a set of KBs that achieved relatively high performance for most parts of the entire simulation of all training scenarios. Comparing the corresponding episode scores of the final KB of the red force against the blue force rule base used during training revealed that the trained red force was better in at least 82% of all cases, as can be seen in Figure 11. The simulation results demonstrate the feasibility of our approach for multi-agent KB learning and the ability to converge to a stable action selection.

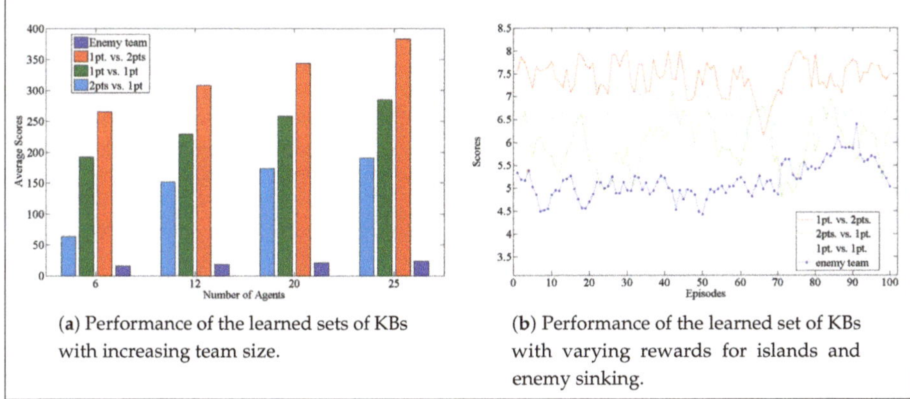

(**a**) Performance of the learned sets of KBs with increasing team size.

(**b**) Performance of the learned set of KBs with varying rewards for islands and enemy sinking.

Figure 11. Performance validation of the best-performing rules of the red force after training. Blue represents the highest performance of the blue force. The red is the performance when two points are assigned to destroying enemy boats and a point for conquering an island. The green is the scores obtained when the value of an island is two points and the destroying opponent boat fetches a point; while the light blue is the case when both conquering an island and destroying the opponent is worth a point.

5. Conclusions and Future Work

In this study, we have presented an unmanned multi-surface vehicle training approach for complex control. We used team learning where a central learner controls the agents in an environment modeled as a partially-observable stochastic game. We modeled the control problem as a decomposed FIS and have provided practical ways for constructing and learning the FIS KB. Our proposed framework enables the usage of different FIS decompositions for a complex control problem with minimal or no modification to the reasoner implementation. It also enables the incorporation of agent-induced non-stationarity awareness in the learning process and a resource model for gathering agents' local observations for off-line learning tasks. Our contributions enable multi-agent control to be performed in domains where no historic data are unavailable for training, but the desired system behavior can be specified as a function of the agents' performance. The multi-surface vehicle island-conquering case study demonstrates the feasibility and convergence properties of this study. Our future works will focus on improving the performance of our approach using multi-agent deep reinforcement learning techniques with the gathered agents' observations.

Author Contributions: Conceptualization, B.Z.; Investigation, W.H.; Methodology, J.L.; Software, W.R.; Writing—original draft, Q.W. and Y.X.; Writing—review & editing, J.L.

Funding: This research received no external funding.

Conflicts of Interest: The authors declare no conflict of interest.

References

1. Weyns, D.; Van Dyke Parunak, H.; Michel, F.; Holvoet, T.; Ferber, J. Environments for Multiagent Systems State-of-the-Art and Research Challenges. In *Environments for Multi-Agent Systems*; Weyns, D., Van Dyke Parunak, H., Michel, F., Eds.; Springer: Berlin/Heidelberg, Germany, 2005; pp. 1–47.
2. Ernest, N.D. Genetic Fuzzy Trees for Intelligent Control of Unmanned Combat Aerial Vehicles. Ph.D. Thesis, University of Cincinnati, Cincinnati, OH, USA, 2015.
3. Magessi, N.T.; Antunes, L. Modelling Agents' Perception: Issues and Challenges in Multi-agents Based Systems. In *Progress in Artificial Intelligence*; Pereira, F., Machado, P., Costa, E., Cardoso, A., Eds.; Springer International Publishing: Cham, Switzerland, 2015; pp. 687–695.

4. Weyns, D.; Michel, F.; Parunak, V.; Boissier, O.; Schumacher, M.; Ricci, A.; Brandao, A.; Carrascosa, C.; Dikenelli, O.; Galland, S.; et al. Agent Environments for Multi-agent Systems—A Research Roadmap. In *Introduction and Challenges of Environment Architectures for Collective Intelligence Systems*; Springer-Verlag: New York, NY, USA, 2015, doi:10.1007/978-3-319-23850-0_1.
5. Pipattanasomporn, M.; Feroze, H.; Rahman, S. Multi-agent systems in a distributed smart grid: Design and implementation. In Proceedings of the 2009 IEEE/PES Power Systems Conference and Exposition, Seattle, WA, USA, 15–18 March 2009; pp. 1–8, doi:10.1109/PSCE.2009.4840087. [CrossRef]
6. Valogianni, K.; Ketter, W.; Collins, J. A Multiagent Approach to Variable-Rate Electric Vehicle Charging Coordination. In Proceedings of the 2015 International Conference on Autonomous Agents and Multiagent Systems (AAMAS '15), Istanbul, Turkey, 4–8 May 2015; pp. 1131–1139.
7. Riedmiller, M.; Moore, A.; Schneider, J. Reinforcement Learning for Cooperating and Communicating Reactive Agents in Electrical Power Grids. In *Balancing Reactivity and Social Deliberation in Multi-Agent Systems*; Springer: Berlin/Heidelberg, Germany, 2001; pp. 137–149.
8. Pita, J.; Jain, M.; Ordóñez, F.; Portway, C.; Tambe, M.; Western, C.; Paruchuri, P.; Kraus, S. Using Game Theory for Los Angeles Airport Security. *AI Mag.* **2009**, *30*, 43–57. [CrossRef]
9. Stone, P.; Veloso, M. Multiagent Systems: A Survey from a Machine Learning Perspective. *Auton. Robots* **2000**, *8*, 345–383, doi:10.1023/A:1008942012299. [CrossRef]
10. Crites, R.H.; Barto, A.G. Elevator Group Control Using Multiple Reinforcement Learning Agents. *Mach. Learn.* **1998**, *33*, 235–262, doi:10.1023/A:1007518724497. [CrossRef]
11. Busoniu, L.; Babuska, R.; Schutter, B.D. A Comprehensive Survey of Multiagent Reinforcement Learning. *IEEE Trans. Syst. Man Cybern. Part C* **2008**, *38*, 156–172, doi:10.1109/TSMCC.2007.913919. [CrossRef]
12. Bibuli, M.; Singh, Y.; Sharma, S.; Sutton, R.; Hatton, D.; Khan, A. A Two Layered Optimal Approach towards Cooperative Motion Planning of Unmanned Surface Vehicles in a Constrained Maritime Environment. *IFAC-PapersOnLine* **2018**, *51*, 378–383, doi:10.1016/j.ifacol.2018.09.458. [CrossRef]
13. Polvara, R.; Patacchiola, M.; Sharma, S.; Wan, J.; Manning, A.; Sutton, R.; Cangelosi, A. Toward End-to-End Control for UAV Autonomous Landing via Deep Reinforcement Learning. In Proceedings of the 2018 International Conference on Unmanned Aircraft Systems (ICUAS), Dallas, TX, USA, 12–15 June 2018; pp. 115–123, doi:10.1109/ICUAS.2018.8453449. [CrossRef]
14. Obst, O. Using a Planner for Coordination of Multiagent Team Behavior. In *Programming Multi-Agent Systems*; Bordini, R.H., Dastani, M.M., Dix, J., El Fallah Seghrouchni, A., Eds.; Springer: Berlin/Heidelberg, Germany, 2006; pp. 90–100.
15. Nalepka, P.; Kallen, R.W.; Chemero, A.; Saltzman, E.; Richardson, M.J. Herd Those Sheep: Emergent Multiagent Coordination and Behavioral-Mode Switching. *Psychol. Sci.* **2017**, *28*, 630–650, doi:10.1177/0956797617692107. [CrossRef] [PubMed]
16. Foerster, J.; Assael, I.A.; de Freitas, N.; Whiteson, S. Learning to Communicate with Deep Multi-Agent Reinforcement Learning. In *Advances in Neural Information Processing Systems 29*; Lee, D.D., Sugiyama, M., Luxburg, U.V., Guyon, I., Garnett, R., Eds.; Curran Associates, Inc.: Red Hook, NY, USA, 2016; pp. 2137–2145.
17. Heins, P.H.; Jones, B.L.; Taunton, D.J. Design and validation of an unmanned surface vehicle simulation model. *Appl. Math. Model.* **2017**, *48*, 749–774, doi:10.1016/j.apm.2017.02.028. [CrossRef]
18. Sonnenburg, C.R.; Woolsey, C. Modeling, Identification, and Control of an Unmanned Surface Vehicle. *J. Field Robot.* **2013**, *30*, 371–398, doi:10.1002/rob.21452. [CrossRef]
19. Yue, J.; Ren, G.; Liang, X.; Qi, X.W.; Li, G.T. Motion Modeling and Simulation of High-Speed Unmanned Surface Vehicle. Frontiers of Manufacturing and Design Science. *Appl. Mech. Mater.* **2011**, *44–47*, 1588–1592, doi:10.4028/www.scientific.net/AMM.44-47.1588. [CrossRef]
20. Liu, Z.; Zhang, Y.; Yu, X.; Yuan, C. Unmanned surface vehicles: An overview of developments and challenges. *Ann. Rev. Control* **2016**, *41*, 71–93, doi:10.1016/j.arcontrol.2016.04.018. [CrossRef]
21. Yan, R.j.; Pang, S.; Sun, H.b.; Pang, Y.j. Development and missions of unmanned surface vehicle. *J. Mar. Sci. Appl.* **2010**, *9*, 451–457, doi:10.1007/s11804-010-1033-2. [CrossRef]
22. IsraelDefense. Rafael's Protector USV Conducts Successful Missile Firing Demo for NATO. 2018. Available online: https://www.israeldefense.co.il/en/node/34530 (accessed on 14 March 2019).
23. Opall-Rome, B. Israel's Elbit Unveils USV for Anti-Sub, Anti-Mine Missions. 2016. Available online: https://www.defensenews.com/naval/2016/02/08/israels-elbit-unveils-usv-for-anti-sub-anti-mine-missions/ (accessed on 14 March 2019).

24. Schnoor, R.T. Modularized unmanned vehicle packages for the littoral combat ship mine countermeasures missions. In Proceedings of the Oceans 2003: Celebrating the Past ... Teaming Toward the Future (IEEE Cat. No. 03CH37492), San Diego, CA, USA, 22–26 September 2003; Volume 3, pp. 1437–1439, doi:10.1109/OCEANS.2003.178073. [CrossRef]
25. Manley, J. Unmanned surface vehicles, 15 years of development. In Proceedings of the OCEANS 2008, Quebec City, QC, Canada, 15–18 September 2008; pp. 1–4, doi:10.1109/OCEANS.2008.5152052. [CrossRef]
26. Wen, G.X. Neural-network-based adaptive leader-following consensus control for second-order non-linear multi-agent systems. *IET Control Theory Appl.* **2015**, *9*, 1927–1934. [CrossRef]
27. Cordón, Ó.; Herrera, F.; Hoffmann, F.; Magdalena, L. *Genetic Fuzzy Systems: Evolutionary Tuning and Learning Of Fuzzy Knowledge Bases;* World Scientific: London, UK, 2004; Volume 141, pp. 161–162, doi:10.1016/S0165-0114(03)00262-8.
28. Holland, J.H.; Reitman, J.S. Cognitive Systems Based on Adaptive Algorithms. *SIGART Bull.* **1977**, 49, doi:10.1145/1045343.1045373. [CrossRef]
29. Smith, S.F. A Learning System Based on Genetic Adaptive Algorithms. Ph.D. Thesis, University of Pittsburgh, Pittsburgh, PA, USA, 1980.
30. Herrera, F.; Lozano, M.; Verdegay, J. A learning process for fuzzy control rules using genetic algorithms. *Fuzzy Sets Syst.* **1998**, *100*, 143–158, doi:10.1016/S0165-0114(97)00043-2. [CrossRef]
31. Herrera, F.; Lozano, M.; Verdegay, J.L. Tackling Real-Coded Genetic Algorithms: Operators and Tools for Behavioural Analysis. *Artif. Intell. Rev.* **1998**, *12*, 265–319, doi:10.1023/A:1006504901164. [CrossRef]
32. Hernandez-Leal, P.; Kaisers, M.; Baarslag, T.; de Cote, E.M. A Survey of Learning in Multiagent Environments: Dealing with Non-Stationarity. *arXiv* **2017**, arXiv:1707.09183.
33. Claus, C.; Boutilier, C. The Dynamics of Reinforcement Learning in Cooperative Multiagent Systems. In Proceedings of the Fifteenth National/Tenth Conference on Artificial Intelligence/Innovative Applications of Artificial Intelligence (AAAI '98/IAAI '98), Madison, WI, USA, 26–30 July 1998; American Association for Artificial Intelligence: Menlo Park, CA, USA, 1998; pp. 746–752.
34. Rada-Vilela, J. Fuzzylite: A Fuzzy Logic Control Library in C++. 2017. Available online: https://pdfs.semanticscholar.org/ec93/4e26ea2950d0f3ab30d31eb8ac239373b4e8.pdf (accessed on 14 March 2019).

MDPI
St. Alban-Anlage 66
4052 Basel
Switzerland
Tel. +41 61 683 77 34
Fax +41 61 302 89 18
www.mdpi.com

Applied Sciences Editorial Office
E-mail: applsci@mdpi.com
www.mdpi.com/journal/applsci

www.ingramcontent.com/pod-product-compliance
Lightning Source LLC
LaVergne TN
LVHW070149170726
843515LV00007B/1878

9783039430468